D1236567

APPLIED CHAOS

APPLIED CHAOS

Edited by
Jong Hyun Kim
John Stringer
Electric Power Research Institute
Palo Alto, California

A Wiley-Interscience Publication
JOHN WILEY & SONS, INC.
New York / Chichester / Brisbane / Toronto / Singapore

Library of Congress Cataloging in Publication Data:

Applied chaos/[edited by] Jong Hyun Kim, John Stringer.
    p.   cm.
    Lectures and discussion from the International Workshop on
Applications of Chaos, sponsored by the Electric Power Research
Institute and held in San Francisco on December 4–7, 1990.
    Includes index.
    ISBN 0-471-54453-1 (cloth)
    1. Chaotic behavior in systems.   2. Chaotic behavior in systems-
Industrial applications.   3. Dynamics.   I. Kim, Jong Hyun.
II. Stringer, John, 1934–   .   III. Electric Power Research
Institute.

Q172.5.C45A66   1992
003'.7—dc20   92-7566
                                                                      CIP

CONTRIBUTORS

MARTIN CASDAGLI, Theoretical Division, Los Alamos National Laboratory, Los Alamos, NM 87545

HILDA A. CERDEIRA, International Centre for Theoretical Physics, 34100 Trieste, Italy

H.-CHIA CHANG, Department of Chemical Engineering, University of Notre Dame, Notre Dame, IN 46556

GARY T. CHAPMAN, Department of Mechanical Engineering, University of California, Berkeley, CA 94720

A. A. COLAVITA, Microprocessor Laboratory, International Centre for Theoretical Physics, 34100 Trieste, Italy

PREDRAG CVITANOVIĆ, Niels Bohr Institute, Blegdamsvej 17, DK-2100, Copenhagen Ø, Denmark

W. J. DECKER, Department of Nuclear Engineering, University of Virginia, Charlottesville, VA 22903-2442

ROBERT DE PAOLA, Division of Cardiothoracic Surgery, Children's Hospital of Philadelphia, Philadelphia, PA 19104

DEIRDRE DES JARDINS, Theoretical Division, Los Alamos National Laboratory, Los Alamos, NM 87545

M. K. DEVINE, Chemical Technology Division, Argonne National Laboratory, Argonne, IL 60439

J. J. DORNING, Department of Nuclear Engineering, University of Virginia, Charlottesville, VA 22903-2442

T. P. EGGARTER, Microprocessor Laboratory, International Centre for Theoretical Physics, 34100 Trieste, Italy

T. ENOMOTO, Department of Electrical Engineering, Kyoto University, Kyoto 606, Japan

STEPHEN EUBANK, Theoretical Division, Los Alamos National Laboratory, Los Alamos, NM 87545

J. DOYNE FARMER,* Theoretical Division, Los Alamos National Laboratory, Los Alamos, NM 87545

JOHN GIBSON, Theoretical Division, Los Alamos National Laboratory, Los Alamos, NM 87545

LEON GLASS, Department of Physiology, McGill University, Montreal, Canada

ARY L. GOLDBERGER, Associate Professor of Medicine, Harvard Medical School, Director of Electrocardiography, Beth Israel Hospital, Boston, MA 02215

MICHAEL GORMAN, Department of Physics, University of Houston, Houston, TX 77204-5504

CELSO GREBOGI, Laboratory for Plasma Research, University of Maryland, College Park, MD 20742

A. HARRISON,† Department of Mechanical Engineering, The University of Newcastle, NSW Australia 2308

JAMES PAUL HOLLOWAY, Department of Electrical and Computer Science Engineering, University of Michigan, Ann Arbor, MI 48100

NORMAN HUNTER, WX Division, Los Alamos National Laboratory, Los Alamos, NM 87545

J. H. KIM, Electric Power Research Institute, Palo Alto, CA 94304

YING-CHENG LAI, Laboratory for Plasma Research, University of Maryland, College Park, MD 20742

A. N. LANSBURY, Mathematical Sciences Group, Department of Applied Science, Brookhaven National Laboratory, Upton, NY 11973

HUAN LIN, Department of Civil Engineering, Oregon State University, Corvallis, OR 97331-2302

ALAN J. MARKWORTH, Metals and Ceramics Department, Battelle Memorial Institute, Columbus, OH 43201-2693

J. KEVIN McCOY, Metals and Ceramics Department, Battelle Memorial Institute, Columbus, OH 43201-2693

---

*Present address: Prediction Co., 234 Griffin Avenue, Santa Fe, NM 87501

†Present address: Winders, Barlow & Morrison, Inc., 14 Inverness Drive East, Englewood, CO 80112.

F. A. McRobie, Centre for Nonlinear Dynamics, Department of Civil Engineering, University College London, London WC1E 6BT, United Kingdom

William I. Norwood, Division of Cardiothoracic Surgery, Children's Hospital of Philadelphia, Philadelphia, PA 19104

J. M. Ottino, Department of Chemical Engineering, Northwestern University, Evanston, IL 60208-3100

Punit Parmananda, Department of Physics and Astronomy, Condensed Matter and Surface Science Program, Ohio University, Athens, OH 45701-2979

Kay A. Robbins, Division of Mathematics, Computer Science, and Statistics, The University of Texas at San Antonio, San Antonio, TX 78249

Roger W. Rollins, Department of Physics and Astronomy, Condensed Matter and Surface Science Program, Ohio University, Athens, OH 45701-2979

O. E. Rössler, Division of Theoretical Chemistry, University of Tübingen, 7400 Tübingen, Germany

Mihir Sen, Department of Aerospace and Mechanical Engineering, University of Notre Dame, Notre Dame, IN 46556

H. B. Stewart, Mathematical Sciences Group, Department of Applied Science, Brookhaven National Laboratory, Upton, NY 11973

John Stringer, Electric Power Research Institute, 3412 Hillview Avenue, Palo Alto, CA 94304

Michael J. Szady, Department of Chemical Engineering, Princeton University, Princeton, NJ 08540

S. W. Tam, Chemical Technology Division, Argonne National Laboratory, Argonne, IL 60439

James Theiler, Theoretical Division, Los Alamos National Laboratory, Los Alamos, NM 87545

J. M. T. Thompson, Centre for Nonlinear Dynamics, Department of Civil Engineering, University College London, London WC1E 6BT, United Kingdom

Y. Ueda, Department of Electrical Engineering, Kyoto University, Kyoto 606, Japan

Leslie Yates, Eloret Institute, Palo Alto, CA 94303

Solomon C. S. Yim, Department of Civil Engineering, Oregon State University, Corvallis, OR 97331-2302

James A. Yorke, Institute for Physical Science and Technology, University of Maryland, College Park, MD 20742

CONTENTS

PREFACE

Chaos umpire sits, ..., Chance governs all.

John Milton, *Paradise Lost*

Over the past few decades, many exciting and interesting ideas have been developed in nonlinear dynamics. In particular, the birth of the science of chaotic dynamics has been a source of great excitement in the scientific community. Chaos has been a subject of intense curiosity, but the activity has been largely confined to a small community of academicians and research scientists. Surely, some popular books on chaos (such as *Chaos*, by James Gleick, Viking Press, 1987) have enlightened the laity about the science of chaos, but chaos has been largely a fertile garden for theoreticians and natural philosophers. Their main interest has been to identify and describe chaotic phenomena or to investigate the fundamental nature of chaos in terms of their own familiar languages. Practical implications of these ideas in engineering, biology, medicine, and other technological fields have not been widely perceived or appreciated. We have often heard many practical-minded people expressing a curt question about chaos: "So what?" This volume is an attempt to answer partially the mundane but nontrivial question, What can we say about applications of chaos? What opportunities does it present to the engineers and applied scientists to understand better the way systems work—or fail—and, ultimately, what can they do about it?

As a first step toward enhancing the awareness of the potential applications of chaotic dynamics and related subjects, the Electric Power Research Institute (EPRI) sponsored the International Workshop on Applications of Chaos, held in San Francisco, December 4–7, 1990. This volume contains the lectures presented at the workshop, as well as some of the very vigorous discussion that the presentations engendered. Although EPRI's long-term objective is to apply chaotic dynamics to problems in the electric utility industry and other energy-related areas, the scope of the workshop was not confined to these topics. Our theme was applications, regardless of the areas. Our strategy was to make the meeting as chaotic as possible—but with order! Experts from all disciplines of chaos were invited—physicists, chemists, mathematicians, engineers (electrical, mechanical, chemical, nuclear, and civil), physiologists, information and computer scientists, material scientists, and others—and were encouraged to speak freely, but about applications of chaos (occasional gentle reminders and guidance were in order to keep them on track). The only constraint—that they should speak about the applications —served rather well to achieve order within chaos! The result was revealing and rewarding. Many participants, who never thought of chaos in terms of practical applications, were presenting specific examples of applications and expounding them with great fascination and enthusiasm. It became clear that applied chaotic dynamics would play an increasingly important role in many branches of science and technology precisely because of its ubiquitous nature. Transition from regular behavior to chaotic behavior is not an exception but rather a norm in many real systems. It was also interesting to see that although in many practical systems a transition to chaos is undesirable, in others it is the preferred mode of operation—this result was even surprising to some of the active investigators in the field. The workshop has also shown that many seemingly disparate fields are connected through the common concepts and language of chaos, proving again the interdisciplinary nature of chaotic phenomena.

The topics contained in this volume are as diverse as the participants, running the gamut from the dynamics of electrocardiograph data and the instability of conveyor belts to the time series modeling and control of chaos. The common thread among them, however, is that, with a few exceptions, they all address some applications aspect of chaos, be it practical or theoretical. We believe this is a main contribution of the workshop to the technical and scientific community.

We thank the authors, speakers, and all participants of the workshop for making the meeting a worthy event. Useful suggestions for the workshop were provided by Bruce Stewart. The workshop was sponsored by the Office of Exploratory Research of EPRI, and we thank Fritz Kalhammer, John Maulbetsch, and Walter Esselman for support and advice.

Jong Hyun Kim
John Stringer

APPLIED CHAOS

PART I

CHAOS IN ENGINEERING AND
TECHNOLOGICAL APPLICATIONS

CHAPTER 1

BRIDGING THE GAP BETWEEN THE SCIENCE OF CHAOS AND ITS TECHNOLOGICAL APPLICATIONS

J. J. DORNING
Department of Nuclear Engineering
University of Virginia
Charlottesville, VA 22903-2442

J. H. KIM
Electric Power Research Institute
Palo Alto, CA 94304

This chapter first reviews selected early scientific studies on nonlinear dynamics and chaos that have fairly obvious potential implications for practical engineering problems arising in technological applications; the emphasis in this review is on these implications. A summary also is given of some related recent examples of applications of the techniques of modern bifurcation theory, nonlinear dynamical systems, and deterministic chaos to the analysis of engineering problems—especially those relevant to electric power generation and distribution technologies. Then some engineering needs related to the control of the dynamics of complex nonlinear technological systems, principally chaotic systems, are discussed, and some results are described briefly on a simple example of modification of periodic and chaotic dynamics of a simple forced Lorenz convection-cooling model of heat-generating components.

1.1. INTRODUCTION

Interestingly, many of the early scientific studies of chaotic dynamics, especially the early experimental studies in which chaotic behavior was observed, were carried out in the contexts of physical problems that are not unrelated

Applied Chaos, Edited by Jong Hyun Kim and John Stringer.
ISBN 0-471-54453-1 © 1992 John Wiley & Sons, Inc.

to important practical engineering problems arising in complex technological systems. In fact, many of the "simple" clean physical problems are related directly to "first-cut" generic models of phenomena and components that are nearly all-pervasive in modern technology—examples abound.

A few such examples of obvious historical importance—and of equally obvious technological relevance—are reviewed in Section 1.2. Section 1.3 summarizes related recent applications to engineering problems, especially engineering problems arising in technologies related to electric power generation and distribution.

Section 1.4 discusses some clear engineering needs related to the modification of the nonlinear dynamical behavior of complex technological systems, the obvious ultimate need being the boundary control of such systems so that they can be operated optimally in equilibrium, periodic, quasiperiodic, or chaotic states—depending on the efficiencies of these various operational modes toward achieving an engineering objective. Section 1.5 describes an example of the modification of the nonlinear dynamical behavior of a simple forced Lorenz convection cooling model for heat-generating electrical or electronic components, such as electrical transformers and electronic control system boards. Some results are discussed briefly on modifying such a system (which otherwise would be chaotic) so that it is periodic and then bringing it through a period-doubling sequence back to chaos.

Finally, Section 1.6 proffers some thoughts—which are certainly preliminary because this area of science and technology probably has not yet begun its infancy—about future directions and possible applications of nonlinear dynamics and chaotic systems analysis to practical engineering systems that might bring to bear the insights of the "new science of chaos" and the power of its continuously developing techniques on the many long-known, but unsolved, nonlinear problems arising in technology, as well as on many of the newly identified important nonlinear problems arising in engineering applications.

1.2. SOME EARLY SCIENTIFIC STUDIES OF CHAOS WITH DIRECT RELATIONSHIPS TO TECHNOLOGICAL APPLICATIONS

Probably one of the earliest reported and best-known experimental examples of chaos is the pioneering work of Gollub and Swinney [1], who showed that the route to turbulence via an infinite sequence of Hopf bifurcations hypothesized by Landau [2] was, in fact, not followed in their Taylor–Couette experiment; rather, the Ruelle–Takens–Newhouse route, three Hopf bifurcations to a structurally unstable torus that decays into a strange attractor [3, 4], explained their measured data. The geometry of their experiment, an annulus of water between a rotating interior stainless steel cylinder and a fixed exterior glass cylinder, certainly is not uncommon in engineering applications; rather, it is almost ubiquitous in rotating machinery used in electrical power generation—journal bearings, steam and gas turbines, electrical generators and motors, centrifuges, pumps, and so on.

Another early experimental example of chaotic dynamics is provided by the works of Linsay [5] and of Testa, Pérez, and Jeffries [6], who studied a simple controlled oscillator-driven *LCR* circuit that included a nonlinear silicone varactor diode,

$$L\ddot{q}(t) + R\dot{q}(t) + V_C(q(t)) = V_{\text{drive}}(t) = V_0 \sin(2\pi ft)$$

By varying the amplitude V_0 of the sinusoidal voltage imposed at fixed frequency on the simple circuit, they showed that the voltage V_C across the silicon varactor became chaotic after following a period-doubling sequence from a period-1 oscillation, with structure analogous to that observed by Feigenbaum [7, 8] for the simple iterated logistic map

$$x_{n+1} = cx_n(1 - x_n).$$

In fact, they even observed period-3, period-5, and period-7 windows in addition to the period-2 sequence in the bifurcation diagram constructed from a window in V_C versus V_0. Later experimental studies by Chua, Yao, and Yang [9] on a slightly more complicated sinusoidally driven electrical circuit—an *RCL* circuit with a negative resistance device synthesized by a two-transistor subcircuit—showed that this still relatively simple circuit behaved chaotically and that as the driver frequency was varied it followed a devil's staircase route to chaos via the period-adding law observed by Kaneko [10] in a one-dimensional discrete map. In fact, van der Pol himself along with van der Mark apparently observed chaos as early as 1927 in an electrical circuit (a simple sinusoidally forced neon bulb circuit) by listening to subharmonics—and the bands of noise (probably chaotic signals) that separated some of them—on a telephone receiver coupled to the circuit [11]. Chaos also has been observed experimentally in a simple autonomous electrical circuit by Matsumoto, Chua, and Komuro [12]. Since then, chaos has been observed in a number of electrical [13] and electronic [14] circuits. Such behavior also has been observed more recently in circuits of practical interest in engineering systems, including even some commercially available products [15]. Clearly then, nonlinear dynamics and chaos must be relevant to the systems involved in the generation and distribution of electrical power (with their numerous electrical and electronic control systems, etc.) and possibly even to the power grid and subgrid distribution systems themselves.

Also endemic to the technologies of electric power generation and distribution are structural elements and other components that undergo flow-induced vibrations, such as turbine blades, condenser tubes, transmission wires, and so on. Analyses have shown that such elements can vibrate chaotically [16, 17], and there are early experimental examples of chaotic vibrations of rods subjected to various forces [18, 19]. In practical engineering systems, heat-exchanger tube arrays subjected to cross flow often experience flow-induced vibration [20–22], sometimes leading to fluid-elastic instabilities that can cause severe fatigue and fretting wear on materials. The richness and complexity of phenomena induced by the interaction between solid materials and fluid flow fields suggests the likelihood of chaotic vibrations in many practical engineering systems.

One of the fundamental physical phenomena that is almost all-pervasive in many technologies is heat removal by convective cooling, both forced and natural thermal convection. It is widespread in the technology of electric power generation, both in single-phase systems and more-complicated two-phase systems. Not surprisingly, some of the exciting early physics experiments that reported chaotic phenomena were done in the context of natural convection of a simple single fluid driven by a temperature gradient—classic Rayleigh–Bénard convection. The period-doubling path to chaos was observed by Maurer and Libchaber [23] in a pioneering experiment in a tiny box (1.25 × 1.5 × 3 mm) of liquid helium, and by Libchaber, Laroche, and Fauve [24] in a cleaner experiment using liquid mercury exposed to a magnetic field in a 7 × 7 × 28-mm box. The experimental temperature time-series data are shown in Figure 1.1, which is adapted from reference [24]. It shows the period-doubling bifurcations from period-2 to period-4 to period-8 to period-16 on the path to chaos as the Rayleigh number—proportional to the amount by which the temperature at the bottom surface of the box exceeds that at the top surface—is increased, moving downward in the figure. The data recorded in this experiment were used to obtain an estimate of one of the Feigenbaum ratios generated from studies of the logistic map [7, 8]. The experimental value, based on the first four bifurcations only, is $\delta = 4.4 \pm 0.1$, in fairly good agreement with the value that corresponds to uninodal maps with quadratic maxima $\delta = 4.6692\ldots$. In another experiment done at about the same time, Giglio, Musazzi, and Perine [25], using a 7.9 × 15 × 25-mm Rayleigh–Bénard cell filled with water, reported an estimate of $\delta = 4.3$, also based on the first four period-doubling bifurcations.

In yet another Rayleigh–Bénard experiment done at about the same time, using a 2 × 2.4 × 4-cm cell filled with silicone oil, Bergé, Dubois, Manneville, and Pomeau [26] provided a lovely example of the intermittency route to chaos, showing that in addition to following the period-doubling path to chaos such cells (depending upon the Prandtl number, etc.) could exhibit intermittency. Moreover, using Baker's transform [27, 28] and the Lorenz equations [29, 30]

$$\dot{x}(t) = \sigma(y - x)$$

$$\dot{y}(t) = \rho x - y - xz$$

$$\dot{z}(t) = xy - bz$$

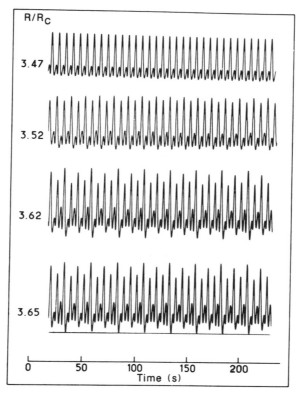

Figure 1.1. Experimental time series data for the temperature reading at a point on the top of a horizontal Bénard cell of aspect ratio 4. The period-2, -4, -8, and -16 oscillations measured at increasing Rayleigh number ratios R/R_C on the path to chaos are shown. (Adapted from reference [24].)

as a simple model, they provided an insightful explanation of their experimental observations, relating them to the theoretical work of Manneville and Pomeau [31] and type-I intermittency.

Thus, although a very oversimplified model of natural convection in a Rayleigh–Bénard cell [29], which in turn usually is a substantial simplification of related thermal convection problems arising in technology, the Lorenz equations exhibit both the period-doubling [7, 8] and intermittency [31] paths to chaos observed in early Rayleigh–Bénard experiments on chaotic dynamics [23–26]. Perhaps then, in some fairly general way they also may bear relationships to some of the vast array of thermal convection problems that arise in technology. Moreover, the Lorenz equations are directly related [32–34] to a simple circular convection loop or closed-loop thermosyphon—a thin torus containing a single fluid, standing vertically on end and heated along its lower portion [35, 36]. In fact, these equations provide a fairly accurate description of such a thermosyphon. Experiments by Gorman, Widmann, and Robbins [37, 38] in a circular glass loop 76 cm in major diameter and 2.1 cm inner minor diameter, using water and in some cases glycerol–water mixtures as working fluids, have shown that the nonlinear dynamics of a simple convection loop of this type are well described both qualitatively and quantitatively over a wide range of parameters by the Lorenz equations. More recently they have shown that experimental departures from the Lorenz model dynamics in the form of two subcritical Hopf bifurcations occur at large values of the driving parameter, but that these can be replicated by an extended four-mode Lorenz model [39]. Thermosyphon cooling of this type provides a simple generic model of thermally driven cooling loops in many engineering systems that depend upon passive heat removal under certain conditions for their reliability and safety. These include various loop cooling components in both conventional and advanced nuclear reactors [40]. In conventional nuclear reactors, natural convection cooling of the reactor loop is extremely important in postaccident long-term cooling; in so-called advanced nuclear reactors it will be essential to their "inherent safety," which, following emergency shutdown with complete loss of electrical power, will depend entirely upon natural thermal convection in the loop.

The so-called passive light water reactors (SBWR and AP-600), currently being designed by the U.S. utilities with international cooperation, and the advanced liquid metal cooled reactors rely heavily on natural convection cooling in loops. In the AP-600, for example, the buoyancy-driven loop flow substantially supplements the forced flow during normal operation, whereas it becomes the primary heat-removal mechanism during an accident. In the simplfied boiling water reactor (SBWR), cooling by the buoyancy-driven loop flow is the sole mechanism of heat removal under all conditions, that is, during normal operation as well as during an accident, and the design includes no pumps. In the advanced liquid metal cooled reactors, safety is ensured during emergency shutdown with complete electrical power loss by natural convection in a large sodium pool and natural convection in several closed and open loops. Thus, understanding the dynamics of natural convection in loops and pools is not only important, it is essential to the reliable and flexible operation of these proposed advanced nuclear reactors.

Closed natural and forced convection loops are widespread in other engineering applications as well, because technology tends to generate heat, which must be transported away from the source efficiently. Frequently, the flow dynamics are further complicated by boiling—resulting in two-phase loops such as those in two-phase flow heat exchangers, refrigeration systems, and boiling water nuclear reactors, all of which present stability problems in various operating and design parameter regimes. Thus, of imminent interest in technology are not only forced convection problems, but also thermally driven convection problems, which are broadly classified as natural circulation problems in technological contexts [41]. These include problems in single- and two-phase flow in open and closed pools (which are generalizations of Rayleigh–Bénard cells) and in open and closed loops (which are generalizations of thermosyphons. If the simple "scientific" experimental systems just discussed can exhibit interesting chaotic dynamics, what then can be expected for the nonlinear dynamical behavior of complex real engineering systems for which Rayleigh–Bénard cells and thermosyphons are grossly simplified models? The nonlinear dynamics of many such engineering systems have been examined in varying degrees of detail and, in many cases, have been found to be very rich indeed!

For example, chaotic solutions have been shown to be possible in a thermosyphon [42]. Experimental observations and theoretical studies on flow in thermosyphons under various conditions demonstrated the existence of periodic oscillations, quasiperiodic oscillations, bursts of chaos, and persistent chaos [43, 44]. A rich variety of dynamic behavior also was shown to exist by direct numerical simulation of two-dimensional natural convection in a vertical cavity [45]. A period-doubling sequence to chaos also was exhibited very recently in a thermal convection loop subjected to time modulation in a case where chaos would not be expected otherwise [46]. Transitions and bifurcations in laminar buoyant flows in confined enclosures of engineering interest also have been discussed extensively [47]. Chaotic behavior also can be observed in isothermal flows in channels and pipes. For example, calculations have indicated that both periodic and nonperiodic oscillations can exist in a simple two-phase gas–liquid flow [48]. Recently, pulsating flow through curved pipes (e.g., blood flow through the aortic arch) was shown to undergo a period-tripling sequence [49].

1.3. SOME RECENT APPLICATIONS TO ENGINEERING PROBLEMS RELATED TO ELECTRIC POWER GENERATION TECHNOLOGY

In the very limited survey of some of the early scientific studies of chaotic dynamics given in the preceding section, connections are suggested between early experiments and technologies related to electrical power generation and distribution. These included rotating machinery, electrical circuits and grids, mechanical vibrations, and natural and forced convection. Clearly, numerous recent examples are available in each of these, as well as in many other areas of engineering.

Voltage collapse and abnormally high and low voltages with severe consequences have been observed a number of times on large interconnected power systems. These phenomena are due mainly to the following: the increased use of transmission systems for the transfer of energy from remote generation sites to local distribution centers; the loading of transmission systems by transferrals of large amounts of energy from region to region; and the use of more devices such as special protection schemes and shunt capacitors to increase capabilities for the transferral of larger amounts of energy. Voltage collapse problems are considered the principal threat to stability, integrity, and reliability of power grid systems, and understanding the causes and solutions of the voltage collapse problem is thus one of the highest priorities in electrical power transmission research [50]. Yet, the problem is not well understood because the complexity of power grid systems makes the phenomenon not amenable to simple analysis and leads to differing views on modeling techniques and analytical methods. Some well-publicized voltage collapses and related events include the following: the cascading voltage collapse experienced in the Memphis, Tennessee, area (August 22, 1987); the near voltage collapse in England (May 20, 1986); and the well-documented power failure in Tokyo on July 23, 1987 [51].

In the past, power system transitions have been interpreted in terms of bifurcations and nonlinear dynamics. Typical examples include the interpretation of *P–V* and *Q–V* curves—used for power system planning and operation—in terms of a saddle-node bifurcation. Examples of a bifurcation due to variation of reactive power demand at a bus can be found [52]; loss of steady-state stability and voltage collapse have been associated with a local bifurcation [53]; more recently, a center manifold model based on a saddle-node bifurcation also was proposed to explain voltage collapse [54]. Power system failures such as the one that occurred in Tokyo in 1987 might be explained within the framework of bifurcation theory and chaotic dynamics. A preliminary study sponsored by EPRI, based on a simple nonlinear model for a three-bus system with reactive demand as a varying parameter, showed that a system could display a sequence of period-doubling bifurcations, subcritical Hopf bifurcation, more period-doubling bifurcations, supercritical Hopf bifurcation, and then a saddle-node bifurcation as the reactive demand is increased [55]. A tentative conclusion on voltage instability was that saddle-node bifurcation leads to voltage collapse, Hopf bifurcation leads to voltage collapse, and different kinds of bifurcations give rise to different modes of voltage instability that eventually lead to voltage collapse [55].

In the past, transient stability analyses of electric power systems have been carried out using models based on the swing equation [56, 57]. According to that model, the generator dynamics for a single generator supplying a *PQ* load can be expressed as

$$\dot{\Theta} = \omega$$
$$\dot{\omega} = -\alpha\omega - \sin(\Theta) + F$$

where Θ is the angle of the generator voltage relative to the bus voltage, α represents normalized damping, and F corresponds to the mechanical power input. In the case of a periodically varying mechanical power input ($F = k + \beta \cos \omega_0 t$), one can write

$$\dot{\Theta} = \omega$$
$$\dot{\omega} = -\alpha\omega - \sin(\Theta) + k + \beta \cos(\omega_0 t)$$

By varying the parameters α, β, k, and ω_0, a rich variety of dynamical behavior, including quasiperiodic behavior and chaotic behavior, can be shown to exist [55]. Clearly, much remains to be done to the application of nonlinear dynamics to the analysis of voltage instabilities and voltage collapse.

Because this survey is by no means meant to be comprehensive, the examples that follow will be limited to a few convection problems that are of considerable engineering interest. The survey will be limited further by omitting a discussion of fluidized beds, which are of great practical interest in the context of burning pulverized coal in fossil-fuel central power stations, and by merely noting the possibility that they might behave under certain practical operating conditions as chaotic dynamical systems [58]. Also omitted are discussions of mixing in chaotic flows induced by simple forcing motions [59], notwithstanding its considerable relevance to combustion and other technological applications based on chemically reacting systems. Rather, the examples emphasized will be from forced, two-phase flow in closed convection loops.

Forced flow heat removal loops in which boiling and condensation occur are in widespread use in electrical power generation. They are present in the form of heated channels—with single-phase and two-phase flow regions—in two-phase heat exchangers, fossil-fueled steam generators, and boiling water nuclear power reactors. Moreover, the problem of the stability of operating steady-flow equilibria or fixed points of such heated channels (even in isolation from a closed loop with a condenser, pump, etc.) has long been recognized by pragmatic design engineers who, decades ago, introduced orifices or nozzles at the inlets of such channels to increase the total frictional pressure drop, thereby making desirable system operating points (such as those with high heat fluxes) stable fixed points at the expense of continuously supplying the additional pump work necessary to overcome the extra pressure drops caused by the orifices. The basic mechanism that can make the flow in these heated channels unstable is the interaction of the height at which boiling begins in the vertical channel with the nonlinear effect of the surface friction in the two-phase flow region, which is larger than that in the single-phase flow region. For a given coolant mean flow rate and a given inward heat flux through the channel walls, this simple mechanism can cause perturbations of an unstable operating flow equilibrium to evolve to growing nonlinear oscillations.

Although a very simple model (of an already simplified engineering model) of a closed two-phase forced convection loop (Figure 1.2) is substantially more complicated than a single-phase closed convection loop of the type discussed in the previous section, such a two-phase loop can be analyzed and has been shown to have fairly rich nonlinear dynamics [60]. Actually, the closed loop represents only a fairly small difference in the sense of a nonlinear dynamical system from the isolated heated two-phase flow channel subjected to a constant externally imposed pressure drop [60]. Such an

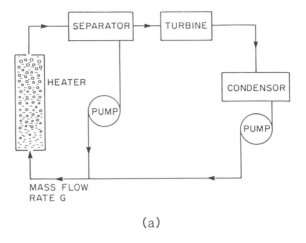

(a)

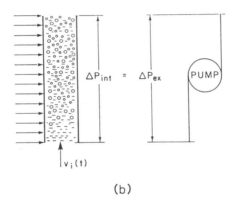

(b)

Figure 1.2. (*a*) Schematic diagram of a heated loop with two-phase flow. Fluid enters bottom of heater in liquid phase and begins to boil part of the way up the heater, as indicated schematically by the appearance of bubbles. (Adapted from reference [60].) (*b*) Schematic diagram of a simplified heated loop where all "pressure drops" are lumped in the heated channel. (Adapted from reference [60].)

isolated channel, or one in parallel with many others, has long been known to have two saddle-node bifurcation points [61, 62]. These two turning points form an S-shaped, or hysteresis, curve in the bifurcation diagram, with the upper and lower branches usually being stable and the middle branch being unstable. This unstable middle branch, a saddle point, corresponds to the well-known Ledinegg, or excursive, instability [61]. It is possible to convert exactly the channel cross section averaged nonlinear partial differential equations in axial position and time [for the single-phase flow in the region between the inlet at the channel bottom and the moving boiling boundary

$\lambda(t)$ and for the two-phase in the region between that boiling boundary and the exit at the top] to nonlinear functional ordinary differential equations with complicated delay integrals. This is done by first integrating the single-phase flow equations along characteristics and then along the channel from the inlet to the boiling boundary, integrating the more-complicated two-phase flow equations along characteristics and then along the channel from the moving boiling boundary to the exit, and combining the results to obtain an expression for the pressure drop over the channel in terms of the dynamical variables. The result is a set of nonlinear functional ordinary differential equations (FDEs) for the channel inlet velocity $v_i(t)$, the boiling boundary $\lambda(t)$, and the two-phase residence time $\tau(t)$. Although still an infinite-dimensional dynamical system, these FDEs can be analyzed—notwithstanding the fact that they are extremely complicated. Thus, recently it has been shown, first using the homogeneous equilibrium model for the two-phase flow [63] and then using the more realistic drift flux model [64], that for most parameter values of practical interest the lower-branch fixed point on the hysteresis curve also can lose stability through a supercritical Hopf bifurcation in which nonlinear density-wave oscillations (stable limit cycles) are

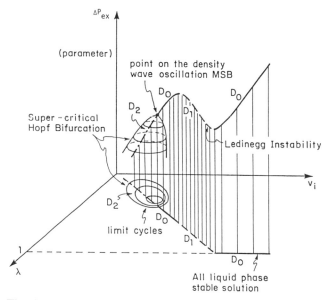

Figure 1.3. The heated channel as an autonomous dynamical system: Schematic bifurcation diagram for the phase-space projection (λ, v_i) of the S-shaped fixed-point curve and of the limit cycle as functions of the parameter ΔP_{ex}, with the Hopf bifurcation point on the marginal stability boundary (MSB) indicated [64]. The index on D_n indicates the number of eigenvalues (with positive real part) of the linearization of the functional differential equations about the indicated fixed-point branch. (Adapted from reference [65].)

born. Physically, an observer located above the oscillating boiling boundary in this case sees a periodic void fraction or mixture density that is equivalent to waves of density passing vertically by—hence the name density wave. Figure 1.3, adapted from reference [65], shows the bifurcation diagram in the three-dimensional space comprised of the parameter ΔP_{ex}, the fixed external pressure drop applied to the channel, and the phase variables λ and v_i. Figure 1.4, also adapted from reference [65], shows the time evolution of v_i

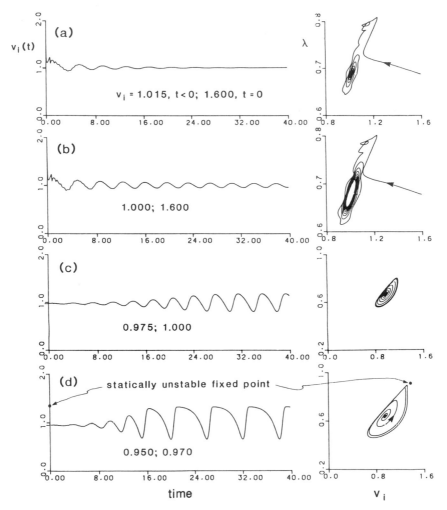

Figure 1.4. Time evolution of the coolant inlet velocity in the heated channel after perturbations in v_i at $t = 0$, and the projection of the corresponding phase-space trajectory onto the v_i–λ plane for a sequence of values of the externally imposed pressure drop ΔP_{ex}. The sequence corresponds to movement through the supercritical Hopf bifurcation point between (b) and (c). (Adapted from reference [65].)

following a small initial perturbation about the fixed point in a sequence of four systems as the Hopf bifurcation point is crossed.

In practical heat exchangers it often is advantageous to enhance net system heat transfer by intentionally oscillating the flow via a periodic externally imposed pressure drop. If the heated channel exhibits limit cycle behavior with natural frequency ω as an autonomous dynamical system, when the same system is forced periodically, depending on the forcing frequency ω_f, one can expect richer dynamics. This is indeed the case as shown in numerical simulations of the forced systems [65]. For various forcing frequencies these heated-channel dynamical systems evolve onto various interesting attracting sets: higher-order limit cycles, invariant tori, and strange attractors. Figure 1.5 adapted from reference [65], shows the computed time series, power spectra, and phase portraits for three forced systems: the first, exhibiting a period-6 limit cycle; the second, quasiperiodic flow onto a torus; the third, chaotic flow onto a strange attractor with one positive Lyapunov exponent [66] and a correlation dimension calculated [65, 67] to be 2.048 $\pm$ 0.003. Analogous dynamical behaviors were found when the system was driven by periodic channel heat fluxes and periodic channel inlet fluid temperatures [68].

Because a boiling water nuclear power reactor has a core that is, to a large extent, an assembly of parallel single- and two-phase flow channels heated by the heat flux resulting from the energy released in the fission process, such reactors inherit aspects of instability phenomena inherent in two-phase flow heated channels. In fact, the desire to understand and ensure the stability of boiling water reactors has been one of the important motivations for the study of the stability—and general nonlinear dynamics—of heated two-phase flow channels [62–65, 68].

Interestingly, fairly recently a boiling water power reactor (LaSalle-2) operating near Chicago in the United States apparently underwent a transition from a stable operating equilibrium to an oscillatory behavior, probably via a Hopf bifurcation, following a valving error made by one of the plant's operating engineers. Although the incident never presented a public hazard or any hazard to the plant employees, it did revitalize interest in the stability and nonlinear dynamics of boiling water reactors. Careful experiments performed on a boiling water reactor in Norway (Forsmark-1) during start-up after a planned shutdown period clearly showed that limit cycle behavior existed at certain operating parameters and raised questions concerning the monitoring of spatially nonuniform power oscillations [69].

When the complex nonlinear FDEs of the two-phase heated channel are coupled to the equations that describe the neutron kinetics in a boiling water reactor (which determine the heat source via the fission process and which are affected by the thermohydraulics through the fuel temperature and, more significantly, through the water density changes due to boiling), dynamics similar to those exhibited by the isolated heated channel are observed. Supercritical Hopf bifurcation from a stable fixed point to a stable limit cycle

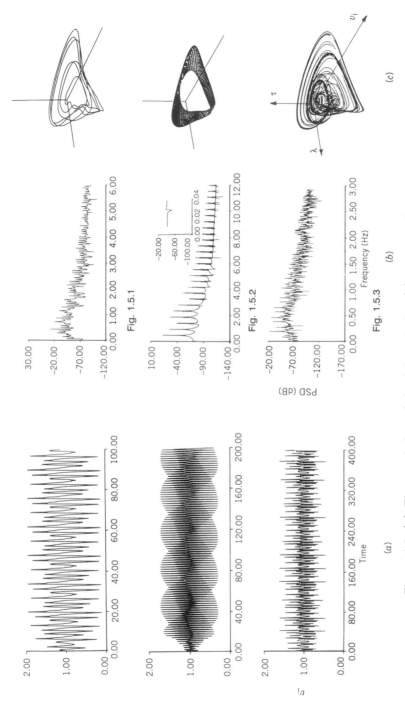

Figure 1.5. (*a*) Time evolution of the inlet velocity; (*b*) corresponding power spectral density calculated from the posttransient behavior; (*c*) the trajectory in the v_i–Λ–τ phase space after the transients have died away, for $\omega_f = 4.500$ (in radians per dimensionless time) (Fig. 1.5.1—limit cycle), $\omega_f = 11.636$ (Fig. 1.5.2—torus), and $\omega_f = 1.200$ (Fig. 1.5.3—strange attractor with dimension 2.048 ± 0.003). (Adapted from reference [65].)

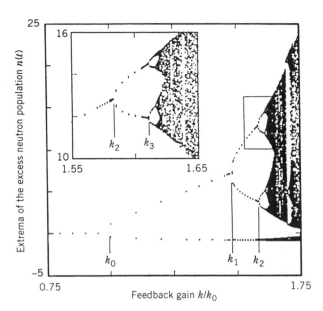

Figure 1.6. Bifurcation diagram for a simple boiling water reactor model [72–74], constructed from the extrema of the time-dependent neutron population as a function of the simple thermal feedback model gain coefficient k/k_0. (Adapted from reference [72].)

results both when a simple "lumped" point kinetics model is used for the neutron kinetics [70] and when the more precise space-dependent model is used [71]; the latter, of course, yielding the spatial changes during the limit cycle oscillations. In fact, earlier work using the simple point kinetics model coupled to an extremely simple model for the complex two-phase heated channel dynamics—a simple forced damped linear oscillator—yielded a supercritical Hopf bifurcation followed by a period-doubling sequence to chaos as a feedback parameter was varied and even gave rather good estimates of the Feigenbaum ratios $\delta = 4.545 \pm 0.648$ and $\alpha = 2.517 \pm 0.094$ based on bifurcations through period-64 [72–74] versus values of $\delta = 4.6692\ldots$ and $\alpha = 2.5029\ldots$ based on the logistic map [7, 8]. An adapted version of the bifurcation diagram generated in reference [72] is shown in Figure 1.6. It was produced by plotting the extreme of the computed time-dependent neutron population versus the heat transfer feedback coefficient of the time-dependent temperature that drove the damped linear oscillator used to model the thermohydraulic feedback.

Because much of the nonlinear dynamics that arise in complex engineering systems—such as those exhibited by the boiling channels relevant to two-phase heat exchangers, refrigeration systems, and boiling water reactors as just discussed—can be explained by the chaotic dynamics paradigms based on simple nonlinear dynamical systems like the logistic map and the Lorenz equations, and because it is clear that the nonlinear dynamics of an engineering system can be made more complicated by nonautonomous forcing (see Figure 1.5, for example), it is natural to ask two questions. Can the behavior of simple nonlinear dynamical systems be modified to achieve desired behavior, complex or simple, when the behavior of the original system is simple or complex? Can the behavior of complex nonlinear dynamical systems be modified to achieve simple behavior when the autonomous system exhibits complex behavior, for example, chaotic behavior? Affirmative answers to questions like these could lead to procedures for the optimal operation of complex nonlinear engineering systems (enhanced heat transfer in simple natural convection pools, in thermosyphons, in two-phase heated channels, etc.) without damaging the systems.

1.4. SOME ENGINEERING NEEDS: MODIFICATION OF THE DYNAMICAL BEHAVIOR OF COMPLEX NONLINEAR TECHNOLOGICAL SYSTEMS

Of course the most desirable situation would be one in which complete control of the nonlinear dynamical behavior of complex technological systems is possible so that they could be operated optimally [75] and in which this control could be exercised as boundary control [76] so that intervention in the interior of the operating system would not be necessary. Short of the ultimate practical engineering objective of optimal boundary control of infinite-dimensional nonlinear dynamical systems with chaotic behavior that are described by nonlinear partial differential equations, for example, the modification (in some systematic way) of the behavior of a simple nonlinear dynamical system is a reasonable shorter-term goal. In fact, such systematic modification of simple dynamical systems, such as the logistic map, has been achieved, and the modification has been more than simple empirical modification. Rather, it has been possible to entrain the system to evolve to specific prespecified dynamical behavior [77]. Entrainment, specifically frequency entrainment, has been studied for some time and has been applied successfully to continuous nonlinear dynamical systems described by ordinary differential equations [78]. Recent work on the logistic map, however, shows that chaotic orbits can be entrained exactly, that is, an otherwise chaotic orbit of the logistic map can be "entrained" so that it asymptotically tends in both amplitude and phase to a prespecified governing set, for example, a periodic orbit [77]. The conditions for entrainment also are established, as are the basins of attraction of initial conditions that evolve to the governing set. Conditions and basins of attraction, for frequency entrainment only, also are established [77]. The logistic map is made nonautonomous by adding to it a forcing function that depends upon the prespecified set $\{g_n\}$, to form

$$x_{n+1} = cx_n(1 - x_n) + g_{n+1} - cg_n(1 - g_n)$$

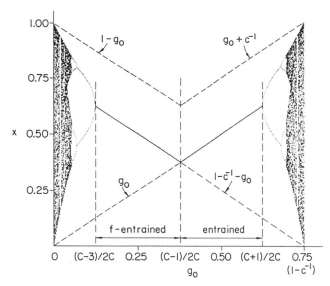

Figure 1.7. The asymptotic behavior of the system for period-1 governing, $g_n = g_0$, for all possible values $(c - 4)/2c < g_0 < (1 + c)/2c$. The figure is drawn for the chaotic system, $c = 4$, but is labeled for general c. The straight lines at the top and bottom are the boundaries of the basin of attraction to the indicated asymptotic motion. Entrainment is limited to $(c - 1)/2c < g_0 < (1 + c)/2c$, required for stability. A frequency-entrained region exists for $(c - 3)/2c < g_0 < (c - 1)/2c$. (Adapted from reference [77].)

which clearly has as a solution $x_n = g_n$ whose stability and basin of attraction are then studied. For the special case of a fixed point or period-1 orbit $g_{n+1} = g_n = g_0(c^*) = 1 - (1/c^*)$, $1 < c^* < 3$, the system for $c = 4$, which would otherwise be chaotic, was entrained so that it evolved to the fixed point $x = g_0$ under the conditions $(c - 1)/2c = 3/8 < g_0 < (c + 1)/2c = 5/8$ for initial conditions in the basin $1 - (1/c) - g_0 < x < g_0 + (1/c)$. These results are shown in Figure 1.7, adapted from reference [77], in which the entrainment conditions, frequency entrainment conditions, and basin boundaries are indicated. If the entrainment had not been induced by the nonautonomous forcing function $g_{n+1} - cg_n(1 - g_n) = g_0 - cg_0(1 - g_0)$, the figure would be a solid black square (for all values of g_0) corresponding to the chaotic orbit of the logistic map for $c = 4$.

Although many large steps are likely to be required to make the transition from entrainment of the logistic map to optimal boundary control of a complex nonlinear engineering system, the basic ideas used in entraining the logistic map are being extended to chaotic continuous dynamical systems. Thus, it is natural to think about boundary control, or entrainment, or at least modification of the nonlinear dynamics of complex engineering systems, and also of the modification of the dynamics of very simple model representations of such systems. To this end, Section 1.5 describes some preliminary ideas that start from boundary control of Rayleigh–Bénard convection—motivated by control of natural convection in complex engineering systems—to arrive straightforwardly at parametrically forced Lorenz equations that are driven from chaotic behavior to simple periodic behavior and then led to chaotic behavior again in simple computer experiments.

1.5. MODIFICATION OF CHAOTIC AND PERIODIC CONVECTION COOLING OF HEAT-GENERATING COMPONENTS

Many engineering systems utilized in the generation and distribution of electrical power, as well as in a vast array of other technologies, rely on convection cooling for heat removal during their normal operation and, in many cases, especially for their emergency operation. Examples include electrical transformers, certain types of nuclear reactors, and electronic components. Frequently, it is necessary, or at least desirable, to maximize or possibly even to control precisely the convective flow that removes heat from such components. The simplest general model for the heat flow from the top of such a component to a cooler surface above it is precisely the one used by Lorenz [29]: an incompressible fluid between two parallel horizontal boundaries, with the lower boundary at a higher temperature than the upper boundary and the fluid motion described by the horizontal and vertical components of the Navier–Stokes momentum conservation equations in the Boussinesq approximation, the mass conservation equation, and the temperature or energy equation. Lorenz, after subtracting the no-flow steady-state solution for the temperature, which is just the solid thermal conduction solution that is linear in the distance between the plates, recast the equations in terms of the stream function and vorticity and, following Saltzman [79], introduced Fourier expansions of the stream function and temperature. Motivated by Saltzman's observation that all but three expansion coefficients eventually tended to zero, Lorenz introduced an extremely low-order truncation of these expansions, one term in the expansion of the stream function and two in that of the temperature, to arrive at the now-celebrated three autonomous nonlinear ordinary differential equations of the Lorenz model.

In most practical convection cooling problems associated with heat-generating components, the top of the component—lower boundary of the convection layer—does not remain at a constant temperature. Even when it does, it might be prudent to vary its temperature intentionally or, more likely in practice, to vary the temperature of the cooler upper boundary in order to exert boundary control over the rate at which heat is being transferred to that boundary. If this temperature is varied periodically, the no-flow solution to the temperature equation, which becomes linear in this zero-velocity case, can be computed (although it is no longer simply constant in time and linear in height) and subtracted from the solution to the original nonlinear partial differential equations. Finally, the Fourier expansion and truncation procedure followed by Lorenz can be repeated, to arrive at parametrically forced Lorenz equations

$$\dot{x}(t) = \sigma[y(t) - x(t)]$$

$$\dot{y}(t) = \rho(t)x(t) - y(t) - x(t)z(t)$$

$$\dot{z}(t) = x(t)y(t) - bz(t)$$

where the parametric forcing function $\rho(t)$, which is proportional to the temperature difference between the bottom boundary and the top boundary, is a periodic Rayleigh number $\rho(t) = \rho_0 + \rho_1 \cos(\omega t)$ that replaces the usual constant value. Thus the boundary control $T(z = 1, t) = T_0 + T_1 \cos(\omega t)$ of the original nonlinear partial differential equations becomes a parametric control of the Lorenz equations.

An even simpler development starting from the equations for the convective motion of the fluid in a thermosyphon [32–39], heated periodically at its bottom or cooled periodically at its top, not surprisingly leads to the same parametrically forced Lorenz equations. This physical problem also is of interest, of course, because of its connection to engineering systems as a simple generic model for closed-loop single-phase convection cooling.

Some simple preliminary computer experiments were performed to explore the possibility of parametrically modifying the behavior of this driven Lorenz system so that a specific system that otherwise would behave chaotically could be made to behave periodically. This was done on personal computers and workstations using the software package available from Yorke [80], which was modified to include the parametric forcing function—or time-dependent Rayleigh number [81].

The parameter values were set at $\sigma = 10.0$ and $b = \frac{8}{3}$ with $\rho_0 = 26.5$, where the well-known Lorenz butterfly, shown in Figure 1.8, exists when $\rho_1 = 0$. The frequency of the forcing function then was set at $\omega = 7.62$ and ρ_1 was increased from 0 to 5, causing the chaotic attractor to collapse into the simple stable single orbit shown in Figure 1.9. The mean value of the Rayleigh number ρ_0 then was increased, resulting in a period-doubling bifurcation to the period-2 orbit displayed in Figure 1.10. Subsequent increases in ρ_0 led to a sequence of period-doubling bifurcations to period-4 (shown in Figure 1.11), . . . , to period-64 (shown in Figure 1.12), and eventually through the period-doubling path to the chaotic attractor shown in Figure 1.13. Thus, the initially complex chaotic dynamics of the system were modified by the introduction of simple parametric forcing so that the system evolved to simple periodic motion; further, the system was then smoothly led, by simply varying the mean value of the forcing function, through a period-doubling sequence back to chaos, but a more controlled chaos. Analogous control also can be achieved in the presence of noise [82].

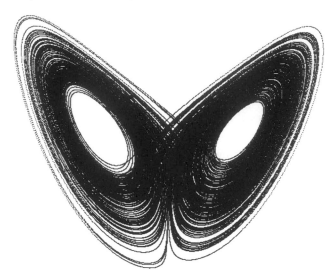

Figure 1.8. The well-known chaotic attractor for the autonomous Lorenz system for $\sigma = 10.0$, $b = \frac{8}{3}$, with $\rho = \rho_0 = 26.5$, above both the value at which the first homoclinic explosion occurs and the value at which the two secondary fixed points become unstable. The numerical calculations that resulted in the attractors shown in Figures 1.8–1.13 were done using a fourth-order Runge–Kutta method with a time step of 0.001.

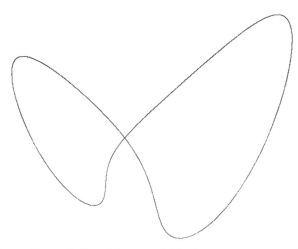

Figure 1.9. The stable periodic orbit onto which the nonautonomous Lorenz system, with $\sigma = 10.0$, $b = \frac{8}{3}$, and $\rho_0 = 26.5$, is forced by the parametric driving function $\rho(t) = \rho_0 + \rho_1 \cos(\omega t)$ with $\rho_1 = 5.0$ and $\omega = 7.62$.

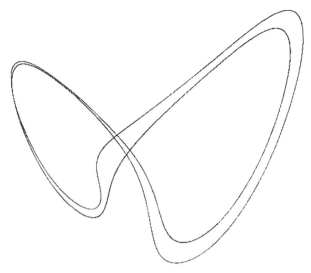

Figure 1.10. The period-doubled stable periodic orbit of the parametrically driven Lorenz system with $\sigma = 10.0$, $b = \frac{8}{3}$, $\rho_1 = 5.0$, $\omega = 7.62$, and ρ_0 increased to 27.5.

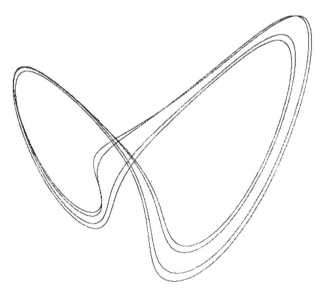

Figure 1.11. The period-4 stable orbit of the parametrically driven Lorenz system with $\sigma = 10.0$, $b = \frac{8}{3}$, $\rho_1 = 5.0$, $\omega = 7.62$, and ρ_0 increased to 27.9.

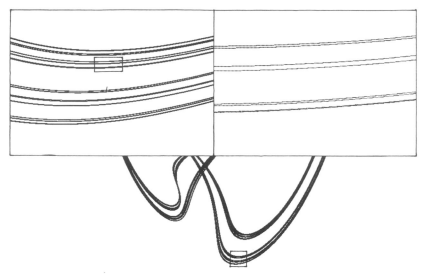

Figure 1.12. The period-64 stable orbit of the parametrically driven Lorenz system with $\sigma = 10.0$, $b = \frac{8}{3}$, $\rho_1 = 5.0$, $\omega = 7.62$, and ρ_0 increased to 27.987.

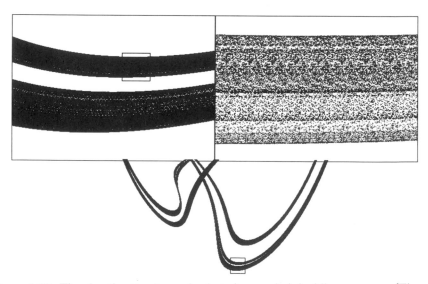

Figure 1.13. The chaotic attractor arrived at via a period-doubling sequence (Figures 1.9–1.12, etc.) for the parametrically driven Lorenz system with $\sigma = 10.0$, $b = \frac{8}{3}$, $\rho_1 = 5.0$, $\omega = 7.62$, and ρ_0 increased to 28.0.

1.6. BRIDGING THE GAP

Just as empirically modifying the behavior of the Lorenz system from chaotic dynamics to simple periodic dynamics is a long way from boundary control of chaotic natural convection in a complex engineering system, the recent advances in nonlinear science and our resulting understanding of nonlinear phenomena, although extremely impressive and in many ways even revolutionary, are a long way from straightforward implementation of these advances in technological applications. Nevertheless, the stone is moving. Slowly, to be sure, and in what direction, no one is sure. Some useful insight into the nonlinear dynamics of many complex engineering systems already has been obtained; some understanding has been developed of the dynamics of bearings and other rotating machinery, of electrical circuits, of single-phase natural convection, of flow-induced structural vibrations, of two-phase heat exchangers, of boiling water reactors, and so on [for exapmle, see 83–85]. Of course, much remains to be done to gain a full or even satisfactory understanding of the dynamics of complex systems such as these, and far more remains to be done even to begin to understand whether the developing techniques of analysis of chaotic dynamical systems will be at all useful in analyzing many of the more-complex engineering systems that arise in practical technological applications. There has been progress, however, and hopefully it will accelerate. This, in fact, is likely to occur when the gap between the science of chaos and its technological applications is bridged—when nonlinear scientists explore technological applications, finding both practical motivation and new needs and as a result developing ideas that are rich in intellectual content and that aid in the solution of basic research problems, and when engineers and scientists oriented toward applied technologies probe into nonlinear science to gain understanding and learn techniques that they then might generalize and use in practical applications.

ACKNOWLEDGMENTS

We gratefully acknowledge valuable discussions with William J. Decker, James Paul Holloway, Rizwan-uddin, and John Stringer.

REFERENCES

*1. J. P. Gollub and H. L. Swinney, Onset of turbulence in a rotating fluid. *Phys. Rev. Lett.* **35** 927–930 (1975).

*References marked with an asterisk are also in the following reprint selection: P. Cvitanović, *Universality in Chaos*, Adam Hilger, Bristol, U.K., 1984.

2. L. D. Landau, On the problem of turbulence. *Dokl. Akad. Nauk SSSR* **44**(8) 339–343 (1944). See also E. Hopf, A mathematical example displaying the features of turbulence. *Comm. Pure and Appl. Math.* **1** 303–322 (1948).

3. D. Ruelle and F. Takens, On the nature of turbulence. *Comm. Math. Phys.* **20** 167–192 (1971).

4. S. Newhouse, D. Ruelle, and F. Takens, Occurrence of strange axiom A_m attractors near quasiperiodic flows on T, $m \geq 3$. *Comm. Math. Phys.* **64** 35–40 (1978).

5. P. S. Linsay, Period doubling and chaotic behavior in a driven, anharmonic oscillator. *Phys. Rev. Lett.* **47** 1345–1352 (1981).

*6. J. Testa, J. Pérez, and C. Jeffries, Evidence of universal chaotic behavior of a driven nonlinear oscillator. *Phys. Rev. Lett.* **48** 714–717 (1982).

7. M. J. Feigenbaum, Qualitative universality for a class of nonlinear transformations. *J. Statist. Phys.* **19** 25–52 (1978).

*8. M. J. Feigenbaum, Universal behavior in nonlinear systems. *Los Alamos Sci.* 4–27 (1980).

9. L. O. Chua, Y. Yao, and Q. Yang, Devil's staircase route to chaos in a non-linear circuit. *Int. J. Circuit Theory Appl.* **14** 315–329 (1986).

10. K. Kaneko, On the period-adding phenomena at the frequency locking in a one-dimensional mapping. *Prog. Theoret. Phys.* **68** 669–672 (1982).

11. B. van der Pol and J. van der Mark, Frequency demultiplication. *Nature* **120** 363–364 (1927).

12. T. Matsumoto, L. O. Chua, and M. Komuro, The double scroll. *IEEE Trans. Circuits and Systems* **CAS-32** 797–818 (1985).

13. M. J. Hasler, Electrical circuits with chaotic behavior. *Proc. IEEE* **75** 1009–1021 (1987).

14. M. Matsumoto, Chaos in electronic circuits. *Proc. IEEE* **75** 1033–1057 (1987).

15. M. I. Sobhy, N. A. Butcher, and A. A. A. Nasser, Chaotic behaviour of microwave circuits. Electronic Engineering Laboratories, University of Kent, Canterbury, Kent CT2 7NT, U.K. (Preprint for Nineteenth European Microwave Conference, 1989.)

16. P. J. Holmes and J. E. Marsden, Bifurcations to divergence and flutter in flow-induced oscillations: An infinite-dimensional analysis. *Automatica* **14** 367–384 (1978).

17. P. J. Holmes and J. E. Marsden, A partial differential equation with infinitely many periodic orbits: Chaotic oscillations of a forced beam. *Arch. Ration. Mech. Anal.* **76** 135–166 (1981).

18. F. C. Moon and P. J. Holmes, A magnetoelastic strange attractor. *J. Sound Vib.* **65** 273–296 (1979). A magnetoelastic strange attractor. *J. Sound Vib.* **69** 339 (1979).

19. F. C. Moon, Experiments on chaotic motions of a forced nonlinear oscillator: Strange attractors. *ASME J. Appl. Mech.* **47** 638–644 (1980).

20. M. P. Paidoussis, Flow-induced vibrations in nuclear reactors and heat exchangers: Practical experiences and state of knowledge. In *Practical Experiences with*

Flow-Induced Vibrations (E. Naudascher and D. Rockwell, eds.) pp. 1–81. Springer-Verlag, Berlin, 1980.

21. S. S. Chen and J. A. Jendrzejczyk, Characteristics of fluidelastic instability of tube rows in crossflow. *J. Pressure Vessel Tech.* **110** 1–5 (1988).

22. C. M. Cheng and P. M. Moretti, Lock-in phenomena on a single cylinder with forced transverse vibration. In *Flow-Induced Vibration* (M. K. Au-Yang and F. Hara, Eds.) PVP vol. 206, pp. 129–133. ASME, New York, 1991.

23. J. Maurer and A. Libchaber, Rayleigh–Bénard experiment in liquid helium: Frequency locking and the onset of turbulence. *J. Phys. Lett.* **40** L419–L423 (1979).

*24. A. Libchaber, C. Laroche, and S. Fauve, Period doubling cascade in mercury: A quantitative measurement. *J. Phys. Lett.* **43** L211–L216 (1982).

*25. M. Giglio, S. Musazzi, and V. Perine, Transition to chaotic behaviour via a reproducible sequence of period-doubling bifurcations. *Phys. Rev. Lett.* **47** 243–246 (1981).

*26. P. Bergé, M. Dubois, P. Manneville, and Y. Pomeau, Intermittency in Rayleigh–Bénard convection. *J. Phys. Lett.* **41** L341–L345 (1980).

27. J. Guckenheimer and P. Holmes, *Nonlinear Oscillations, Dynamical Systems, and Bifurcations of Vector Fields.* Springer, New York, 1983.

28. J. M. T. Thompson and H. B. Stewart, *Nonlinear Dynamics and Chaos*, Wiley, Chichester, U.K., 1986.

*29. E. N. Lorenz, Deterministic nonperiodic flow. *J. Atmos. Sci.* **20** 130–141 (1963).

30. C. Sparrow, *The Lorenz Equations: Bifurcations, Chaos and Strange Attractors*, Springer, New York, 1982.

31. P. Manneville and Y. Pomeau, Intermittency and the Lorenz model. *Phys. Lett. A.* **75** 1 (1979).

32. J. A. Yorke and E. D. Yorke, Chaotic behavior and fluid dynamics. In *Hydrodynamic Instabilities and the Transition to Turbulence* (H. L. Swinney and J. P. Gollub, eds.) pp. 77–95. Springer, New York, 1981.

33. J. Miles, Strange attractors in fluid dynamics. *Adv. in Appl. Mech.* **24** 189–214 (1984).

34. J. E. Hart, A new analysis of the closed loop thermosyphon. *Int. J. Heat Mass Transfer* **27** 125–136 (1984).

35. P. Welander, On the oscillatory instability of a differentially heated fluid loop. *J. Fluid Mech.* **29** 17–30 (1967).

36. W. V. R. Malkus, Non-periodic convection at high and low Prandt number. *Mem. Soc. R. Sci. Liège* **4** 125–128 (1972).

37. M. Gorman, P. J. Widmann, and K. A. Robbins, Chaotic flow regimes in a convection loop. *Phys. Rev. Lett.* **52** 2241–2244 (1984).

38. M. Gorman, P. J. Widmann, and K. A. Robbins, Nonlinear dynamics of a convection loop: A quantitative comparison of experiment with theory. *Physica D* **19** 255–267 (1986).

39. P. J. Widmann, M. Gorman, and K. A. Robbins, Nonlinear dynamics of a convection loop II: Chaos in laminar and turbulent flows. *Physica D* **36** 157–166 (1989).

40. J. H. Kim, Heat removal by natural circulation in light water reactors. In *Proceedings of the Fourth International Topical Meeting on Nuclear Reactor Thermal-Hydraulics* (U. Müller, K. Rehme, and K. Rust, eds.), vol. 1, pp. 430–447, G. Braun, Karlsruhe, Germany, 1989.

41. J. H. Kim and Y. A. Hassan (eds.), *Natural Circulation*. FED vol. 61, ASME, New York, 1987.

42. M. Sen, E. Ramos, and C. Trevino, The toroidal thermosyphon with known heat flux. *Int. J. Heat Mass Transfer* **28** 219–233 (1985).

43. J. E. Hart, Observations of complex oscillations in a closed thermosyphon. *J. Heat Transfer* **107** 833–839 (1985).

44. J. E. Hart, A model of flow in a closed-loop thermosyphon including the Soret effect. *J. Heat Transfer* **107** 840–849 (1985).

45. S. Paolucci and D. R. Chenoweth, Transition to chaos in a differentially heated vertical cavity. *J. Fluid Mech.* **201** 379–410 (1989).

46. Y. Wang and H. H. Bau, Period doubling and chaos in a thermal convection loop with time periodic wall temperature variation. In *Heat Transfer 1990* (G. Hetsroni, ed.), pp. 357–362. Hemisphere, New York, 1990.

47. K. T. Yang, Transitions and bifurcations in laminar buoyant flows in confined enclosures. *J. Heat Transfer* **110** 1191–1204 (1988).

48. H. B. Stewart, A two-fluid instability. *AIChE Symposium Series, No. 236,* **80** 187–193 (1984).

49. C. C. Hamakiotes and S. A. Berger, Period tripling in periodic flows through curved pipes. *Phys. Rev. Lett.* **62** 1270–1273 (1989).

50. R. A. Schlueter, A. G. Costi, J. E. Sekerke, and H. L. Forgey, Voltage stability and security assessment. EPRI Report EL-5917, Electric Power Research Institute, August 1988.

51. A. Kurita and T. Sakurai, The power system failure on July 23, 1987 in Tokyo. In *Proceedings of the 27th IEEE Conference on Decision and Control, Austin, Texas, December 1988*, pp. 2093–2097. IEEE, New York, 1988.

52. Y. Tamura, H. Mori, and S. Iwamoto, Relationships between voltage stability and multiple load flow solutions in electric power systems. *IEEE Trans. Power Apparatus and Systems* **PAS-102** 1115–1125 (1983).

53. H. G. Kwatny, A. K. Pasrija, and L. Y. Bahar, Static bifurcations in electric power networks: Loss of steady-state stability and voltage collapse. *IEEE Trans. Circuits and Systems* **CAS-33** 981–991 (1986).

54. I. Dobson and H. D. Chiang, Towards a theory of voltage collapse in electric power systems. *Systems and Control Lett.* **13** 253–262 (1989).

55. F. Wu, personal communications, 1991.

56. P. M. Anderson and A. A. Fouad, *Power System Control and Stability*, Iowa State University Press, Ames, IA, 1977.

57. N. Kopell and R. B. Washburn, Jr., Chaotic motions in the two-degree-of-freedom swing equations. *IEEE Trans. Circuits and Systems* **CAS-29** 738–746 (1982).

58. J. Stringer, Is a fluidized bed a chaotic dynamical system? In *Proceedings of the Tenth International Conference on Fluidized Bed Combustion: FBC Technology for Today* (A. M. Manaker, ed.), pp. 265–272. ASME, New York, 1989.

59. J. M. Ottino, C. W. Leong, H. Rising, and P. D. Swanson, Morphological structures produced by mixing in chaotic flows. *Nature* **333** 419–425 (1988).

60. Rizwan-uddin and J. J. Dorning, The nonlinear dynamics of two-phase flow in a simple heated loop. *Chem. Engr. Comm.* **87** 1–19 (1990).

61. M. Ledinegg, Instability of flow during natural and forced circulation. *Die Wärme* **61** 8 (1938).

62. R. T. Lahey, Jr., and F. J. Moody, *The Thermal-Hydraulics of a Boiling Water Nuclear Reactor.* American Nuclear Society, LaGrange Park, IL, 1977.

63. J.-L. Achard, D. A. Drew, and R. T. Lahey, Jr., The analysis of nonlinear density-wave oscillations in boiling channels. *J. Fluid Mech.* **155** 213 (1985).

64. Rizwan-uddin and J. J. Dorning, Some nonlinear dynamics of a heated channel. *Nucl. Eng. Des.* **93** 1–14 (1986).

65. Rizwan-uddin and J. J. Dorning, A chaotic attractor in a periodically forced two-phase flow system. *Nucl. Sci. Engr.* **100** 393–404 (1988).

66. A. Wolf, J. B. Swift, H. L. Swinney, and J. Vastano, Determining Lyapunov exponents from a time series. *Physica D* **16** 285–317 (1985).

67. P. Grassberger and I. Procaccia, Dimensions and entropies of strange attractors from a fluctuating dynamics approach. *Physica D* **13** 34–54 (1984).

68. Rizwan-uddin and J. J. Dorning, Chaotic dynamics of a triply-forced two-phase flow system. *Nucl. Sci. Engr.* **105** 123–135 (1990).

69. B. G. Bergdahl, F. Reisch, R. Oguma, J. Lorenzen, and F. Akerhielm, BWR stability investigation at Forsmark 1, *Ann. Nucl. Energy* **16** 509–520 (1989).

70. Rizwan-uddin and J. J. Dorning, Nonlinear dynamics of nuclear coupled density-wave oscillations. *Trans. Am. Nucl. Soc.* **59** 165 (1989). See also Rizwan-uddin and J. J. Dorning, Periodicity, quasi-periodicity and chaos in two-phase fluid flow. In *Forum on Chaotic Dynamics in Fluid Mechanics*, (K. Ghia, Ed.) ASME, New York, 1992. To appear.

71. Rizwan-uddin and J. J. Dorning, Nodal nonlinear BWR dynamics with nodal two-phase thermal hydraulics. *Trans. Am. Nucl. Soc.* **62** 294 (1990).

72. J. March-Leuba, D. G. Cacuci, and R. B. Perez, Universality and aperiodic behavior of nuclear reactors. *Nucl. Sci. Eng.* **86** 401 (1984).

73. J. March-Leuba, D. G. Cacuci, and R. B. Perez, Nonlinear dynamics of boiling water reactors: Part 1—Qualitative analysis. *Nucl. Sci. Eng.* **93** 111 (1986).

74. J. March-Leuba, D. G. Cacuci, and R. B. Perez, Nonlinear dynamics of boiling water reactors: Part 2: Quantitative analysis, *Nucl. Sci. Eng.* **93** 124 (1986).

75. E. B. Lee and L. Markus, *Foundations of Optimal Control Theory*, Wiley, New York, 1967.

76. I. Lasiecka and R. Triggiani, Exact controllability of the Euler–Bernoulli equation with controls in the Dirichlet and Neumann boundary conditions: A nonconservative case. *SIAM J. Control and Optim.* **27** 330–373 (1989).

77. E. A. Jackson and A. Hübler, Periodic entrainment of chaotic logistic map dynamics. *Physica D* **44** 407–420 (1990).

78. C. Hayashi, *Nonlinear Oscillations in Physical Systems*, Princeton University Press, 1985. (Previously published by McGraw-Hill, New York, 1964.)

79. B. Saltzman, Finite amplitude free convection as an initial value problem—I. *J. Atmos. Sci.* **19** 329–341 (1962).

80. J. A. Yorke, *Dynamics* (*Version: March 1990*). Institute for Physical Science and Technology, University of Maryland, College Park, MD, 1990.

81. W. J. Decker, Personal communication (University of Virginia, Charlottesville, VA), 1990.

82. J. Dorning, Controlling chaos in the presence of noise. *Trans. Am. Nucl. Soc.* **64** 295 (1991).

83. J. Dorning, Nonlinear dynamics and chaos in heat transfer and fluid flow. *AIChE Symposium Series, No. 269*, **85** 13–29 (1989).

84. Rizwan-uddin, Strange attractors in forced two-phase flow heated channels. *AIChE Symposium Series, No. 269*, **85** 249–255 (1989).

85. J. Dorning, An introduction to chaotic dynamics in two-phase flow. *AIChE Symposium Series, No. 269*, **85** 241–248 (1989).

CHAPTER 2

GLOBAL INTEGRITY IN ENGINEERING DYNAMICS — METHODS AND APPLICATIONS

F. A. McROBIE and J. M. T. THOMPSON
Centre for Nonlinear Dynamics
Department of Civil Engineering
University College London
London WC1E 6BT
United Kingdom

Rather than focusing on the long-term (possibly chaotic) steady-state solutions of engineering dynamic systems, a method is discussed aimed at quantifying the overall safety of the transient motions that arise from short-term forcing excitations. Topological methods of geometric nonlinear dynamics are applied to the phase space of simple driven oscillators that exhibit softening. An organization of the bifurcational behavior of steady-state solutions is perceived and related to the evolution of a homoclinic tangle. An associated fractal basin boundary circumscribing the global basin of all safe long-term motions underlies the dynamics in parameter regions that are safe under short-term excitation. Proposed applications in ship stability analysis and structural engineering dynamics are discussed.

2.1. INTRODUCTION

In engineering dynamics, a fundamental enterprise involves establishing whether or not the motions of the system under consideration will become unacceptably large—in particular, so large that the system fails. An application of "chaos theory" or geometric nonlinear dynamics is discussed for

Applied Chaos, Edited by Jong Hyun Kim and John Stringer.
ISBN 0-471-54453-1 © 1992 John Wiley & Sons, Inc.

engineering systems that exhibit softening, where system failure can be formulated in terms of escape from a potential well. Dissipative single-degree-of-freedom driven oscillators are considered, with potential functions permitting escape in one direction only.

Parameter regimes where chaotic solutions exist have been identified for many periodically forced nonlinear oscillators. However, such regimes are often restricted to extremely narrow parameters ranges, particularly in softening systems. In order for a physical system to achieve a steady-state chaotic response in such circumstances, extremely long-term noise-free periodic forcing is required and forcing parameters must vary only very gradually and very smoothly. There is thus an understandable skepticism among practicing engineers about the applicability of chaos theory to real engineering systems where inevitable noise and nonsmooth, nongradual parameter variation will prevent the development of pure chaotic motions.

The application of chaos theory discussed here, however, rather than focusing on the delicate long-term or steady-state behavior after the application of possibly hundreds of forcing cycles, provides information about the system safety under forcing histories of only a few cycles. Some of the techniques of this approach, referred to as the *fractal basin boundary method*, are covered briefly in this chapter. A more comprehensive and formal presentation is in preparation.

2.2. THE DYNAMICS OF SOFTENING SYSTEMS

Under periodic forcing, the analysis of the possible asymptotic behaviors (stable steady-state oscillations, or *attractors*) of a nonlinear oscillator is most readily achieved using Poincaré sections through the full phase space (see reference [18]). In engineering terms, the displacement and velocity of the system are stroboscopically sampled at successive periods of the applied forcing. Starting from a range of initial conditions, the possible long-term behaviors are elicited, together with their corresponding *basins of attraction*. In mathematical terms, the system motion is represented by a parametrized area-contracting diffeomorphism of the plane.

The nature of significant attractors and their basins can be established by a variety of numerical techniques (e.g., reference [2]). At a particular value of forcing, observation of attractor amplitudes at a range of different frequencies leads to the frequency-response diagram (Figure 2.1).

For small values of forcing, the behavior is substantially linear, giving the familiar single resonance peak at a ratio of forcing to natural frequency close to unity (Figure 2.1a). For larger forcing amplitudes the resonance peaks of softening systems lean toward the left (lower forcing frequency ranges) creating a zone of resonant hysteresis where there exist two possible long-term stable motions. Under gradual increase of frequency, an evolving system can follow the lower curve (Figure 2.1b) to its termination (at a *fold* or *saddle-*

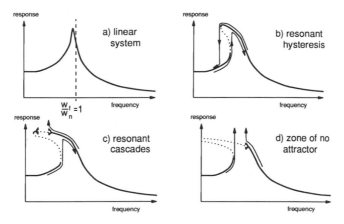

Figure 2.1. Frequency-response curves of a softening oscillator showing nonlinear behavior. (Arrows indicate likely paths of evolving systems.)

node bifurcation), where it undergoes a *catastrophe*, jumping to the resonant limb before further evolution. On subsequent reduction of forcing frequency, the system returns by a different route (see Figure 2.1*b*).

At higher forcing values the resonant attractor undergoes a period-doubling cascade leading to an apparently chaotic attractor that exists over a narrow frequency band (Figure 2.1*c*) [14]. This chaotic solution terminates at a *blue sky catastrophe* or *boundary crisis*, and any system following the resonant attractor across this catastrophe is liable to escape [13].

At even higher forcing values there is a frequency range where there exists no robust stable solution of the system (Figure 2.1*d*). (Strictly, there are some short-lived, high period attractors here of little practical importance.) Any evolving system that enters this regime must inevitably escape because there is no robust attractor available on which to stabilize.

Combining all frequency-response curves at all forcing amplitudes leads to the *response surface* (see Figure 2.2), on which the paths of systems may be followed as either forcing or frequency are gradually varied. The locus of the catastrophes and other bifurcations of the response surface can be projected onto the parameter plane to give the bifurcation diagram. In the nonlinear regime, there can be many (indeed, infinitely many) coexistent stable solutions, and their location and bifurcational behavior can be plotted onto the bifurcation diagram. Figure 2.3 shows the bifurcation diagram for two nonlinear oscillators. Figure 2.3*a* corresponds to a "generic" escape potential [14, 16]; Figure 2.3*b* shows the bifurcation behavior in a mathematically cumbersome but realistic potential based on the softening characteristics of annular reinforced concrete sections under combined flexure and compressive axial load effects. Despite the mathematical dissimilarities in the stiffness functions (Figures 2.3*c* and 2.3*d*), the potential functions possess the same

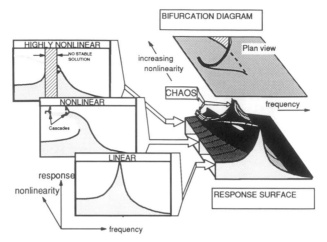

Figure 2.2. Schematic sketch of response surface and bifurcation diagram for a softening oscillator [4].

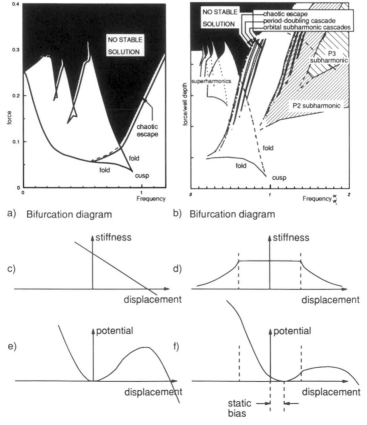

Figure 2.3. Bifurcation diagrams, stiffness, and potential functions for two softening oscillators.

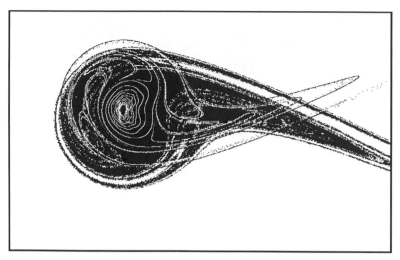

Figure 2.4. A homoclinic tangle showing basins of coexistent attractors. The period-2 subharmonic orbiting the central period-1 attractor is approaching chaotic escape associated with an incipient change in the period-2 Birkhoff signature of the manifolds of the hilltop saddle.

broad features. Broad similarities in the resulting bifurcation diagrams are evident. This is in line with and reinforces the similarities already established between the escape scenarios of the single- and twin-well Duffing oscillators [3, 19].

Around the local potential maximum, there usually exists an unstable periodic solution referred to as the *hilltop saddle* cycle. Consideration of the *stable and unstable manifolds* of this solution by topological methods is one of the approaches of geometric nonlinear dynamics that will be pursued here. The stable manifold comprises those orbits that converge to the saddle, and the unstable manifold can be described best as comprising those orbits that converge to the saddle in reverse time. On the Poincaré section, the hilltop saddle cycle is represented by a single point, and the stable and unstable manifolds are curves intersecting at that point.

Above certain forcing values these manifolds intersect at points other than at the saddle itself, and each manifold then becomes geometrically complicated, such that under any magnification an infinitely striated nature of nested "stripes-within-stripes" is revealed. The overlap of these two manifold structures is referred to as a *homoclinic tangle* (see Figure 2.4). The closure of the stable manifold forms the *global basin boundary*. The stable manifold partitions the Poincaré plane $(x(\phi_0), \dot{x}(\phi_0))$ at any forcing phase ϕ_0 of sampling into two sets $\mathscr{B}$ "inside" and $\mathscr{E}$ "outside." Those initial conditions $\mathscr{E}$ outside necessarily escape and, likewise, those points $\mathscr{B}$ inside cannot escape and together with the stable manifold form the *global basin*. It immediately follows from the area-contracting property of the map that if there exists any attractor, then the global basin has infinite area.

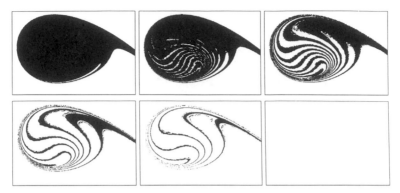

Figure 2.5. Basin erosion: Incursion of the fractal boundary into the central regions of the potential well under increasing forcing amplitude.

The closure of the unstable manifold makes no such partition of the plane. Loosely speaking, the unstable manifold bites out its own inside and is the infinitely long and smooth perimeter around a set of no area. This trapped set of zero measure contains all attractors, all other saddles and all other unstable manifolds of all other saddles. In the zone of no attractor, the stable manifolds likewise bite out their own inside. There is no attractor and yet because the stable manifolds still exist into this zone, they must then trap no area. The evolution of the manifolds from the occurrence of the first homoclinic intersection to the zone of no attractor is referred to as *basin erosion* (see Figure 2.5). At all stages in this erosion process prior to the final escape, the global basin has infinite area, this dropping instantaneously to zero as the zone of no attractor is entered. Various measures of the degree of erosion (or rather its complement, the basin integrity) are discussed later.

As the homoclinic tangle evolves, there are infinite sequences of inner tangencies (referred to as *Birkhoff signature changes*) between the stable and unstable manifolds. Associated with a subset of these tangencies are infinite cascades of subharmonic saddle-node bifurcations, creating attractors that orbit the peripheral regions of the global basin before undergoing rapid period-doubling cascades to crises. Research at University College London, using cell-to-cell mapping techniques, has located a large number of such orbital subharmonics in order to understand their involvement in the full dynamical erosion process [3]. There is a necessary partial order on subsets of the Birkhoff signatures: Some cannot be reached without passing through others.

Plotting the locus of the major signature changes of the hilltop manifolds on the bifurcation diagram, a strong correlation between a signature change of some periodicity and the escape of an attractor of equivalent period is observed [7].

The escape of the main resonant attractor is observed to be strongly associated with the first change of the period-1 Birkhoff signature of the hilltop manifolds. The coexistence of a nonresonant attractor (resonant hysteresis) appears to be strongly dependent upon the existence of an inner eyelike region within the period-2 (and all higher-period) signatures.

Chaotic escape of the first resonant superharmonic is likewise strongly related to a signature change in the inner eye, with the coexistence of nonresonant attractors being likewise dependent upon the existence of an *inner* inner eye. The sequence of superharmonic zones of resonant hysteresis to the left of the main resonance (Figures 2.3*a* and 2.3*b*) is thus related to the existence of nested sequences of inner eyes within inner eyes. This general superharmonic activity appears to be a feature of an evolving spiral horseshoe configuration, which, having three stripes near the primary resonance, accrues two extra internal stripes at each superharmonic resonance. Cascades of subharmonics orbiting the superharmonics are likewise implied by the inner signature changes, and many have been observed numerically.

The association of homoclinic tangencies with the escape of attractors and cascades of both saddle-node and flip bifurcations is strongly related to the work of Newhouse [8], Yorke and Alligood [20], and the mathematicians Gavrilov and Silnikov [1], among others. Loosely speaking, the passage through a tangency creates *Smale horseshoes*, and the creation process implies associated sequences of bifurcations.

Immediately following a tangency, the manifold structures are amenable to analysis by *symbolic dynamics*. Using elementary binary logic on bi-infinite strings of two symbols (0 and 1, say) containing a decimal point, for example,

$$\dots 01110100101100110010 1.000111100011101010010 \dots$$

the existence and location of important saddle solutions can be rapidly estimated relative to the invariant manifolds. Corresponding to any such sequence there exists a point on the Poincaré plane, and shifting the decimal point one place to the left obtains the image of that point under one iteration of the map. The trajectories of saddles and associated homoclinic and heteroclinic orbits can then be deduced readily.

Returning to the preordering of subsets of signature changes, it follows that many such saddles deducible by the symbol shift were in existence at earlier signature changes. Such considerations of allowable sequences and necessary orderings and precedences provide detailed insight into the organizational mechanisms of the full dynamical evolution. Returning to the wider perspective, the system evolution along some path leading to the zone of no attractor (e.g., path A–G of Figure 2.9 in Section 2.5) can be broadly summarized by the following topological relations.

At all parameter values:

$$\Omega \subset \overline{W}_2^u(\mathbf{D}^1) \tag{2.1}$$

$$\forall \, \Omega_r \quad \overline{W}^u(\Omega_r) \subset \overline{W}_2^u(\mathbf{D}^1) \tag{2.2}$$

$$\forall \, \Omega_r \quad \overline{W}^s(\mathbf{D}^1) \subset \overline{W}_j^s(\Omega_r) \tag{2.3}$$

In the zone of no attractor:

$$\Omega \subset \overline{W}_2^s(\mathbf{D}^1) \tag{2.4}$$

$$\forall \, \Omega_r \quad \overline{W}^s(\Omega_r) \subset \overline{W}_2^s(\mathbf{D}^1) \tag{2.5}$$

$$\forall \, \Omega_r \quad \overline{W}^u(\mathbf{D}^1) \subset \overline{W}_j^u(\Omega_r) \tag{2.6}$$

where Ω is the nonwandering set, Ω_r is any invariant component of $\{\Omega - \mathbf{D}^1\}$, W^s and W^u are, respectively, the stable and unstable sets of any set, $\mathbf{D}^1$ is the hilltop saddle, and the subscript 2 or j denotes particular or arbitrary branches of manifolds as appropriate (see [6] for proofs).

Further results readily follow from these strikingly symmetric relations. Broadly speaking, the erosion process involves filling up the last three relations. The tangencies create saddle-node bifurcations, the attractors period-double to escape and the remaining saddles, the residue of these cascades, are additional elements of the nonwandering set that become contained in the incursive global basin boundary. On entering the zone of no attractor, relation (2.4) holds that all nonwandering elements are so contained and it follows that the closure of all stable manifolds are identical and the closures of all unstable manifolds likewise. In the zone of no attractor, the nonwandering set is thus indecomposable. In summary there is thus a strong organization to the bifurcational behavior and the mechanism of basic erosion that can be discerned by topological and geometric analysis of all invariant sets of the plane.

2.3. SOME ENGINEERING CONSIDERATIONS

In order for an engineering system to adhere to the path of an attractor across the response surface, there are the underlying assumptions that the forcing has a long-term consistent periodic nature and that variation of forcing parameters is gradual and smooth. Such conditions may be well approximated in many mechanical or rotodynamic systems, for structures supporting turbomachinery, or in, say, laser systems. In many engineering applications, however, the presence of noise prevents any such rigorous path-following. A particular naval architecture scenario involves the analysis of the safety against capsize of a craft subjected to, say, a protracted sequence of small waves followed by a short sequence of abnormally large waves. Capsizing is necessarily a transient phenomenon and the relevance of any concept of a steady-state motion is questionable in such conditions. There is insufficient time for the vessel to converge to some stable periodic motion, let alone undergo a detailed period-doubling cascade. The fundamental issue is whether the vessel will capsize. How many of the transients lead to escape? In such applications it is arguable that attractor behavior is of less significance than the global integrity measure, which, defined in the following section, in some sense quantifies that proportion of starts that do not escape. In Figure 2.6*a*, the asymptotic behavior of a craft under a wave force F of period T is shown, and the plane of initial conditions exhibits the familiar fractally eroded basin. The associated transient behavior is summarized in Fig. 2.6*b*. Assuming the vessel is initially stable (but not necessarily at rest), those sets of initial conditions are plotted from which starts led to eventual escape after the incidence of from one to five large waves, all of force F and period T. Over the region of interest, the fractally bounded basin is a lower bound on the regions safe to transient capsize.

The parameter locations of sharp decrease in global integrity, such as the reported *Dover Cliff effect*, represent regions where capsize is more likely to occur [15, 9].

2.4. LOBE DYNAMICS, BIRKHOFF SIGNATURES, AND INTEGRITIES

Various *integrity measures* have been proposed to quantify the intuitive engineering implication that an attractor well-distanced from the boundary in a substantially uneroded basin is safe to realistic perturbation [11].

An integrity is defined here using *lobe dynamics*, which is a systematic method of labeling segments of the overlapping manifold structures and monitoring the evolution of sets enclosed by these trellises.

Given the homoclinically tangled manifolds of the hilltop saddle $\mathbf{D}^1$, it is always possible to extract from the set of all homoclinic points a primary intersection point (or *pip*) (see, e.g., reference [10]), such that the forward-directed stable manifold segment $S(\mathbf{x}_0, \mathbf{D}^1)$ connecting the pip ($\mathbf{x}_0$, say) to the saddle $\mathbf{D}^1$ does not cross the unstable manifold segment $U(\mathbf{D}^1, \mathbf{x}_0)$ connecting the pip back to the saddle (see Figure 2.7). The closed set bounded by this closed loop is referred to as an *eye* denoted Ξ_0.

For the tangles of the hilltop manifolds considered here, there are two distinct forms of pip. This results from the shape of the potential well, which is such that immediately following the initial homoclinic tangency there are only two distinct primary homoclinic orbits that traverse the well once. Those pips where slight further extensions of the manifold segments beyond the pip are not part of the associated eye are referred to as *main pips*. All images and preimages of a pip are also pips, each with a corresponding eye. Connecting any main pip $\mathbf{x}_0$ to its image $\mathbf{x}_1$ by fundamental domains of the invariant manifolds, then these domains necessarily intersect at a pip $\mathbf{x}_{1/2}$ of the other form (referred to as an *intermediate pip*) and possibly at many other nonprimary intersections. The configuration of the *lobes* enclosed by the manifold segments connecting a main pip to its image is referred to here as the *period-1 Birkhoff signature* of the tangle. The lobe configuration between a main pip and its image after n forcing cycles is similarly referred to as the *period-n Birkhoff signature* (see Figure 2.8).

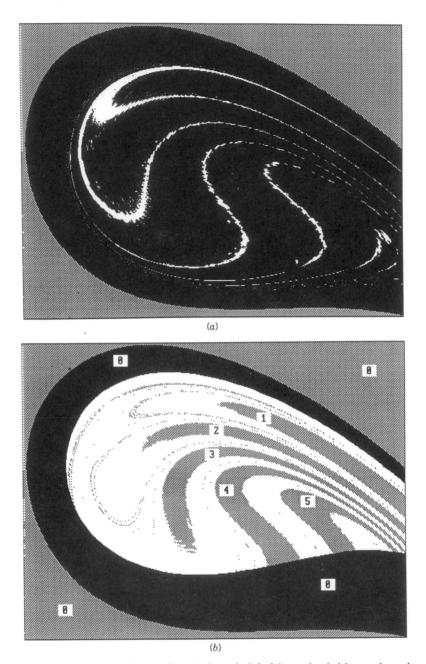

Figure 2.6. Asymptotic and transient basins of global integrity (white regions do not escape): (*a*) Full asymptotic behavior under long-term forcing; (*b*) Transient integrity after application of between one and five cycles.

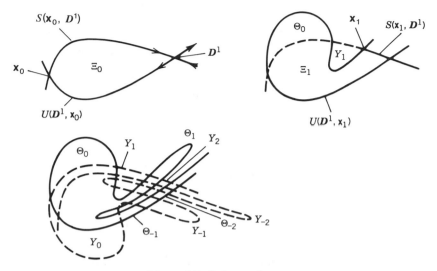

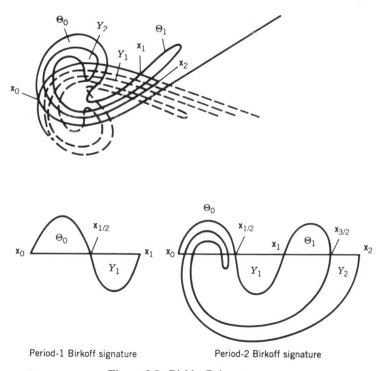

Figure 2.7. Lobes and eyes.

Period-1 Birkoff signature   Period-2 Birkoff signature

Figure 2.8. Birkhoff signatures.

The lobe circumscribed by the manifold segments from x_0 to $x_{1/2}$ is denoted Θ_0 and is said to be of outgoing variety, because all points inside the lobe must necessarily escape. Conversely, the lobe circumscribed by the manifold segments from $x_{1/2}$ to x_0 is denoted Y_0 and is said to be of ingoing variety, because at least some points inside the lobe do not escape.

For the tangles of the hilltop saddle in the one-sided escape equations under consideration, these lobes Θ_0 and Y_0 are outside the corresponding eye Ξ_0. Furthermore, no outgoing lobe intersects any preceding ingoing lobe. It follows that

1. no outgoing lobe can contain any stable set of any nonwandering element and, likewise, no ingoing lobe can contain any unstable set of any nonwandering element;
2. no attractors or saddles (indeed, no elements of the nonwandering set) are contained in any lobe;
3. all attractors and all saddles (indeed, all elements of the nonwandering set) are contained in all eyes.

Given even a limited description of the tangled invariant manifolds of the hilltop saddle, the possible regions of the plane where attractors or saddles can exist is thus considerably restricted. An association between the invariant manifold structures and the location of attractors is thus apparent. The more traditional approaches of engineering mathematics to nonlinear equations often aim directly for an important known *stable* solution and then examine its bifurcation behavior under parameter variation. The approach here is to find an important known *unstable* solution, the hilltop saddle, and then extract information about the location and behavior of all other possible stable (and unstable) solutions by considering its invariant manifolds.

Taking any eye and successively adding and subtracting future outgoing and ingoing lobes, respectively, it can be seen how the unstable manifold of the hilltop saddle "bites out its own inside," as previously stated, leaving a set of no area inside. Formally, the closure of the unstable manifold of the hilltop saddle has zero two-dimensional Lebesgue measure (i.e., no area) and this closed set of no area contains all elements and all unstable sets of all components of the nonwandering set [i.e., relations (2.1) and (2.2)].

Assuming the area-contracting properties of the map are independent of velocity and displacement such that an area $|A|$ of the Poincaré plane has an area $k|A|$ after one forcing cycle, it readily follows that

$$(1 - k)|\Xi_0| = |Y_1| - |\Theta_0|$$

It further follows that an ingoing lobe is always larger than the preceding outgoing lobe. As previously stated, if there are attractors present, then the area of the global basin is necessarily infinite. However, a measure of the degree of erosion (or its complement, the integrity) naturally follows from the lobe configurations.

We define the *global integrity* ρ_G as

$$\rho_G = 1 - e_G = 1 - \sum_{i=-1}^{-\infty} \frac{|\Xi_0 \cap \Theta_i|}{|\Xi_0|}$$

Loosely speaking, e_G measures the eroded fraction of any eye, that is, that proportion of the eye that is contained in some outgoing lobe Θ_i and thus destined to escape.

Additional indicators of erosion may be defined likewise:

Lobe Integrity ρ_L. That fraction of any ingoing lobe that does not escape, related to global integrity via

$$\rho_G(1-k)|\Xi_0| = \rho_L|Y_1|$$

Mean Lobe Number N_G. For that area of any eye that escapes, the average (with respect to area) of the number of cycles required to leave the eye

Integrity Diameter. A metrical rather than topological concept, but loosely the largest-diameter set that can be constructed within the global basin that does not intersect any outgoing lobe. Various definitions of diameter are possible.

The integrities and the mean lobe number so defined possess the beneficial property of being independent of the forcing phase of the Poincaré sampling.

Analogous to the response surface (Figure 2.2), a global integrity surface can be constructed above the parameter plane. Contours of global integrity likewise can be superimposed on the parameter plane for comparison with the attractor bifurcation diagram. It is argued in this chapter and elsewhere that this integrity diagram, rather than the bifurcation diagram alone, contains information pertinent to a wide class of engineering systems subject to noise and nonsmooth parameter variation. One aim of present research is to identify the mathematical methods with which to analyze the "slope" of the integrity surface and to describe its continuity. By relating integrity to perceived organization underlying the dynamics, the intention is that any significant discontinuities or steep decreases in integrity may be better understood and thus anticipated.

2.5. TANGLED SADDLE-NODE BIFURCATIONS

As stated before, at small values of forcing, systems on the lower-limb attractor in the main zone of resonant hysteresis (Figure 2.1*b*) will jump to

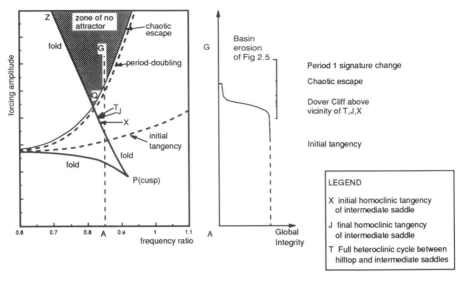

Figure 2.9. Schematic bifurcation diagram and global integrity near optimal escape for a softening oscillator.

the resonant attractor path on passing through the saddle-node bifurcation (arc PT of Figure 2.9). Where the bifurcation is adjacent to the zone of no attractor it follows that any such evolving system must necessarily escape on crossing the bifurcation (arc QZ of Figure 2.9).

For intermediate values (arc TQ), the outcome of such a jump is indeterminate (as shown by Thompson and Soliman [17]), even under mathematically ideal conditions of gradual and smooth parameter variation. The system may jump to resonance or escape. This indeterminacy follows from the fact that the ω-limit set of the *center-unstable manifold* of the saddle node contains more than one attracting element, namely, the resonant attractor and the "attractor at infinity." Other outcomes are necessarily possible if there are other attractors in this set.

From the geometry of the manifolds, it immediately follows that any such saddle-node bifurcation permitting jumps to escape must be homoclinic. In such a case the invariant manifolds of the saddle node constitute an interesting example of a "nonhyperbolic" homoclinic tangle, involving transverse intersection of the (linelike) center-unstable manifold with the nonunique center-stable manifolds whose union constitutes the (area-like) residual basin of the saddle node. A saddle node possessing the property of homoclinicity is referred to as a *tangled saddle node* (or TSN) [17].

The point T on the bifurcation arc corresponds to the lowest possible point at which jumps to escape occur; thus it coincides with the first heteroclinic intersection between the center-unstable manifold of the TSN and the stable manifolds of the hilltop saddle. Below this point T, the saddle node may still possess the property of homoclinicity. There is necessarily a first homoclinic intersection (at the point X) of the center-unstable manifold with a single outermost center-stable manifold. Moving up the bifurcation arc, progressively more of the center-stable manifolds are intersected until at point J all center-stable manifolds are intersected. Necessarily, not before this final homoclinic tangency can jumps to escape occur. The point J thus necessarily precedes or is coincident with the point T.

The engineering implications of such indeterminant jumps are significant in their own right. However, a different consequence of the presence of such a bifurcation within the global basin is the strong conditions that it places upon the configurations of the invariant manifolds of the hilltop saddle, upon the allowable sequences of Birkhoff signature changes that can occur in its vicinity, and the constraints that these impose upon the process of global basin erosion.

At the final homoclinic tangency of the TSN manifolds (point J), lobe-dynamics considerations reveal the necessary presence of a substantial safe global basin. At slightly different parameter values (to the right of the bifurcation arc immediately above the point T), further invariant manifold analysis reveals that the global basin boundary, previously restricted to the peripheral regions of the tangle, now penetrates along paths bounded by the former outer center-stable manifolds of the TSN, striating the central regions of the potential well. Indeed, this incursion is so pervasive that the diameter of the largest sphere that can be constructed in the parameterized vicinity of the residual basin of the TSN that does not contain some wispy outgoing lobe permitting eventual escape shrinks arbitrarily close to zero alongside the bifurcation arc above T. However, starts in this vicinity, although necessarily unsafe, may require the application of arbitrarily large numbers of forcing cycles before escaping over the hilltop. This is a consequence of both the retardation of orbits that occurs beyond any saddle node and the reinjection properties implied by the homoclinicity of the TSN. Loosely speaking, there is a slow channel of escape beginning along the lines of the former center-stable manifolds and exiting through the lines of the former center-unstable manifolds of the TSN.

By such analysis, the global integrity behavior for any system evolving under increased forcing in this parameter regime may be anticipated to possess substantial integrity until the region adjacent to point T is encountered. Beyond this point, fractal incursion of the boundary into the more central regions will occur. Much information about the shape of such incursion follows from consideration of the configuration of the invariant manifolds of the TSN at point T. Combination of this information with the associated knowledge of the organization of the hilltop manifolds near a TSN and the sequences of Birkhoff signatures that must necessarily occur between the point T and the period-1 signature change in the zone of no attractor is leading to some understanding of the location of the Dover Cliff.

2.6. SUMMARY AND APPLICATIONS

Adopting topological methods, strong organization can be discerned in the possible behavior of nonlinear engineering oscillators. These organizations persist over wide classes of oscillator, depending on broad features and general shapes of potential functions and so on, rather than requiring precise evaluation of physical coefficients or exact algebraic formulation of constitutive relations. Such robustness is essential in engineering applications where a priori knowledge of parameter values is necessarily imprecise and incomplete.

Although many of the detailed internal fractal structures and intricate bifurcational cascades predicted by such analysis may be destroyed in the presence of inevitable noise or nongradual variation for forcing parameters, it is contended that consideration of global integrities can provide meaningful indications of regimes where system failure is of increased likelihood because the integrity of the underlying fractal structures provides lower bounds on regions safe to transient excitation.

Because most engineering systems do not undergo detailed period-doubling cascades to failure, the applicability of the methods of geometric nonlinear dynamics to real engineering systems may not be apparent at first to practicing engineers. The often-unappreciated advantage of these techniques is twofold: Not only do they utilize broad rather than detailed system features, but they can provide information about *transient* behavior and integrity unavailable by many alternative approaches. Shifting the analytical emphasis slightly away from the bifurcational behavior of attractors toward global integrities, the applicability of the methods is extended to a wider set of engineering systems subject to realistic forcing histories.

In summary, the particular approach we shall refer to as the fractal basin boundary method comprises the following:

1. Assuming short-term excitation, directly construct the transient integrity diagram by straightforward numerical time-integration [15].
2. Assuming long-term noise-free periodic forcing, apply topological methods to all invariant sets of the Poincaré plane, to identify organizations of the attractor bifurcation, boundary incursion, and global integrity behaviors. Identify parameter regimes of severe boundary incursion and infer integrity information regarding short-term transient behavior in order to verify and supplement the purely numerical results.

Good engineering involves an understanding of system behavior, rather than an unthinking reliance on numerical computer output. Although the transient integrity diagram involves only simple time-integration of a differential equation and readily provides the information necessary to establish the transient integrity of the system, the complex possibilities of nonlinear behavior may be unfamiliar to linear dynamicists, and the intuitive engineering feel for linear dynamic response is harder to replicate in nonlinear systems. It is argued here, however, that the application of topological methods can provide significant insight into the global behavior and the short-term transient possibilities and can lead at least some of the way toward the necessary understanding of the motions and safety of nonlinear systems. Current research is directed in the fields of ship stability, offshore structural dynamics, and more-general structural engineering applications.

In ship stability studies, analysis of escape from potential wells resulting from the GZ curves (or righting characteristics) of real vessels has led to the proposal [15] of an *index of capsizability* based upon the first incursion of the fractal boundary into central regions of the formerly safe basin. One particular case study involves the *Gaul*, a large-stern trawler from Hull, England, which disappeared in heavy seas off the coast of Norway in February 1974 with the loss of 36 lives. The official inquiry into the disaster indicated that the design of the vessel conformed to all required stability design criteria against capsize, leading various naval architects to question whether those criteria (based on static considerations) were always conservative given the dynamic and transient nature of the capsize. In other marine and offshore structural applications, research has been conducted into irregular behavioral effects that result from nonlinearities in the stiffness characteristics of various mooring systems. Mooring cables exhibit displacement-dependent (or geometric) nonlinear stiffness. Vessels tied alongside rigid wharves experience restoring forces strongly bilinear or impacting in nature. A study is being conducted regarding a particular offshore construction procedure where various unanticipated dynamic phenomena have been observed while making the connection of a rigid-leg platform to a fixed seabed foundation. In such studies the concept of escape from a potential well is less applicable and the considerations are focused more toward the long-term dynamic amplitudes rather than ultimate integrity. Nonescape phenomena such as jumps to resonance, subharmonic and superharmonic resonances, and, in the impacting systems, "chattering" instabilities can lead to fatigue, fracture, or generally unacceptable vibrational amplitudes. Although ultimate global integrity is not at issue, such systems generally contain multiplicities of coexisting steady-state solutions, and analysis of the relative dominance of competing attractors involves internal basin boundaries that may be fractal.

In general structural dynamics, a particular system under study arose from the design of a reinforced concrete chimney at a power station, where under the action of mean along-wind forces the flexural stiffness of the tower enters a dramatically nonlinear regime. The standard design methods involve comprehensive random dynamic analysis of the excitation resulting from the turbulence of the wind, but have an underlying unproven assumption that flexural restoring forces may be adequately modeled by linear elastic behavior. The concept of escape is applicable in such an application, and, although there are many complications resulting from the highly stochastic nature of the wind, the simultaneity of crosswind responses due to vortex-shedding and the many-degree-of-freedom nature of the tower, the possible application of the techniques of the fractal basin boundary method is being investigated.

It is difficult to foresee any future introduction of invariant manifold-following algorithms as standard packages on the PCs of practicing engineers. However, it is envisaged that not only can geometric and topological methods help to explain unusual and unexpected dynamical phenomena encountered in existing systems, but that by consideration of the integrity of *transient* motions rather than the amplitude of long-term motions, it is suggested that the techniques of the fractal basin boundary method may prove useful in ultimate limit state design by identifying any regimes where the linearity assumptions underlying existing dynamic design methodologies may not always lead to prudent engineering conservatism.

REFERENCES

1. N. K. Gavrilov and L. P. Silnikov, On three-dimensional dynamical systems close to systems with a structurally unstable homoclinic curve. *Math. USSR Sb.* **88** 467–485 (1972); **90** 139–156 (1973).

2. C. S. Hsu, *Cell-to-Cell Mapping*, vol. AMS 64. Springer, New York, 1987.

3. A. N. Lansbury and J. M. T. Thompson, Incursive fractals: A robust mechanism of basin erosion preceding the optimal escape from a potential well. *Phys. Lett.* **150** 8, 9 355–361 (1990).

4. F. A. McRobie and J. M. T. Thompson, Chaos, catastrophes and engineering. *New Scientist* **126** (1720) 41–46 (9 June 1990).

5. F. A. McRobie and J. M. T. Thompson, Lobe dynamics and the escape from a potential well. *Proc. Roy. Soc. London Soc. A* **435** 659–672 (1991).

6. F. A. McRobie and J. M. T. Thompson, Invariant sets of planar diffeomorphisms in nonlinear vibrations. *Proc. Roy. Soc. London A* **436**, 427–448 (1992).

7. F. A. McRobie, Birkhoff signature change: A criterion for the instability of chaotic resonance. *Philos. Trans. Roy. Soc. London Ser. A* **338** 1651, 557–568 (1992).

8. S.E. Newhouse, Lectures on dynamical systems. In *Dynamical Systems, CIME Lectures, Bressanone, Italy, June 1978. Progress in Mathematics*, vol. 8, pp. 1–114. Birkhäuser, Boston, 1980.

9. R. C. T. Rainey, J. M. T. Thompson, G. W. Tam, and P. G. Noble, The transient capsize diagram—a route to soundly-based new stability regulations. Presented at the Fourth International Conference on Stability of Ships and Ocean Vehicles, Naples, September 24–28, 1990 (Department of Naval Engineering, University of Naples).

10. V. Rom-Kedar and S. Wiggins, Transport in two-dimensional maps. *Arch. Ration. Mech. and Anal.* **109** (3) 239–298 (1990).

11. M. S. Soliman and J. M. T. Thompson, Integrity measures quantifying the erosion of smooth and fractal basins of attraction. *J. Sound and Vibration* **135** 453–475 (1989).

12. H. B. Stewart, A chaotic saddle catastrophe in forced oscillators. In *Proceedings of the Conference on Qualitative Methods for the Analysis of Nonlinear Dynamics, Henniker, New Hampshire, June 1986* (F. Salam and M. Levi, eds.). SIAM, Philadelphia, 1987.

13. H. B. Stewart and Y. Ueda, Catastrophes with indeterminate outcome. *Proc. Roy. Soc. London A* **432** 113–123 (1991).

14. J. M. T. Thompson, Chaotic phenomena triggering the escape from a potential well, *Proc. Roy. Soc. London A* **421** 195–225 (1989).

15. J. M. T. Thompson, R. C. T. Rainey, and M. S. Soliman, Ship stability criteria based on chaotic transients from incursive fractals. *Philos. Trans. Roy. Soc. London A* **332** 149–167 (1990).

16. J. M. T. Thompson and M. S. Soliman, Fractal control boundaries of driven oscillators and their relevance to safe engineering design. *Proc. Roy. Soc. London A* **428** 1–13 (1990).

17. J. M. T. Thompson and M. S. Soliman, Indeterminate jumps to resonance from a tangled saddle-node bifurcation. *Proc. Roy. Soc. London Ser. A* **432** 101–111 (1991).

18. J. M. T. Thompson and H. B. Stewart, *Nonlinear Dynamics and Chaos*. Wiley, Chichester, 1986.

19. Y. Ueda, S. Yoshida, H. B. Stewart, and J. M. T. Thompson, Basin explosions and escape phenomena in the twin-well Duffing oscillator: Compound global bifurcations organising behavior. *Philos. Trans. Roy. Soc. London A* **332** 49–186 (1990).

20. J. A. Yorke and K. T. Alligood, Cascades of period-doubling bifurcations: A prerequisite for horseshoes. *Bull. Amer. Math. Soc.* **9**(3) 319–322 (1983).

CHAPTER 3

DYNAMIC INSTABILITIES AND CHAOS IN RUNNING BELTS AND THEIR CLEANING DEVICES

A. HARRISON*
Department of Mechanical Engineering
The University of Newcastle
NSW Australia 2308

Rubber conveyor belts reinforced with steel cords and fabric ply layers are used extensively in industry to move bulk materials. For the past 10 years, the stability of running belts has been investigated both experimentally and theoretically. Unpredictable transverse vibrations, beats, and drop-out phenomena have been found to be sensitively related to parameters that govern the belt's dynamic motion. Chaos has been observed in time, phase-plane records, and Poincaré maps for transverse belt vibration. In addition, longitudinal resonances in belts subjected to a step change in force have been measured, and some interesting unstable effects are described here with the aid of the phase-plane method. New design methodologies for belt systems are based on controlling these types of nonlinear processes. Frictionally excited plough cleaners are also described as another example of chaotic vibrations in belt conveying systems.

3.1. INTRODUCTION

Vibration and its control in large, unsupported mechanical structures is an emerging field of engineering. The multidisciplinary approach to engineering

*Present address: Winders, Barlow & Morrison, Inc., 14 Inverness Drive East, Englewood, CO 80112.

Applied Chaos, Edited by Jong Hyun Kim and John Stringer.
ISBN 0-471-54453-1   © 1992 John Wiley & Sons, Inc.

problem evaluation being adopted by many of the larger organizations worldwide has brought together electrical and mechanical engineers with complimentary skills in computing, control, and modeling of real physical systems.

In most cases, vibration problems are solved by brute-force methods without understanding the underlying mechanisms that govern the motion. Often, chaotic motion is encountered in mechanical systems such as conveyor belts, and usually large safety factors mask a potential failure although fatigue may occur at a rate faster than linear.

Belt conveyors are used to handle large volumes of bulk material. In the United States, coal conveyors can carry upward of 9 million tonnes per annum; in the former West Germany, open-pit belts carry 30,000 tonnes per hour of overburden. These belts are rubber, reinforced along their length by steel cables embedded inside the belting. The belt may be up to 3 cm thick with steel cords up to 1.3 cm in diameter. Some belts are 3 m wide and contain up to 200 cords.

The size of these belts alone indicates large tensions and high speeds, typically 1000 kN and 6 m/s, respectively. The belts are supported on rollers, spaced typically 1.5 and 5 m on the carrying and return sides, respectively. They are driven by up to four large electric motors with total installed power of over 6 MW.

On shutdown, the differential stretch around the belt (due to distributed friction) relaxes at the speed of sound in the belt material. This relaxation causes elastic or stress waves to propagate along the belt, initiating at the drive location. The forces oscillate from high to low in a complex manner that can cause very substantial sag to be generated on the carry side. Experiments indicate this process is a possible cause of belt-snap on start-up, with the precise location of high sag being difficult to predict because of sensitivity to initial conditions of load, friction, and temperature.

Another operational problem with conveyors is transverse belt vibration, or flap. Flap leads to dynamic impact forces of over twice the static load, and this leads to idler bearing failure by reducing bearing life by a factor of 8. Transverse belt vibration has been studied and the large deflections introduce nonlinear effects that lead to unpredictable and chaotic behavior of belting in multispan belt sections.

The phase-plane trajectories of belt particle displacement are examined in this chapter, with the view to developing design strategies to eliminate resonances or to improve the stability prediction of the moving belt.

As a final example, a simple v-plough is described that exhibits self-excited oscillations due to belt–plough interactions. Although the vibration of the plough is excited by slip–stick mechanisms, other nonfriction factors such as belt vibration, belt thickness changes, dirt, and water affect the motions of the plough cleaner in the vertical direction. The efficiency of the plough scraper depends on its stability in contact with the belt surface. The chaotic vibrations have been overcome by varying the mass of the plough and modifying the interface friction using a water spray. This is a practical solution that has a basis in theory.

This chapter describes some of the stability investigations for belts exhibiting chaotic phenomena in their transverse and longitudinal vibration, as well as looking at the chaos in plough cleaning devices.

3.2. CHAOTIC TRANSVERSE VIBRATION IN BELTS

Considerable research has been conducted on the subject of dynamic stability related to strings, band-saw blades, and moving thin tapes [1–4]. Thick belts that are either flat or troughed need to be analyzed using plate mechanics in which particular attention is payed to the development of boundary conditions [5]. One of the more interesting dynamic aspects of moving flat belts is the instability of highly tensioned belts on multiple roller or idler supports. The flexural interaction between spans, when one span is resonant, results in tension-pulling on adjoining spans at twice the flap frequency. This leads to large deflections, lift-off from the supports and drop-out behavior of adjoining spans.

The instantaneous transverse deflections of four spans in a multiply supported belt under tension have been investigated for chaotic motions.

Figure 3.1 illustrates the measured deflection history of four consecutive spans during a 20-s period. At any instant it is difficult to determine the exact level of interaction between spans. The signals shown in Figure 3.1 were obtained from noncontact CBM magnetic reluctance probes [6, 8] mounted at the midpoint edge of the belt in each span. Figure 3.2 pictorially shows the vibrations of these four spans at any instant, and Figure 3.3 shows the experimental setup used to obtain the data in Figure 3.1.

3.2.1. The Motion

The equation of motion of belt can be approximated by a flat beam in which modal deflections across the belt are neglected and effort is concentrated on tension-driven deflections along the belt. In this case the free vibration for support length a is governed by

$$-\frac{EI}{\mu}\frac{\partial^4 w}{\partial x^4} + c_0^2\left[\frac{\partial^2 w}{\partial x^2} - \frac{1}{c_0^2}\frac{\partial^2 w}{\partial t^2}\right] - \frac{v^2 \partial^2 w}{\partial x^2} - \frac{2v \partial^2 w}{\partial x\,\partial t} - \frac{k \partial w}{\partial t} = 0 \quad (3.1)$$

which is the PDE for the problem with damping k, where EI is the stiffness, μ is the belt mass per unit length, $c_0^2 = (aT_0 + EI\pi)/\mu a$ is the wave speed in the belt, T_0 is the tension, w is the deflection, and v is the belt speed. In the case that $\mu c_0^2 \gg (EI)$ and $c_0 \gg v$, we can perturb the belt tension T_0 at

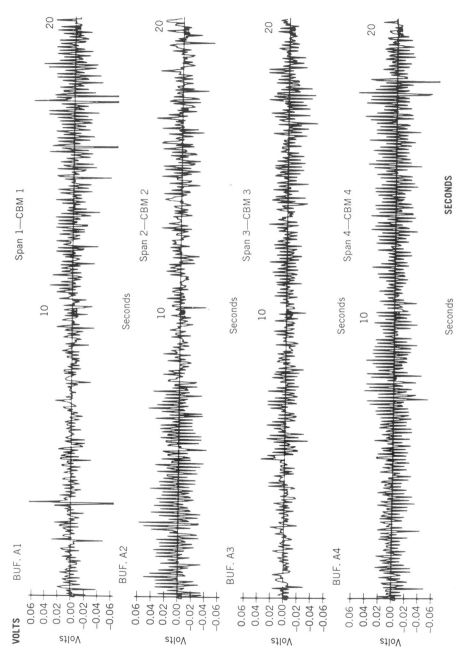

Figure 3.1. The resonant behavior during 20 s of four adjoining v-return spans, showing the unpredictable nature of the vibration of each span with time.

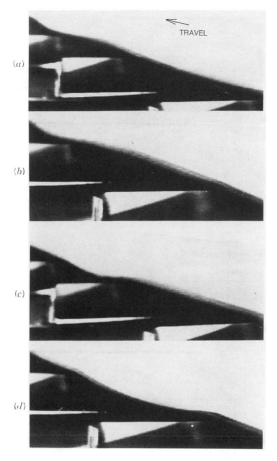

Figure 3.2. Resonating flat belt showing [(a) and (b)] flexural waves near the resonant span and [(c) and (d)] sinusoidal deflections of the belt edge.

twice the natural belt frequency ω,

$$T(t) = T_0 + \Delta T_0 \cos 2\omega t \tag{3.2}$$

With the preceding constraints, substitution of (3.2) into (3.1) with a particular solution for the nonforced case [1, 7],

$$w(x, t) = A_m \sin \lambda x \phi(t) \tag{3.3}$$

leads to the ODE for the motion (let $z = 2\omega t$),

$$\frac{d^2\phi}{dz^2} + (p + q \cos z)\phi + \frac{r\,d\phi}{dz} = 0 \tag{3.4}$$

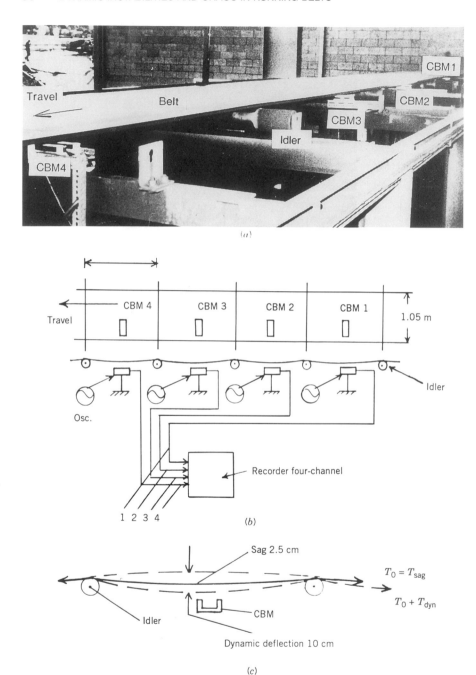

Figure 3.3. (*a*) Photograph showing the location of four CBM transducers for monitoring the flexural behavior of a flat belt. (*b*) Plan view of the monitoring system for recording multispan flexural behavior. (*c*) Dynamic and static deflections in relation to the CBM transducer.

where $\lambda = m\pi a$, a is the support spacing of the span, m is the mode shape along x, and

$$p = \left(\frac{c_0 \lambda}{2\omega}\right)^2, \qquad q = p \frac{\Delta T_0}{T_0}, \qquad r = \frac{k}{(2\omega)^2} \qquad (3.5)$$

It is noteworthy that (3.4) is the Mathieu equation with damping, which also arises directly from the well-known forced Duffing equation with losses

$$\ddot{x} + c\dot{x} + (\alpha x + \beta x^3) = \Gamma \cos \Omega t \qquad (3.6)$$

The fact that a Duffing-type equation with a nonlinear spring also defines this nonlinear problem suggests the possible existence of chaotic trajectories in the phase plane.

3.2.2. Intermittent Chaos in a Flexural Cascade

At least five flexural-mode combinations can occur in the preceding experiment. Figure 3.4 illustrates the predicted flexural combinations for just one type of wave combination, termed a type (IV-A) mode [8], with a time history of each span for 2.5 wave periods. Figure 3.5 shows the actual measurement of four spans using the system shown in Figure 3.3. Square boxes on the

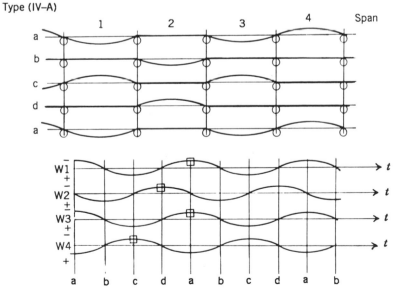

Figure 3.4. A possible flexural cascade and span flexure evolution, termed type (IV-A) mode.

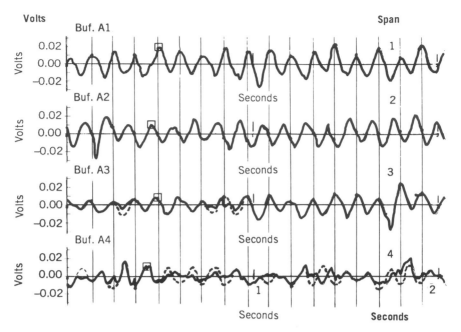

Figure 3.5. Complex flexural behavior with a deflection of the type predicted by Figure 3.4.

waveforms are intended to show phase relationship and the dotted lines show where the deflection would be expected for linear periodic motion.

Span 1 is resonant, the excitation coming from an eccentric idler at the frequency

$$f_i = \frac{v}{\pi D} \tag{3.7}$$

where D is the idler diameter. Its frequency is very close to the belt natural frequency, which for a highly tensioned flat belt, is given by

$$f_b = \left(\frac{c_0^2 - v^2}{c_0}\right) \Big/ 2a \quad \text{for mode } m = 1 \tag{3.8}$$

Phase-plane measurement of the spans reveals attractors that are distinct for each span. The voltage signal $V(t)$ for each span represents the displacement signal $x(t)$. An analog differentiator, with phase preservation above 5 Hz, was used to produce $\dot{x}(t)$. An x–y recorder was used to trace the phase plane $[x(t), \dot{x}(t)]$. Figure 3.6 shows the phase plane for spans 3 and 4, respectively. The orbits are not closed and there are obvious regions of attractions in phase space. Poincaré maps could be drawn using span 1 to

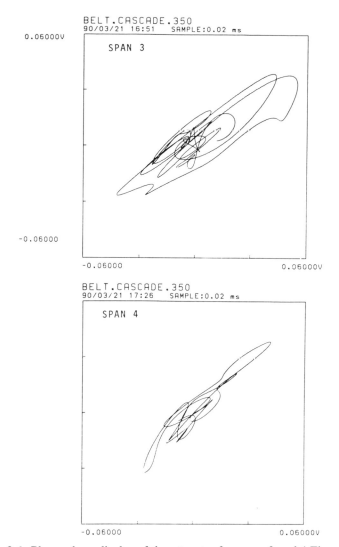

Figure 3.6. Phase-plane display of the attractor for spans 3 and 4 Figure 3.5.

"clock" the signal of span 3 or 4, and this would show the existence of the strange attractors. In a later section, Poincaré maps of forced belt vibration are shown.

This particular data set was recorded at Newvale 1 colliery, NSW, Australia. The resonance problem was solved by changing the idler spacing a so that $f_b > f_i$. The extent to which the belt natural frequency required shifting depends on the mechanical Q of the system. An experiment was designed to determine this amount using an electromagnet shaker [8], described later in this chapter.

Clearly nonlinear effects are occurring in multispan belt vibrations. The periodic coefficient in the Mathieu equation (3.4) generally leads to diverging solutions in the phase plane, the degree of divergence may be tested using Lyapunov coefficients [9, 10], although this has not been done for this example.

From the preceding experiment it is evident that a span near resonance will flex through direct excitation and will then begin to excite adjoining spans by tension fluctuations. The amplitude of vibration may build up to large values, typically 10 cm over a 3-m span. This results in large values of ΔT_0 and it will change the ensuing vibration frequency resulting in dropout and belt lift-off from the support rollers. This is often observed in fiercely resonant belts. In addition, the resonances may move along the belt system to equalize at a point of optimum tension for resonance by eccentric idlers [5]. Parametric variation in belt mass μ also can initiate resonances or dropout. A test on the stability at resonance comes from the Mathieu equation [8, 9] and is approximated by

$$\left(\frac{\omega}{\omega_m}\right)^2 \pm 2\frac{\Delta T_0}{T_0} = 4 \tag{3.9}$$

where ω is the axial forcing frequency, ω_m is the frequency of the span, and $(\Delta T_0/T_0)$ governs the stability as we approach resonance from above or below 2ω or from above 2ω if ω_m is approached from above.

3.2.3. Chaotic Beating between Belt Spans

In another conveyor system, shown in Figure 3.7, two spans were measured that had different natural frequencies, 7.75 and 8.6 Hz. Figure 3.8 illustrates the beating and the bursting of the 2.7-m span. The bursting (upper graph) indicates a parametric variation in one of the governing parameters of the

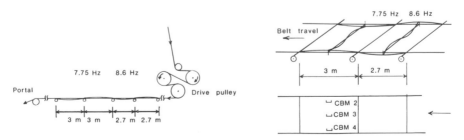

Figure 3.7. Belt system, showing the position of the CBM modal analysis transducers beneath the belt.

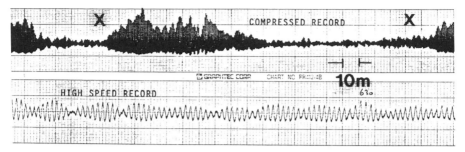

Figure 3.8. Resonance bursting and drop-out behavior of a flat belt system (see Figure 3.7).

vibration, such as belt tension distribution or belt mass. The lower trace is a higher-speed record of the same data and shows dropout and amplitude variations resembling beats in which the interaction is irregular. One of the precursors for chaos is sensitivity to initial conditions; in this case, belt sideways tracking changed at location X, initiating resonance. This is a parametric effect caused by a shift in tension center-line [12].

Figure 3.9 shows both spans vibrating and beating. The lower graph shows the fast Fourier transform (FFT) of the time-domain signal, obtained using an SD 375 Spectral Dynamics FFT analyzer. There is evidence of subharmonic activity in this figure, which could indicate some period-doubling effects, but periodic limit-cycle behavior dominates.

The phase-plane diagrams (Figure 3.10) for sections of the vibration indicate a wandering of orbits, but this in itself is not a guarantee that chaos is present. It may be that the driving frequency ω_1 is close to the limit-cycle frequency ω_2 for resonance of one of the spans and that some entrainment is occurring at the forcing frequency. This would tend to give quasiperiodic vibration if ω_1/ω_2 is irrational. In this case $\omega_1/\omega_2 = 0.90116$ or $\omega_2/\omega_1 = 1.1096$, depending on which span is driven. From the data, both spans are 180° different in phase, which indicates that the excitation is forced from an adjoining span as a flexural cascade pair, not as a classically forced resonance.

Figure 3.10a and 3.10b shows the phase-plane trajectories of the 8.6-Hz span, at different parts of the vibration.

An interesting aspect of these phase-plane results is the similarity to double-egg patterns [11] produced as the forcing frequency Ω and amplitude Γ of the Duffing equation (3.6) are varied. However, orbits are not closed, so there is good indication in the phase-plane maps that the system has nonlinear characteristics and that the span motions are regular but unpredictable. Therefore one can conclude that chaos is present, perhaps continually being modified by geometric effects such as mass, tension, stiffness, and rubber hardness.

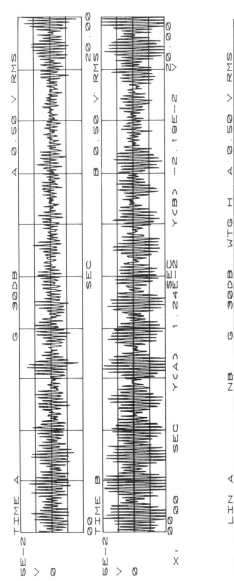

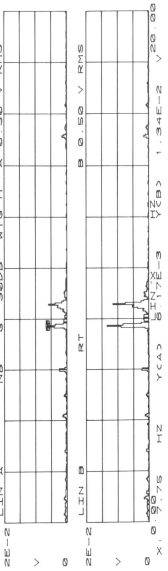

Figure 3.9. Time-domain and frequency-domain data for the beating between two spans vibrating at different frequencies. Notice the build-up and decay of resonance.

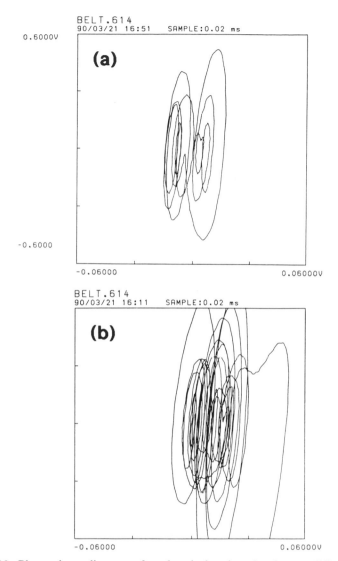

Figure 3.10. Phase-plane diagrams for chaotic beating showing at different times: (*a*) two attractors; (*b*) many orbits and quasiperiodic bursting about two attractors.

3.2.4. Impact FFT Analysis of Belt Vibration

Design of nonresonant belt systems requires a knowledge of damping parameters and displacement related instability. In running belts, the excitation frequency must be sufficiently far away from the natural frequency of the belt to prevent build-up of large amplitudes of vibration that become unstable and chaotic.

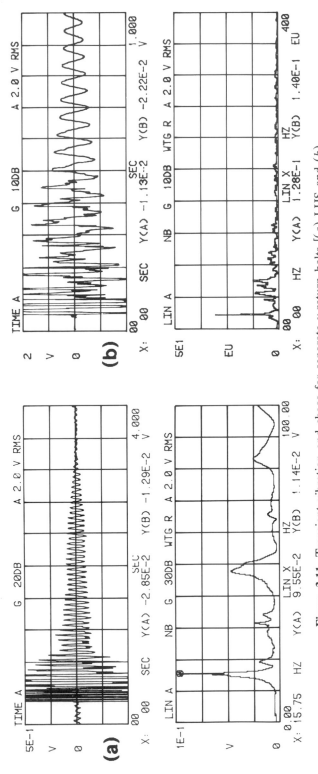

Figure 3.11. Transient vibration and chaos for separate v-return belts [(*a*) LHS and (*b*) RHS] during the initial impact and the FFT response for the respective signals.

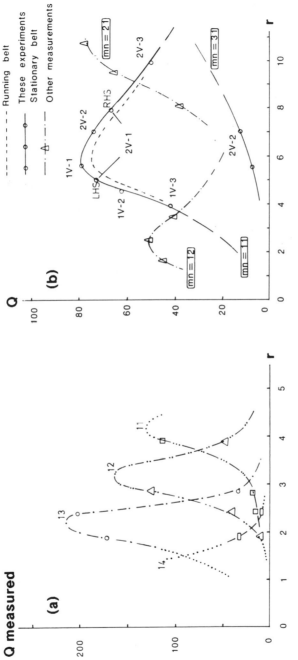

Figure 3.12. Mechanical Q of (*a*) flat and (*b*) v-return belts measured by impact and shaker methods.

Impact methods allow one to investigate chaotic and limit-cycle behavior of tensioned belts using time-domain and FFT analysis. Figure 3.11a shows the impact evolution and its FFT for a v-return belt over a 4-s period. Although the vibration amplitude becomes periodic, it is initially chaotic, which is observed as broad regions of response in the FFT signal. The FFT uses peak averaging to capture all frequencies in the response. Figure 3.11b shows the response over 1 s for another v-return belt. There are initial bifurcations and period doubling, settling down to the steady-state low-amplitude limit cycle. The decay of this signal in the time domain is used to obtain damping from the logarithmic decrement calculation.

When a conveyor belt support system is being designed, the belt tension, idler spacing, and belt material wave speeds need to be determined, as do the loss coefficients and the mechanical Q. It is the latter that will determine the likelihood of chaotic responses, and, in some belt systems that are highly "tuned," large amplitudes will build up naturally and become unstable. The impact experiments and methodology clearly show this effect.

Measurements on resonant systems of v-return and flat belts have been carried out using both impact and shaker experiments to determine the mechanical Q as a function of plate aspect ratio [5]. Figure 3.12a shows the response for flat belts, and Figure 3.12b shows the response for v-return belts. The very high Q values for the various modal deflections of a flat belt are indicative of systems that can become unstable due to large-amplitude resonances that lead to significant values of the βx^3 term in the Duffing equation. The curves in Figures 3.12 are used to design nonresonant systems.

3.2.5. Forced Chaos Experiments

A flat, tensioned steel-cord-reinforced belt has been excited to resonance when in the running state by using a large (70-kg mass) electromagnetic shaker, the system being described by Figure 3.13.

This experiment has some merit in that it allows one to scan the excitation frequency to the point where the electromagnetic force on the embedded steel cords is "pulling" at or near the natural frequency of the belt. This is quite a difficult experiment to perform on large working conveyors; nevertheless, large deflections ($\sim$ 15 cm) were able to be induced and nonlinearities were observed as bifurcations.

Figure 3.14 shows the results of an experiment to observe the onset of chaos in a tensioned span of belting as frequency of excitation increases. Resonance of the span was approached from above, as shown in Figure 3.14a. Amplitude increased at resonance but not before bifurcations were evident, and two attractors became clear on the phase plane. At resonance, period doubling (Figure 3.14c) and instabilities were observed (drop-out phenomena), indicative of multispan interactions discussed earlier. Increased excitation frequency just past resonance (Figure 3.14d) shows smaller-amplitude chaotic activity. The span could not support a proper amplitude of

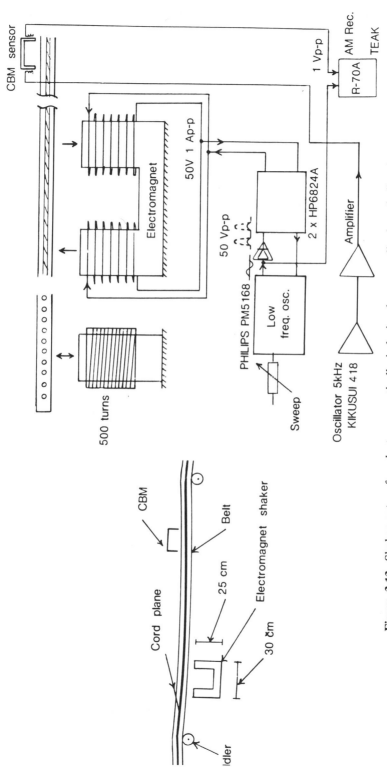

Figure 3.13. Shaker setup for electromagnetically inducing large amplitudes of vibration into a naturally nonresonant moving belt.

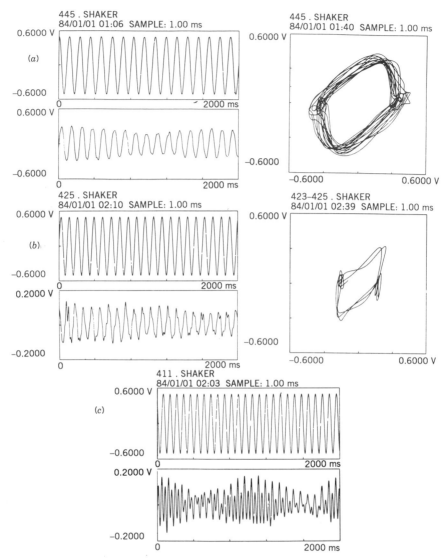

Figure 3.14. Forced excitation of a belt span using a shaker, showing the forcing period (upper time trace) and belt response (lower time trace) with respective phase space plots for the following: (*a*) near-resonance belt response with onset of bifurcations; (*b*) increased excitation frequency, showing two attractors with period-doubling bifurcations; (*c*) period-doubling route to chaos. (*d*) Fully developed chaos and its phase-plane response. (*e*) Poincaré maps of the forced response in the chaotic regions of the waveform shown in (*d*). The upper map is near resonance response, the lower map is at or just beyond natural resonance.

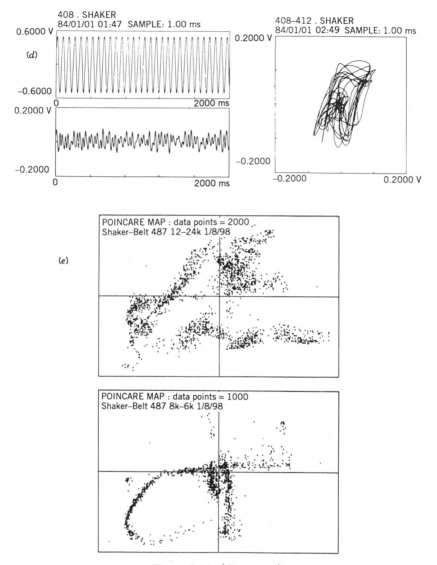

Figure 3.14. *(Continued)*

vibration because of softening nonlinearities as the large deflections collapse and tension in the span reduces rapidly. Adjoining spans would assume resonance while the excited one shimmered chaotically.

In this experiment and others like it, it was determined that for resonances not to be excited by mechanisms of the type (IV-A) mode discussed earlier it is necessary to place the exciting frequencies of the idlers at least 1 Hz below natural resonance of the belt span. However, even this tolerance was found to be inadequate and chaotic responses were excited, due to small changes in initial conditions that govern the vibration. In practical terms, all variables that may sensitively determine the outcome of an excited belt system will have a cumulative or total range of influence on f_b, and this needs to be determined.

Returning to the forced Duffing equation (3.6) and equation (3.7), we may take into account variations in belt mass per unit length, stiffness, tension, and exciting frequency (due to idler build-up from carry-back of fine material) by adopting

$$\ddot{x} + c\dot{x} + R(x)[x + bx^3] = \Gamma(x)\cos\left(\frac{2v}{D(t)}\right)t \qquad (3.10)$$

where c is a viscoelastic loss factor, $D(t)$ is a time variation in idler diameter due to carry-back build-up, and (x) is the position variable. Letting $\delta m(x)$ be the change in belt weight per unit length and letting k_1 (b is a constant) be the equivalent spring stiffness, then, for μ = average belt mass per unit length,

$$M(x) = \mu(x), \qquad R(x) = \frac{k_1}{M(x)}, \qquad \mu(x) = [\mu + \delta m(x)] \quad (3.11)$$

In Figure 3.14e, the trajectory returns for the forced vibration show as regions of strange attractors in the Poincaré map. Although the shaker experiment clearly shows that there are regions of strange attractors in the Poincaré map near resonance, one may determine more about the motion (in terms of period doubling) from the Poincaré map. In the shaker experiment, the forcing frequency is used to strobe the response at the peak of the forcing current in the shaker, and both forcing and response signals are sampled synchronously, stored in a 386 PC file, and then plotted as the Poincaré map. The fractal dimensions and Lyapunov exponents have not been derived at this point, because the drift in the chaotic signal due to the sweep of the forced exciting method will not generate reliable values in this case. The maps do, however, show strange-attractor regions.

Belt mass variation is common, due to manufacturing tolerances and width variations. Splices have mass and stiffness different from the parent belt and they affect resonances in a small way; such effects are observed in the v-plough experiment described later. Given that the tolerances can be quantified, it may be possible to design a stable system for certain belt supports, but only in certain cases.

3.2.6. Design Considerations

In essence, equations (3.7) and (3.8) define the design problem and together with Figure 3.12, the value of f_i can be estimated. In general, parametric variations of the type discussed previously lead to unstable or chaotic variation at some position along the conveyor.

To achieve transverse belt stability and to ensure against chaotic flexure about two strange attractor basins of the type illustrated in the previous shaker experiment, the Δf of belt natural frequency f_b is of the order of ± 2 Hz for a flat belt at a particular tension. The natural bandwidth is ± 0.1 Hz for a flat belt and ± 0.25 Hz for v-return belt structures, so we need to add the extra tolerance of 2 Hz for variability in parameters that govern resonances. In this case, at speed v and belt tension T_0,

$$f_i \leq f_b - \Delta f$$

$$(mc_0 - 4a) \geq \frac{2va}{\pi D}; \qquad m = 1, 2, \ldots \tag{3.12}$$

As an example, a slope belt has $T_0 = 70$ kN, $\mu = 60$ kg/m, $D = 0.15$ m, and $v = 3.4$ m/s. Then, for $m = 1$ stability, the maximum idler support spacing would need to be $a = 1.853$ m. This design will ensure that tension-driven resonances that lead to flexural instabilities are avoided. The underlying assumption is that belt tension is evenly distributed. This is not always the case and may be only 75% of T_0 at the edges due to the way the belt is manufactured. In this case $a < 1.6$ m is required. Build-up on idlers works in favor of resonances as long as $f_b > f_i$. If 5 mm of material builds up on the rollers, D increases to 0.155 m and f_i changes from 7.215 to 6.98 Hz. However, if the construction is such that $f_i = 7.4$ Hz and $f_b = 7$ Hz (nominally), then build-up on the rollers results in belt flap excitation with large and most probably chaotic displacement.

3.3. STRESS WAVES IN BELTS AND DYNAMIC INSTABILITIES

3.3.1. Theoretical Background

Conveyor belts are constructed as a reinforced rubber composite with considerable elasticity in the longitudinal direction. On starting and stopping, a conveyor belt will undergo elastic vibration due to a sudden step-change in drive power. If the vibration can be predicted, the peak stress levels also can be predicted. Dynamic stresses induced in the belt on stopping are well known to be responsible for expensive shutdowns caused when belts and structure (take-up devices) fail.

The potential stress amplitude depends on the distributed mass and friction around the system, and some systems contain locked-up strain energy of the order of 1000 kN of effective tension. This amount of stress is also available for stress-wave generation. The problem requires solving two coupled wave equations for the carry-side and return-side belts, each with different elastic wave velocities V_C and V_R, respectively. Using u^C and u^R to

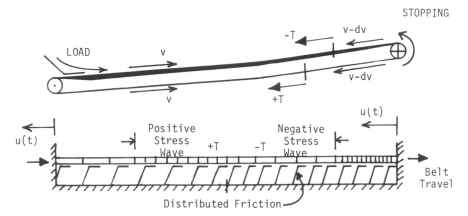

Figure 3.15. A belt conveyor model for elastic wave analysis.

represent carry-side and return-side displacements, the wave equations that require solving are (free-wave case) [4, 8, 11]

$$u_{xx}^C - \frac{1}{V_C^2} u_{tt}^C = 0, \qquad 0 \leq x \leq L$$

$$u_{xx}^R - \frac{1}{V_R^2} u_{tt}^R = 0, \qquad L \leq x \leq 2L \qquad (3.13)$$

where distributed damping of the form $-ku_t$ has been omitted because damping has only a small effect during the transient stopping time in most conveyors. To solve these equations is a straightforward case of applying boundary conditions to the problem (see Figure 3.15). The boundary conditions at $x = L$ are

$$u_t^C = u_t^R$$

$$u_x^C = u_x^R$$

$$u^C = u^R \qquad (3.14)$$

and at $x = 0$ and $x = 2L$, the adopted condition is

$$u_x^C = -u_x^R, \qquad u_x^R > 0 \text{ at } t > 0 \qquad (3.15)$$

(assuming a fixed take-up at drive output T_2).

The reduction of the problem to its solution at $x = 0$ (the return side of the belt) gives a time-dependent function for the first three Fourier terms of the form $u(x, t) = U(x)U(t)$, where

$$U(t) = \frac{1}{(\pi A)^2} \sin(A\pi t) + \frac{1}{(\pi B)^2} \sin(B\pi t) + \frac{1}{(\pi C)^2} \sin(C\pi t) \quad (3.16)$$

where πA, πB, and πC are eigenvalues of the problem's frequency equation

$$\varepsilon \sin \xi \cos \varepsilon + \xi \sin \varepsilon \cos \xi = 0 \quad (3.17)$$

where

$$\xi = \frac{\gamma \pi L}{V_C} \quad \text{and} \quad \varepsilon = \xi \frac{V_C}{V_R} \quad \text{with the variable } \gamma$$

In this analysis, the first three roots of (3.17) become πA, πB, and πC. The values of V_C and V_R have been determined from a lumped-mass model of the belt as a composite and have been verified to be accurate when compared to field experiments [13]. The values for steel belt are [13]

$$V_C = V_{Cu} - \frac{\Delta Q}{Q_D} (V_{Cu} - V_{CL}) \quad (3.18)$$

$$V_{CL} = V_{sc} \sqrt{\frac{\mu_{sc}}{\mu + \mu_{iC}(x) + \mu_m(x)}} \quad (3.19)$$

$$V_{Cu} = V_{sc} \sqrt{\frac{\mu_{sc}}{\mu + \mu_{iC}(x)}} \quad (3.20)$$

$$V_R = V_{sc} \sqrt{\frac{\mu_{sc}}{\mu + \mu_{iR}(x)}} \quad (3.21)$$

where $V_{sc} = 4300$ m/s, μ_{sc} = mass of steel cables (in kilograms per meter), μ_{iR} and μ_{iC} are the masses of idlers on the return and carry sides (in kilograms per meter), and $\mu_m(x)$ is the mass (in kilograms per meter) of bulk material on the belt. In the preceding discussion, $0.3 < \Delta < 1$, depending on the bulk material [13], and Q/Q_D is the ratio of load on the belt to the design load.

3.3.2. Period-Doubling Elastic Instabilities — Example and Application

Irregular behavior has been observed for many years in long belts during shutdown, and it is related to the generation of excessive sag (low tensions) being locked in along the carrying side of the belt. The location of high sag usually occurs near the tail end of the belt and, especially in flat-profile belt systems, the effect can cause belt snap on start-up. Figure 3.16a illustrates

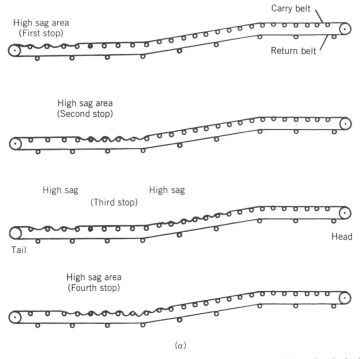

(a)

Figure 3.16. (a) Series of sag formations along the carrying side of a belt following shutdown, showing the variability of the process with each stop. (b) Variation in stopping time for an unloaded belt is attributed to viscoelastic coupling variation along the conveyor.

the effect of unpredictable sag production along a belt on stopping. Figure 3.16b shows the stopping times of a long unloaded belt on five consecutive start-ups and shutdowns. The stopping time of this system is erratic and has been attributed to variations in viscoelastic interactions along the belt.

The phase plane has been used to model and observe instabilities in long belts. Equation (3.16) has been used to generate $x(t)$ and $\dot{x}$ for steel cord belt with $V_R = 2200$ m/s, $L = 2000$ m (flat underground coal mine belt used for longwall coal extraction), and V_C is kept variable. The following discussion shows that the elastic motion sensitively depends on initial conditions of V_C and L. In this example, V_C is used to initiate period doubling.

A belt conveyor with these parameters has been evaluated for stability using the model just given. For this belt, $V_R = 2200$ m/s and V_C is usually 2100 m/s (empty) and about 1000 m/s fully loaded. Phase-plane plots for this conveyor show rapid period doubling and bifurcations as material load on the belt increases [V_C decreases as $\mu_m(x)$ increases—see equation (3.19)].

The period-doubling route to chaos is shown by way of the examples shown in Figure 3.17, where the periods $T \to 2T \to 4T \to 8T \to 16T\ (=2T)$

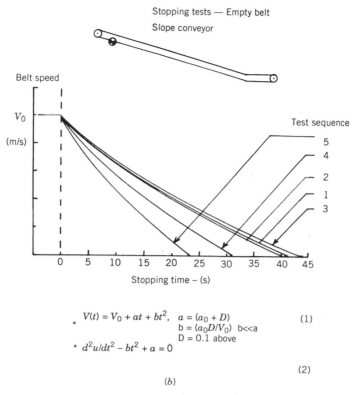

Figure 3.16. (Continued)

The equations below the figure:

$$V(t) = V_0 + at + bt^2, \quad a = (a_0 + D) \tag{1}$$
$$b = (a_0 D / V_0) \quad b \ll a$$
$$D = 0.1 \text{ above}$$

$$d^2 u / dt^2 - bt^2 + a = 0 \tag{2}$$

(b)

occur for carry-side belt wave speeds (critical points) 2200, 2194, 869, and 583 m/s. Although these values represent the points where bifurcation begins, they are not the period-doubled values. The variable parameter V_C that gives rise to exact period doublings occurs at $V_C = 2200$ m/s (P1), 2177 m/s (P2), 869 m/s (P4), 416 m/s (P8), and 217 m/s (P16 = P2). A study of Figure 3.17 shows the various zones of stability and the zones of chaos. It is noteworthy that at the critical points, $(869 - 2194)/(583 - 869) = 4.633$, which is close to Feigenbaum's constant (4.6692...); the error is probably due to root-finding jitter in equation (3.16). At $V_C = 2195$ m/s, bifurcation has started; after $V_C = 2194$ m/2 the period-2 mode shape is "locked" until $V_C = 2177$ m/s, at which time a single loop at double the period occurs as a fully periodic response. Thereafter, as V_C decreases, the motion in phase space is quasiperiodic up to P4.

From a design point of view, the loads on the belt (3.18) that give rise to chaotic orbits in phase space will also produce displacements in the belt on stopping that cannot be predicted. Therefore, design stability and predictability can be set by selecting tonnage on the belt (or another parameter, such as

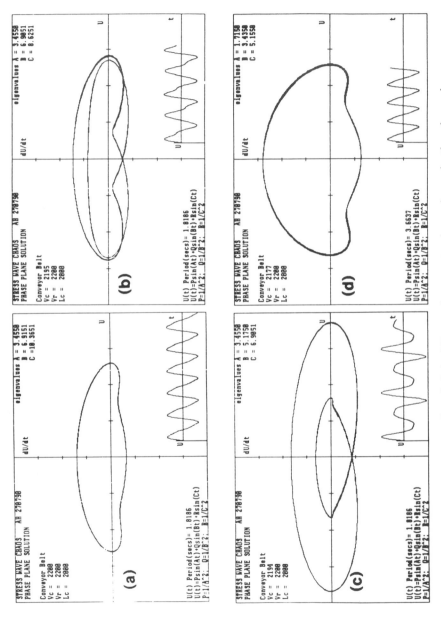

Figure 3.17. Sequence of period doubling and routes to chaos shown in the phase-plane simulation for a belt of length $L = 2000$ m as the load on the belt is varied: (*a*) period-1 (P1); (*b*) and (*c*) period doubling; (*d*) period-2 (P2).

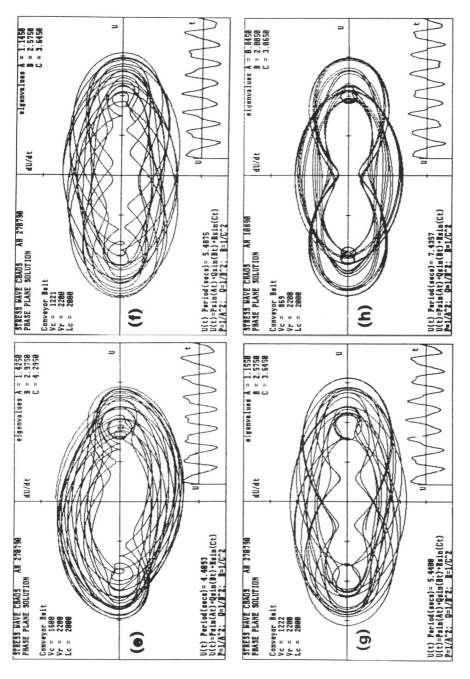

Figure 3.17. (*Continued*). (*e*) chaotic orbits between two strange attractors, quasiperiodic orbits; (*f*) chaos and quasiperiodic orbits near period-3; (*g*) closed orbit (P3); (*h*) closed-orbit period-4 bifurcations (P4).

77

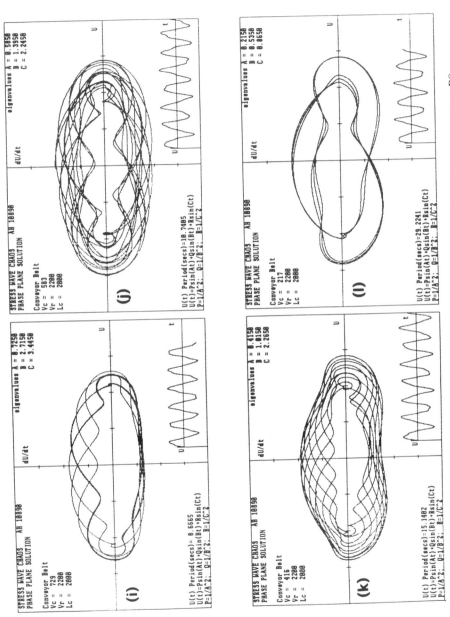

Figure 3.17. (*Continued*). (*i*) and (*j*) quasiperiodic P4 closed orbits and chaos onset to P8; (*k*) quasiperiodic (P8); (*l*) closed-orbit period-16 (P16) equal to P2 orbit.

belt length) to produce P2 stable solutions. A stable solution at P4 would result in severe belt overloading and this is not a practical solution. The prediction of belt displacement stability permits one to compute stress loadings and hence belt safety factors at the initial design stage; however, failure to achieve stable solutions can result in a design with large dynamic belt loads on stopping.

3.4. CHAOTIC BEHAVIOR OF v-PLOUGH BELT CLEANERS

One important aspect of proper belt operation is the control of fine and coarse carry-back material. This material builds up on idlers, piles up along the return run of the system as it is squeezed from the belt by rollers, and generally leads to a dirty and difficult operating environment. There are numerous cleaning devices on the market to remove carry-back material, ranging from water jets to scraper blades and v-ploughs.

The v-plough rides on the return side of the belt and its primary purpose is to prevent spillage and tramp material entering high-tension bend or drive pulleys. Figure 3.18 illustrates the mechanical setup of a v-plough and Figure 3.19 shows an equivalent model of the system.

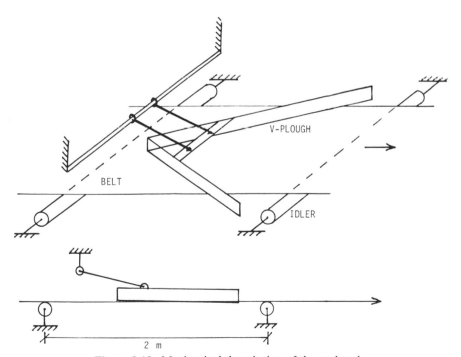

Figure 3.18. Mechanical description of the v-plough.

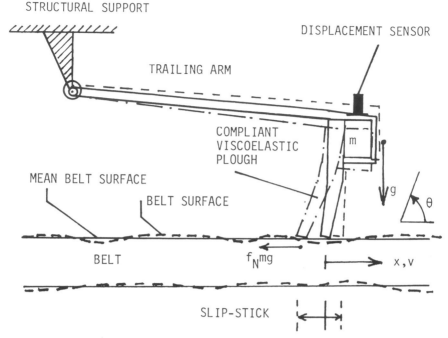

Figure 3.19. Model of the v-plough.

3.4.1. Model Analysis

Analysis of the problem proceeds by assuming that there are no net sideways forces, caused by imbalanced moments, about the v-plough pivot. Coulomb friction is assumed to drive the vibrations through slip–stick mechanisms, although transverse belt-edge vibration also sets up a vibration in the plough arms. The plough has a mass m and rests on the belt at a point near the belt edges. The front point of the v-plough is raised off the belt by about 1 mm in practice, to improve stability, and the blades that are in contact with the moving belt are a polyurethane-type flexible elastomer with nonlinear spring characteristics of the type $(\alpha x - \beta x^3)$. This material's stiffness increases as its deflection increases. The difference between belt speed v and blade velocity $\dot{x}$ is an important factor in the excitation process. The plough blade will deflect by staying in contact with the moving belt until the stiffness increase prevents further motion. At this point Coulomb friction is overcome and the blade snaps back to repeat the process.

The equation of motion is derived using the preceding assumptions and description, with k_1 and k_2 spring constants:

$$\ddot{x} + (v - \dot{x})mc + \left(k_1 x - k_2 x^3\right) = F(x) \tag{3.22}$$

where $(v - \dot{x})$ is the speed-dependent slip velocity of the two interacting surfaces. Equation (3.22) is of the form

$$\ddot{x} = f(x, x^3, \dot{x}) \tag{3.23}$$

and is nonconservative. Such equations are generally expected to produce new phenomena when studying their mechanical behavior. In this case the problem is more complicated because the two plough arms appear to vibrate 180° out of phase, so that slip–stick motion on one side is continually modified by motion of the complementary blade. This problem therefore requires a three-degrees-of-freedom solution, but because of belt sag and cover roughness irregularities, this aspect is largely academic and the problem obeys the general Duffing equation developed earlier [see equation (3.6)].

In terms of (3.23), let the friction be f_N and let θ be the blade angle to the belt. Then the forcing function

$$F(x) = f_N mg \sin \theta \, \text{sgn}(v - \dot{x}) + F_0 \cos \Omega X \tag{3.24}$$

is the switching or exciting force as the blade flexes in and out of slip–stick motion, and F_0 is the amplitude of the forcing frequency of any belt roughness vibration at frequency Ω, which may be nonconstant.

Combining these equations we have $(\alpha = k_1/m, \beta = k_2/m)$

$$\ddot{x} + c\dot{x} + (\alpha x - \beta x^3) + cv = f_N g \sin \theta \, \text{sgn}(v - \dot{x}) + \Gamma \cos \Omega X \tag{3.25}$$

This equation is, once again, a Duffing equation $(\alpha > 0, \beta > 0)$, so strange-attractor motion is expected to be modeled. A value of $f_N = 0.3$, $v = 3$ m/s, $c = 0$, $\Gamma = 0$, and $\beta = 0$ can be used initially to observe the effect of varying α in terms of the phase-plane response $(\psi = f_N g/\alpha, \theta = \pi/2)$ using standard procedures,

$$y = my^2 + \alpha(x + \psi)^2 = C1, \qquad \dot{x} > v \tag{3.26}$$

$$y = my^2 + \alpha(x - \psi)^2 = C2, \qquad \dot{x} < v \tag{3.27}$$

These two equations represent ellipses with centers at the phase-plane points $+\psi$ and $-\psi$. Because of the use of the signum function to approximate the slip–stick, there is a singularity at $(\psi, 0)$ on the phase plane.

The v-plough cleaner exhibits periodic and perhaps chaotic motion and the preceding theory can be taken further if necessary to observe the effect of varying m and c on the model behavior, where damping is expected to produce a strange-attractor motion in phase space.

3.4.2. Phase-Plane Experiments (Real System)

A v-plough, installed on Newstan Colliery slope belt, was measured by attaching displacement sensors to the arms of the v-plough so that vertical displacement was measured. Belt vibration and vertical support loads were also monitored for later Poincaré map analysis in a region on motion where $\Gamma \neq 0$.

Figure 3.20 shows two time waveforms for the v-plough arm motion at counters 400 and 501 on the tape recording of the vibration, together with their respective phase-plane orbits. Figure 3.21 shows a whole series of phase-plane orbits as the belt moves past the v-plough, and the significance of the variations will now be discussed in light of the previous knowledge [equation (3.10)].

From Figure 3.20, the time signal is observed to jump chaotically about upper and lower bifurcations in both cases at tape counters 400 and 501. At tape counter 400 there is evidence of rapid decay in signal at a fixed level on the positive cycle of chaos, consistent with the phase-plane response, which shows a transient forced van der Pol–type relaxation oscillation, with spirals toward a strange attractor followed by chaotic whipping in phase space to another attractor level. The upper level attractor in time (the right-hand one in phase space) is considered to be the airborne state of the instrumented v-plough arm. Although we assumed $\beta = 0$ in the preceding discussion, if its value were positive in (3.25), then it is most likely that jump phenomena would occur, particularly if $c \neq 0$ (damping). In the observed phase-plane response we see these effects and, in practice, stability is considerably improved by using a water spray ($c \approx 0$) and increasing the stiffness of the arms to the structure (α increases, $\beta \approx 0$) to reduce the flexural range of the elastomer blades.

Record 501 describes another section of belting within the same conveyor belt. Its frictional characteristics are clearly different, and reference to the phase plane for record 501 shows distinct bifurcations and two attractor regions where periodic oscillations switch positions in phase space. The shape of the phase-plane response is not unlike the beating experiment described earlier, indicating that some entrainment is occurring due to the existence of belt vibration ($\Gamma > 0$). It is well known, too, that stability and transients for subharmonics of Duffing's equation can be displayed on the van der Pol plane and in this case one can observe periodic and decaying spiral motion. In addition, (3.25) can be made to self-excite if $c > 0$ and $\beta > 0$ without any influence of a forcing function, consistent with what we observe when f_N and Γ are both small.

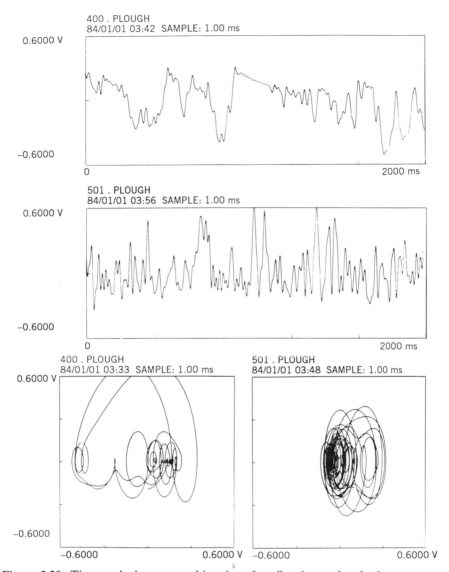

Figure 3.20. Time and phase-space histories of a vibrating v-plough cleaner at two temporal values, showing sensitivity to belt–scraper frictional coupling.

Figure 3.21 is included to illustrate further the variability over time of the v-plough phase-plane orbits. The orbits are all chaotic and serve to show just how sensitive the v-plough motion is to exciting conditions that occur along the belt. This conclusion is consistent with (3.10), which allows for small variations in initial conditions—the outcomes of which are all different but fall consistently about a strange attractor in phase space. Sometimes, the

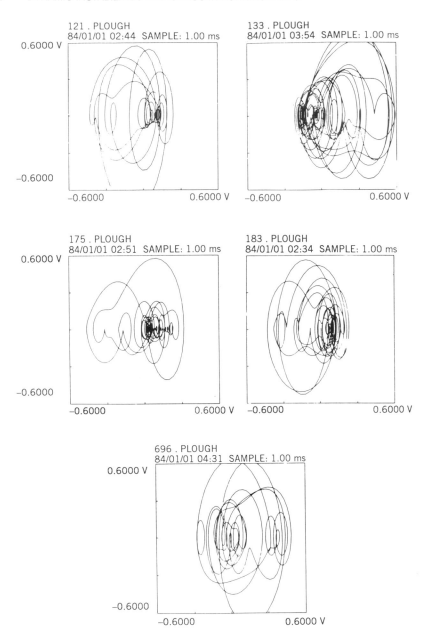

Figure 3.21. Phase-plane responses for a v-plough at different times, showing the effect of variable belt cover friction on the location and type of the strange attractors.

strange attractor occurs for $+\psi$ and other times for $-\psi$; this is believed to be related to an equivalent negative a during v-plough blade spring-back and contact with the belt, giving rise to the signum function in the motion.

3.5. CONCLUDING REMARKS

This chapter describes three systems that are related to belt conveying technology and that exhibit chaotic behavior. Transverse belt vibration is shown to be chaotic under large-amplitude forcing, and various tests are described to elucidate the nature of the vibration. These tests include self-excited hammer impact and shaker response experiments. Modeling of axial tension perturbation to the transverse wave equation clearly shows the existence of mechanisms that are well known to give rise to vibration in Duffing-type oscillators. This chapter explores the effects of varying initial conditions on the response of the belt and applies this new knowledge in chaos technology to belt design. This research is very new and still developing in terms of its application to design.

Elastic vibration in long belts can also be modeled to determine stability in the phase plane, and concepts are developed to improve the stability (on stopping) of long belt conveyor designs. A v-plough cleaning device also is studied using chaos technology and stability theory. It is shown that forced Duffing equations can be derived to describe the motion of the plough and that, by varying friction and mass displacement, the stability can be improved to a level that permits the practical use of such devices in industry.

REFERENCES

1. S. Naguleswaren and C. J. H. Williams, Lateral vibrations of band-saw blades, pulley belts and the like. *Int. J. Mech. Eng. Sci.* **10** 239–250 (1968).
2. F. R. Archibald and A. G. Einslie, The vibration of a string having uniform motion along its length. *J. Appl. Mech.* **24** 347–348 (1957).
3. J. E. Rhodes, Parametric excitation of a belt into transverse vibration. *J. Appl. Mech.* **12** 1055–1060 (1970).
4. W. L. Miranker, The wave equation in a medium in motion. *I.B.M. Journal* **1** 36–42 (1960).
5. A. Harrison, Determination of the natural frequencies of transverse vibrations of conveyor belts with orthotropic properties. *J. Sound and Vibration* **110**(3) 483–493 (1986).
6. A. Harrison, A transducer for testing steelcord deterioration in high-tensile strength conveyor belts. *NDT Int J.* **15**(3) 133–138 (1985).
7. K. Klotter, Non-linear vibrations. In *Engineering and Mechanics Handbook* (H. Flügge, ed.), Chapter 65. McGraw-Hill, New York, 1962.

8. A. Harrison, Dynamic measurement and analysis of steel cord conveyor belts [Chapter 4.3, Non-linear Behaviour (Stability at Resonance)]. Ph.D. dissertation, March 1984, University of Newcastle, Australia.

9. F. C. Moon, *Chaotic Vibrations*. Wiley, New York, 1987.

10. A. V. Holden, *Chaos*. Princeton University Press, 1986.

11. H. T. Davis, *Introduction to Nonlinear Differential and Integral Equations*. Dover Press, New York, 1962.

12. A. Harrison, Predicting the tracking characteristics of steel cord belts. *Int. J. Bulk Solids Handling* **10** 47–53 (1990).

13. A. Harrison, Stress front velocity in elastomer belts with bonded steel cable reinforcement. *Int. J. Bulk Solids Handling* **6** 27–31 (1986).

CHAPTER 4

ATMOSPHERIC FLIGHT DYNAMICS AND CHAOS: SOME ISSUES IN MODELING AND DIMENSIONALITY

GARY T. CHAPMAN
Department of Mechanical Engineering
University of California
Berkeley, California 94720

LESLIE A. YATES
Eloret Institute
Palo Alto, California 94303

MICHAEL J. SZADY
Department of Chemical Engineering
Princeton University
Princeton, New Jersey 08540

Atmospheric flight dynamics is the study of two interacting dynamical systems: the fluid system and the vehicle itself. An overview of this interaction and the possible behaviors it can exhibit is presented. Two widely different sets of issues that occur in studying dynamical systems are also discussed. The first is the modeling of the forces and moments that act on the flight vehicle and are produced by the fluid. Some of the issues that arise here are illustrated using the example of a spring-mounted circular cylinder in cross flow. The second set of issues concerns some practical issues, such as finite data sets and noise, that arise in the determination of the dimensionality of low order chaotic attractors. These will be illustrated with simple geometric shapes, a cantor set, and a low order dynamical system.

4.1. INTRODUCTION

The study of the flight dynamical characteristics of aerodynamic vehicles is important for:

1. The design of vehicles with adequate aerodynamic flight characteristics.

Applied Chaos, Edited by Jong Hyun Kim and John Stringer.
ISBN 0-471-54453-1 © 1992 John Wiley & Sons, Inc.

2. The design of robust control systems.
3. The development of representative mathematical models for flight simulators.

The mathematical study of the flight dynamic characteristics of aerodynamic vehicles dates back to work of Bryan [1] in the early 20th century. Bryan recognized the importance of the interaction between the fluid dynamic system surrounding the vehicle and the dynamics of the vehicle itself. He was the first to introduce modeling of the aerodynamic forces and moments acting on the vehicle. Although his modeling was strictly linear, it is consistent with the idea of a Taylor series expansion of the aerodynamic forces and moments in terms of the vehicular velocity, angular rates, and their derivatives with respect to time (e.g., reference [2]). In 1954, Tobak [3] introduced the concept of indicial response to aerodynamic modeling for linear aerodynamics. Ten years later Tobak and Pearson [4] applied the indicial response concept to modeling nonlinear aerodynamic forces and moments. Up until about 1980 the indicial approach was believed to be generally valid (e.g., reference [5]) provided that the rate of change of the flight conditions with time was slow. However, with the extensive research on nonlinear dynamic systems and the role of bifurcations and chaos in the late 1970s and early 1980s, it became obvious that the indicial response approach, as proposed in the 1950s and 1960s, was not valid if the fluid flow was not steady under steady flight conditions or when serious bifurcations occurred in the flow [6, 7]. It became increasingly clear that flight dynamic behavior of aircraft at high angles of incidence was not predictable with existing mathematical aerodynamic models, and there were also several classical, aerodynamically forced, dynamics problems that defied adequate explanation. Two examples are:

1. Aerodynamic induced oscillations of a circular cylinder in cross flow, the so-called galloping telegraph wire problem (also related to the Tacoma Narrows bridge failure).
2. The oscillating airfoil in dynamic stall.

Since the early 1980s aerodynamic modeling in the presence of bifurcations and chaos has been an active area of research [8, 9].

The introduction of the concepts of chaos and strange attractors provided a method for characterizing the flight dynamics for these types of motion. The dimension of the attracting subspace is one of the commonly noted characteristics, and in the early 1970s several methods for measuring these dimensions were put forward. One widely used method is the Grassberger–Procaccia algorithm [10] for the correlation dimension. The theoretical ideas behind this are excellent but the theory is for an infinite, error-free data base; real data bases are finite and contaminated by errors. Other factors also enter into the practical determination of the dimension, and the accuracy of the calculated dimensions is often not known.

This chapter covers three topic areas and includes:

1. A brief discussion of atmospheric flight dynamics.
2. An introduction to aerodynamic mathematical modeling and some major issues that arise with examples for an oscillating circular cylinder in cross flow.
3. A study of issues associated with determining accurate dimensions using the Grassberger–Procaccia algorithm.

4.2. FLIGHT DYNAMICS

The study of flight dynamics is the study of two interacting systems, the inertial system of the vehicle and the aerodynamic system of the air flow around the vehicle. In most cases this interaction is moderated by a control system. In some cases the inertial body may also be deforming elastically during flight and this must be accounted for in the interaction. The elements of flight dynamic systems are illustrated in Figure 4.1. The coupling of the interacting systems can be complex due to the complex nature of the

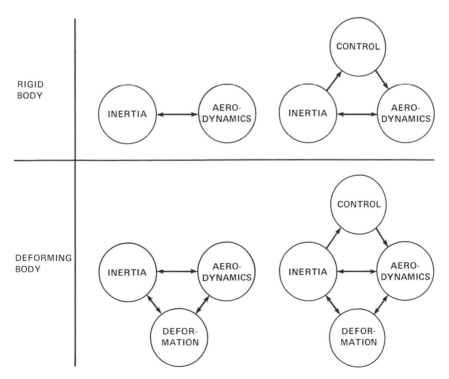

Figure 4.1. Elements of flight-dynamic systems.

aerodynamic forces and moments acting on the vehicle. In this section we will briefly describe this interaction and some of the complexities of the aerodynamic forces and moments.

4.2.1. Coupled Inertial and Aerodynamic System

A block diagram of the coupled system is illustrated in Figure 4.2. The inertial system is defined by the equations of motion for the flight vehicle (Newton's laws for a rigid or elastic body). For a rigid body this is a set of six second-order ordinary differential equations. With appropriate initial conditions and aerodynamic forces and moments, this set of equations provides the motion history of the vehicle and, therefore, the boundary conditions for the aerodynamic system. The aerodynamic system is given by the equations of motion for the fluid flow around the vehicle, a set of nonlinear partial differential equations called the Navier–Stokes equations. With appropriate initial conditions and boundary conditions, the Navier–Stokes equations provide the aerodynamic forces and moments for the vehicle inertial set of equations. The complexity of the interaction, and hence the vehicle motion, results from the complex nature of the aerodynamic forces and moments. *It should be noted that this coupled system has a special characteristic; some of the bifurcation parameters of the fluid dynamic system (e.g., angular orientation) are the dynamic variables of the inertial system describing the vehicle motion.* This can lead to very complex behavior, and at present this interdependency of two systems has received little attention in the study of nonlinear dynamical systems.

At the present time computational methodology and computer power is not adequate to provide time accurate solutions to the Navier–Stokes equations. Even if the methodology existed, the cost of the computations would be prohibitive. Therefore, some form of mathematical modeling of the aerodynamic forces and moments must be used. Before we take up the issue of

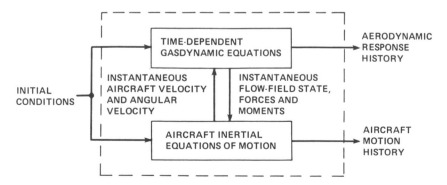

Figure 4.2. Block diagram of a rigid-body flight-dynamic system.

modeling, we will first take a look at the range of aerodynamic characteristics that can arise using simple bodies with quasistatic or forced time dependent boundary conditions and some examples of free interactions.

4.2.2. Quasistatic Aerodynamics

The range of aerodynamic characteristics that occurs when the boundary conditions are not a function of time can be illustrated with a couple of examples. The first is a slender body of revolution being pitched very slowly (quasistatically) through a range of angles of incidence. The types of flow observed around this body are illustrated in Figure 4.3 [11] and the corresponding aerodynamic forces normal to the body are plotted in Figure 4.4 [6] (note this is either the steady or time averaged force; at the higher angles the nature of the unsteady forces is noted). At angles of incidence less than 30° the flow field is steady, and for angles above approximately 5°, it is dominated by a pair of symmetric vortices. The normal force is initially a linear function of the angle of incidence and becomes nonlinear at higher angles. At angles larger than 30°, the vortex system becomes asymmetric and additional vortices are added (see $\alpha = 48°$ in Figure 4.3). The direction of the asymmetry is not predetermined, but the flow is still steady. In this range the normal force is more nonlinear and a side force, out of the plane of the angle of incidence, occurs (not shown). As the angle of incidence increases still further the flow field becomes unsteady, and vortices are shed periodically into the wake (shown in Figure 4.3). At 90° the wake flow is the classic Karman vortex flow behind a circular cylinder in cross flow. There is a range of Reynolds numbers where the wake flow is chaotic at these large angles of incidence (noted in Figure 4.4). This simple body generates a wide range of aerodynamic characteristics in the absence of body dynamics.

Subcritical bifurcations or folds in the solution space can produce hysteresis in the flow and result in another type of aerodynamic characteristic. An example of this is illustrated in Figure 4.5 for the case of a body with a square cross section at an incidence angle of 90° and rotating about the velocity vector at a constant rate [12] (note the boundary conditions on the body are steady in a rotating coordinate system). The side force is well behaved with rotational speed at high Reynolds numbers, but at lower Reynolds numbers there is a pronounced hysteresis loop, and at very low Reynolds numbers the forces change drastically at very low rotation rates. This is the result of complex flow separation on the sides of the body.

The preceding examples for quasistatic conditions illustrate four distinct types of behavior exhibited by aerodynamic forces (the term forces will be used generically to indicate both forces and moments). These, summarized in Figure 4.6, are single valued, periodic, aperiodic/chaotic, and multivalued. We see that the forces acting on the vehicle can be chaotic even before any vehicle dynamics are included.

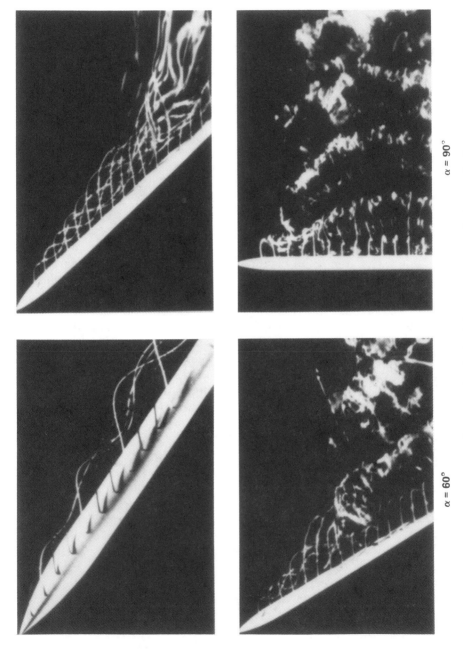

α = 60°

α = 90°

Figure 4.3. Vortex flow on a slender body in a water tunnel (reference [11]).

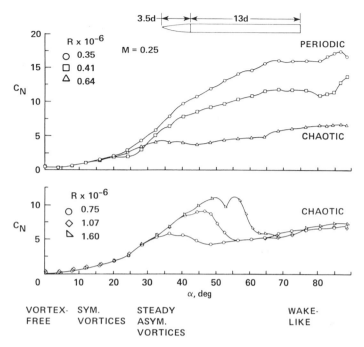

Figure 4.4. Normal-force coefficient for a slender body in a wind tunnel.

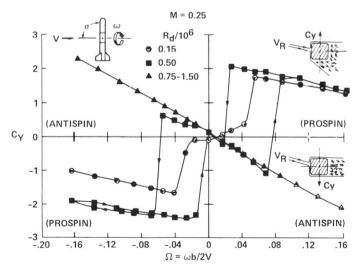

Figure 4.5. Aerodynamic hysteresis: nose side force coefficients on a spinning body with a square cross section (reference [12]).

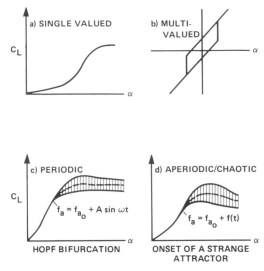

Figure 4.6. Basic types of aerodynamic forces with steady boundary conditions.

4.2.3. Forced Unsteady Aerodynamics

We will now consider the impact of the body dynamics on the aerodynamic forces in the absence of any feedback from the aerodynamic forces. This can be illustrated by considering forced periodic oscillations of the body angle of incidence. From a practical standpoint these oscillations do not have to be very rapid; that is, the reduced frequency $\omega L/U$ can be less than 1. Here, ω is the frequency, L is a characteristic length, and U is the flight velocity. When the aerodynamic forces are single valued under quasistatic conditions, this oscillation normally results in periodic aerodynamic forces with a phase shift relative to the body motion. It is possible that rapid periodic forcing of the body could lead to more complex behavior, such as doubly periodic, quasiperiodic, or even chaotic aerodynamic forces.

When the quasistatic aerodynamic behavior is more complex, the effects of forced oscillations can be much more dramatic. An extreme example of this is dynamic stall of an airfoil. In this case, the oscillations take the airfoil from a condition of steady state, single-valued forces through a Hopf bifurcation, which occurs under quasistatic conditions, and into a periodic domain and back. This results in the very complex flow structure and lift force illustrated in Figure 4.7 [13, 14]. The flow field is periodic during the upstroke but is more complex during the down stroke. At very low reduced frequencies, the lift force during the entire cycle of motion does not deviate very far from the quasistatic results; however, there is a slight hysteresis introduced even at this very low rate of oscillation (note this is only the time averaged lift; the unsteady component at angles above 12° is not shown). As the rate of

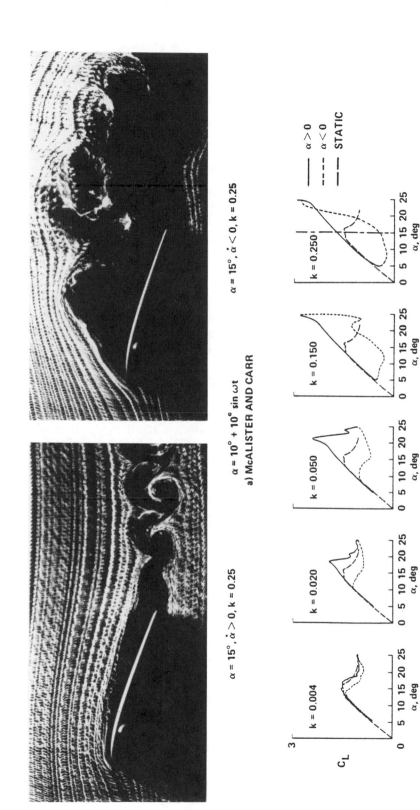

$\alpha = 15°, \dot\alpha > 0, k = 0.25$ $\alpha = 15°, \dot\alpha < 0, k = 0.25$

$\alpha = 10° + 10° \sin \omega t$

a) McALISTER AND CARR

$\alpha = 15° + 10° \sin \omega t$

b) (McALISTER, CARR AND McCROSKEY)

Figure 4.7. Dynamic stall on an oscillating airfoil: (*a*) Flow fields; (*b*) Lift coefficients.

oscillation is increased, the size of the hysteresis loop becomes very large. The flow and its effect on the aerodynamic forces in dynamic stall are not fully understood.

4.2.4. Free Interacting Aerodynamics and Motions

Next, three examples will be described where the body is free to respond to the aerodynamics. The first of these is a delta wing that is free to roll about its longitudinal axis. This problem has been studied extensively by Nelson and his students [15, 16]. At low to moderate angles of incidence a pair of symmetric vortices occurs much in the manner as for the slender body shown in Figure 4.3. At low angles of incidence, small perturbations to the roll angle will decay to zero. At moderate angles small perturbations will grow to a limit cycle (periodic behavior). This appears to be the result of a damping force that is destabilizing to small roll angles but becomes stabilizing at higher roll angles and is a result of the motion of the vortices in response to the wing motion. At still higher angles of incidence an asymmetry occurs in vortex structure (asymmetric vortex bursting) at zero roll angle (see the sketch in Figure 4.8). This leads directly to periodic motion without the introduction of a perturbation.

A second example of interacting systems is an aircraft model in a wind tunnel that is free to roll about the body axis. In this case the flow is

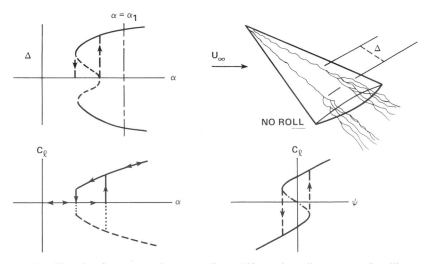

Figure 4.8. Sketch of asymmetric vortex flow, bifurcation diagram, and rolling moment curves for a delta wing.

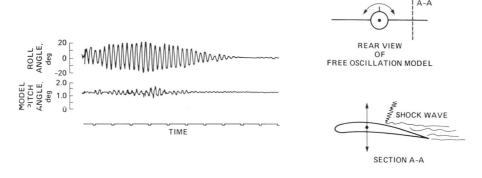

Figure 4.9. Wing rock induced by shock-induced flow separation (reference [17]).

transonic and there are shock waves on the wing. These lead to separation of the flow near the trailing edge of the wing. At low angles of incidence, if perturbed in roll, the wing will return to zero roll angle. At higher angles of incidence a rather curious behavior occurs. The model will begin to oscillate erratically in roll (Figure 4.9 [17]). These oscillations will damp out only to recur in an aperiodic manner. This has some of the characteristics of intermittent chaos and appears to be the result of the shock induced flow separations on each wing panel not being locked onto one another.

A third and much more complex example is that of a free-to-oscillate wire (circular cylinder) in cross flow. This has received considerable attention in the literature, for example, references [18] and [19]. Two examples are shown in Figure 4.10 [18]. In both cases the vortex shedding frequency is near a harmonic frequency of the wire. In the first case, the wire is very heavily damped and shows very little motion. The wake flow, however, exhibits a strong peak in the power spectra at the shedding frequency. For the case where the wire motion is undamped, the wire motion produces a strong peak in the power spectra at the shedding frequency of the vortices. However, now the wake structure appears very chaotic. This chaotic structure may result from the fact that the wire exhibits several standing waves across the wind tunnel and the flow is probably three dimensional and complex.

From these three cases we see that the modeling of aerodynamics is critical for adequate predictions of the vehicle motion. Mathematical modeling of the aerodynamic forces and moments can be difficult even if one considers only quasistatic cases. The cases of dynamic stall, wing rock, and vortex shedding from a circular cylinder pose major problems for mathematical modeling. These are the issues to which we now turn.

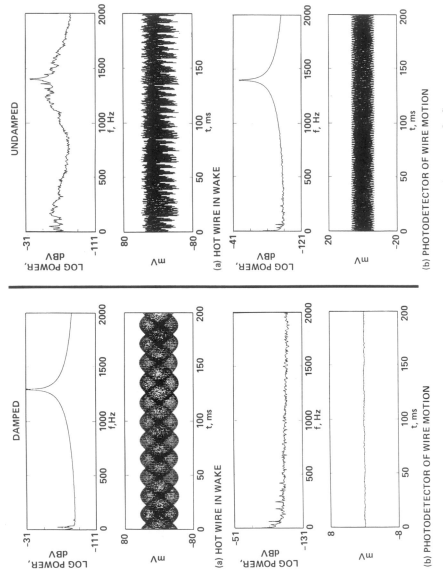

Figure 4.10. Dynamics of a circular cylinder in cross flow (reference [18]).

4.3. AERODYNAMIC MODELING

In order to provide a complete mathematical model for a flight dynamic system that retains its ordinary differential equation form, the aerodynamic forces and moments must be reduced to ordinary differential equations at the most (e.g., reference [20]) or preferably to analytic expressions of motion parameters. The ordinary differential equation approach is not widely used at present. This paper will concentrate on the analytic approach.

4.3.1. General Methodology

Two commonly used approaches will be described briefly: the functional Taylor series expansion approach and the indicial response approach. These two are known to yield the same expressions for single-valued aerodynamics; however, it is not known at the present time if this is the case for more complex aerodynamic behaviors. The first method is more widely used and is easier to apply, whereas the second provides more insight into its limitations and the role of the fluid dynamics. When considering the modeling of aerodynamic forces, one must remember that the aerodynamic forces are functionals. That is, the aerodynamic forces are functions of the vehicle linear and angular velocities, which are themselves functions of time. Hence, for example, in two dimensions the aerodynamic lift force can be written as

$$C_L = C_L[u(t), v(t), q(t)] \tag{4.1a}$$

where C_L is the normalized lift coefficient, $u(t)$ and $v(t)$ are the components of velocity, and $q(t)$ is the angular velocity relative to an inertial coordinate system. In most cases this is rewritten as

$$C_L = C_L[U(t), \alpha(t), q(t)] \tag{4.1b}$$

where $U(t)$ is the resultant velocity and $\alpha(t)$, the angle of incidence, is the arcsine of $v(t)/U(t)$. In most modeling the resultant velocity is assumed constant and C_L becomes

$$C_L = C_L[\alpha(t), q(t)] \tag{4.1c}$$

The functional nature of the aerodynamic forces persists throughout all of these descriptions.

The Taylor series expansion approach to reducing this functional to a function is to expand the flight history functions $\alpha(t)$ and $q(t)$ in Taylor series about some time t_0. For example,

$$\alpha(t) = \alpha(t_0) + \dot{\alpha}(t_0)(t - t_0) + \ddot{\alpha}(t_0)(t - t_0)^2/2 + \cdots$$

Then equation (4.1c) can be written as

$$C_L = C_L(\alpha(t_0), \dot{\alpha}(t_0), \ddot{\alpha}(t_0), \ldots, q(t_0), \dot{q}(t_0), \ddot{q}(t_0), \ldots) \quad (4.2)$$

At first glance the situation does not appear to have improved; a functional has been replaced by a function of an infinite number of variables. In practice, however, most of the derivatives with respect to time are small. For oscillatory motion it can be shown that to first order in reduced frequency this expression becomes

$$C_L = C_L(\alpha) + C_{L_{\dot{\alpha}}}(\alpha)\dot{\alpha} + C_{L_q}(\alpha)q \quad (4.3)$$

Note we have dropped the t_0. The first term on the left is the lift force under quasistatic boundary conditions and the next two are the first order effects of vehicle dynamics. All three coefficients can be nonlinear functions of α. This form is valid for all angles of incidence and low reduced frequencies provided that the quasistatic term is single valued and time independent. In many practical instances this is a very good model, and it is widely used throughout the aircraft industry. Even this rather simple aerodynamic model can result in very complex flight dynamic behavior. Note that the flight dynamic system is nonlinear and of sufficiently high order to allow both periodic and chaotic behaviors. An example is a spring mounted control surface going into periodic motion due to induced shock wave motion on the wing at transonic speeds [21].

The limiting assumption in this approach comes with the Taylor series expansion of the flight history. Both the angle of incidence α and the angular rate q must be continuous because the forces and moments encountered are all finite. However, the coefficients of the time derivatives may not remain small, even at low reduced frequencies of the vehicle, if a rapid change occurs in the structure of the air flow around the vehicle. Even under these conditions the changes will not be instantaneous due to the inertia of the fluid itself, but these can be much more rapid than the vehicle motion and hence the assumption that the higher order terms in the expansion are small may not be valid. This is a little easier to see in the indicial response approach to which we now turn.

In the indicial response approach we consider both the vehicle dynamics and the resulting aerodynamic response. Two vehicle motions are considered. In the first, the base motion, the motion is suddenly frozen at the incidence angle at $\xi = \tau$ as seen in the sketch on the left in Figure 4.11. In the second, the perturbed motion, the motion is the same up to $\xi = \tau$, at which point there is a small incremental step in the angle of incidence. The resulting aerodynamic response is illustrated in the sketch on the right in Figure 4.11. The indicial lift response C_{L_α} is the difference between the lift for these two vehicle motions divided by the incremental change in angle of incidence $\Delta\alpha$ in the limit of $\Delta\alpha \to 0$. The generalized superposition integral for the

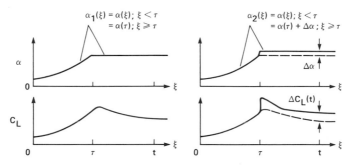

Figure 4.11. Indicial response approach.

response in C_L to an arbitrary angle of incidence variation is given as

$$C_L(t) = \int_{-\infty}^{t} C_{L_\alpha}(\alpha(\xi); t, \tau) \frac{d\alpha}{d\xi} \, d\xi \qquad (4.4)$$

Note that the indicial response depends on the past history and on the values of t and τ explicitly. From this point a series of approximations similar to that of the Taylor series discussed previously is employed to reduce this expression to that of equation (4.3). However, at this point it is already apparent on how the modeling breaks down under conditions where the quasistatic boundary conditions yield unsteady flows or bifurcations or folds occur. They all revolve around the indicial response function C_{L_α}. First, if a subcritical bifurcation or fold in the solution space occurs at $t = \tau$, the indicial response function does not exist for some values of t. Also, if the long term response for the base motion is periodic, then the response function determined with the perturbed motion is also periodic but not unique; the phase of the periodic indicial response function is a function of the history prior to the perturbation. It should be noted that the indicial response approach for single-valued, time-invariant equilibrium states has been derived from the Navier–Stokes equations by Truong and Tobak [22]. We will now consider the modeling and motion response of a simple system to show some of the limitations and behavior manifest in system simulations.

4.3.2. Oscillating Circular Cylinder in Cross Flow

This is a classic problem involving the interaction of an inertial system and a fluid dynamic system [23]. It is often called the galloping telegraph wire problem and is one of the possible causes of the Tacoma Narrows bridge failure. An extensive nonlinear structural analysis of the Tacoma Narrows

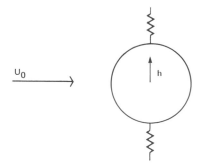

Figure 4.12. Sketch of a circular cylinder in cross flow.

bridge failure problem is presented in reference [24]. However, little attention has been paid to the highly nonlinear nature of the fluid dynamic interaction that is the principal forcing function. The interaction of an elastically mounted circular cylinder in cross flow has been widely studied from both an experimental and mathematical standpoint, and several different levels of modeling have been used. It is normally modeled as a 1-degree-of-freedom oscillation with two-dimensional fluid cross flow as illustrated in Figure 4.12. We will concentrate on the effects of the aerodynamic modeling for the 1-degree-of-freedom system as well as allowing a second degree of freedom in the direction of the free stream.

First, the 1-degree-of-freedom system will be examined. The equation of motion for this system is

$$m\ddot{h} + 2\mu\dot{h} + k^2 h = L(t)$$

where m is the mass of the cylinder, μ is the mechanical damping, k is the spring constant, and $L(t)$ is the lift force per unit length of the cylinder. The simplest form for modeling the aerodynamics is to use the lift from the quasistatic boundary conditions,

$$L(t) = \tfrac{1}{2}\rho \, dU_0^2 \, C_L \sin(\omega t) \tag{4.5}$$

where ρ is the air density, d is the diameter of the cylinder, U_0 is the free-stream velocity of the air, C_L is the amplitude of the unsteady lift coefficient, and ω is the frequency of the periodic lift force generated by the vortices shedding into the wake. Under quasistatic boundary conditions all these terms are constant and $L(t)$ is periodic with frequency ω. The dynamics of this case are those of a classic damped harmonic oscillator forced at the frequency ω, and they are not particularly interesting. It does provide correct dynamics of the system under some limited conditions but not those of interest, such as near resonance.

A slightly more complete quasistatic modeling was proposed in a paper by Tobak, Chapman, and Unal [8]. They proposed modifying the quasistatic model by taking into account the cylinder dynamics directly. This model assumes that the aerodynamic coefficients are constant, the reduced shedding frequency is constant (Strouhal number is constant), but the velocity magnitude and direction must account for the vertical motion of the cylinder. This results in the equation

$$m\ddot{h} + 2\mu\dot{h} + k^2 h = \tfrac{1}{2}\rho\, dU^2\left[C_D\dot{h}/U + C_L(U_0/U)\sin(k_S Ut/d)\right] \quad (4.6)$$

where $U = \sqrt{U_0^2 + \dot{h}^2}$, C_D is the normalized drag coefficient, and k_S is the Strouhal number. This can be simplified by normalizing h by the diameter of the cylinder d and normalizing t by the time it takes a fluid element to move one body diameter U_0/d. With this normalization equation (4.6) becomes

$$\ddot{h} = 2\mu\dot{h} + k^2 h = \beta\sqrt{1 + \dot{h}^2}\left[C_D\dot{h} + C_L\sin\left(k_S t\sqrt{1 + \dot{h}^2}\right)\right] \quad (4.7)$$

Here, β is a function of ρ, d, and m; μ and k have been normalized in a manner as to make equation (4.7) nondimensional.

This equation has been studied extensively by Yates, Unal, Szady, and Chapman [25]. The results are summarized here to illustrate the impact of the modeling. Three sets of cases were considered in reference [25]: a high level of forcing used to demonstrate characteristic long term behavior, a medium level of forcing similar to conditions of a hollow cylinder tested in air [19], and a low level of forcing similar to conditions of a wire tested in water [18].

The first thing that should be noted about this model is that the forcing term (sine) exhibits a very strange behavior that dictates the long term behavior in all cases and in the final analysis dooms this model to failure. However, this model does produce, in the short term, lock-in, harmonics, side bands, and transient chaotic behavior. These are all observed phenomena and could easily be mistaken for correct simulation.

The long term behavior is illustrated for the large forcing and large natural frequency $(k > k_S)$ in Figure 4.13. At early times the motion is sinusoidal, but by $t = 1580$ (body diameters traveled) the behavior is starting to deviate considerably. Note the behavior of the forcing term. As time progresses the motion, although periodic at the Strouhal frequency, is becoming sawtoothed and the forcing function is a narrow spike alternating in sign at the Strouhal frequency. Here we see that the motion has locked to the Strouhal frequency, but the behavior of the forcing is not physically real.

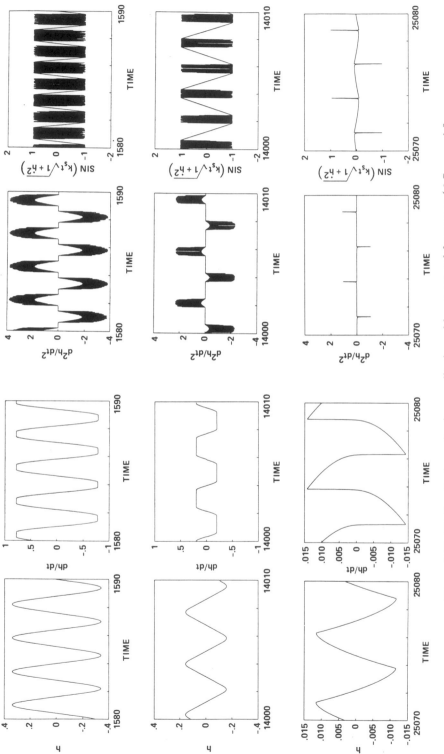

Figure 4.13. Motion for large forcing amplitude and large natural frequency ($\beta C_D = -1.0$, $\beta C_L = 1.0$, $\mu = 0.0054$, and $k = 1.1 k_S$).

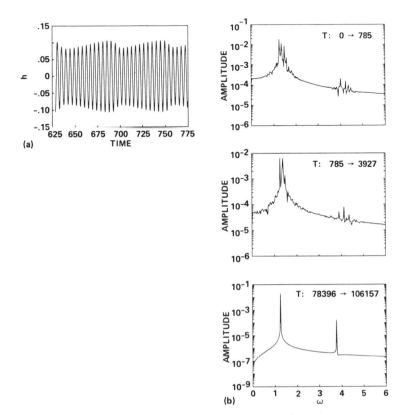

Figure 4.14. Medium forcing and large natural frequency ($\beta C_D = -0.0294$, $\beta C_L = 0.0147$, $\mu = 0.0054$, and $k = 1.1 k_S$): (a) Short term motion; (b) Fourier transforms.

Computed results for the large forcing amplitude and a small natural frequency show the same long term behavior, but the long term behavior is assumed more quickly, and the motion is damped. An analysis of the long term behavior shows that the motion damps as $t^{1/2}$. This rate of damping results directly from the motion minimizing the forcing term, and it agrees very well with the computed motion for the low natural frequency. At the large natural frequency there is evidence of transient chaos (not shown); this was not evident in the case of the low natural frequency.

The moderate forcing cases also showed evidence of going toward this long term behavior, but it took an order of magnitude longer in time to even approach this condition. In the case where the frequencies were matched, there was evidence of transient chaos and the power spectra showed considerable broadening around the natural frequency and at the odd harmonics. This is illustrated in Figure 4.14.

When the forcing is small, the time to reach the long term behavior is even longer. Again when the natural frequency matches or is larger than the

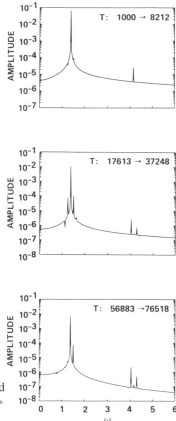

Figure 4.15. Fourier transforms for low forcing and large natural frequency ($\beta C_D = -0.00020$, $\beta C_L = 0.00007$, $\mu = 0.00001$, and $k = 1.1 k_S$).

Strouhal frequency there is transient chaos. The power spectra for the higher frequency case are shown in Figure 4.15.

Although this model does exhibit much of the behavior observed in experiments on oscillating cylinders, the strange long term behavior would appear to doom this model. Before discarding the model the features that give rise to this strange behavior were examined further to see what might be missing in the model. The strange behavior is a result of the forcing term. Analysis of this term shows that it is not invariant under a shift of the time origin. This property also makes the long term behavior sensitive to disturbances, and perturbations may in effect reset the time origin. Hence, any disturbances, such as fluctuations in the air flow or motion of the cylinder in another degree of freedom, might break up the physically unrealistic behavior. To that end, the 1-degree-of-freedom model was extended to 2 degrees of freedom, retaining the same assumptions of an extended quasistatic aerodynamic model. In this case the total velocity must include both components of velocity introduced by the cylinder motion. The equations for the 2-

degrees-of-freedom case are

$$\ddot{h}_1 + 2\mu\dot{h}_1 + k^2 h_1 = \beta\sqrt{\left(1 + \dot{h}_2\right)^2 + \dot{h}_1^2}$$

$$\times\left[C_D\dot{h}_1 + C_L \sin\left(k_S t\sqrt{\left(1 + \dot{h}_2\right)^2 + \dot{h}_1^2}\right)\right]$$

$$\ddot{h}_2 + 2\mu\dot{h}_2 + k^2 h_2 = \beta\sqrt{\left(1 + \dot{h}_2\right)^2 + \dot{h}_1^2}$$

$$\times\left[C_D\left(1 + \dot{h}_2\right) - \dot{h}_1 C_L \sin\left(k_S t\sqrt{\left(1 + \dot{h}_2\right)^2 + \dot{h}_1^2}\right)\right]$$

$$(4.8)$$

where h_1 is the displacement of the cylinder in the free-stream direction and h_2 is the displacement perpendicular to the free stream. As in the one-dimensional case, three different levels of forcing were used to demonstrate the characteristic long term behavior, if any, and to duplicate the experimental conditions for the hollow cylinder in air [19] and the wire in water [18].

The motions for the large forcing cases are shown in Figure 4.16. Unlike the one-dimensional case, the motion does not die out, but a dependency on the time origin remains. The position of the wire perpendicular to the free-stream direction, h_1, changes gradually and almost linearly until it reaches a peak value equal to βC_D. After it reaches this peak value, it suddenly decreases and oscillates about zero. It then again begins its slow, linear increase. This pattern of slow, linear increases and subsequent rapid decreases in h_1 occurs repetitively. However, two different mechanisms prevent the motion from becoming periodic. First, the length of time for each cycle of slow, linear increase and subsequent rapid decrease in h_1 generally increases as time increases. However, this change in period is not monotonic; any given cycle can take a slightly longer or a slightly shorter time than the previous cycle. Second, when h_1 is increasing linearly, the sign of h_1 can be either positive or negative, and there is no regular pattern to the sign changes. This period and phase shifting result in the motion behaving chaotically.

The displacement parallel to the free stream, h_2, is much smaller than the displacement perpendicular to the free-stream direction. It does not exhibit the linear increases nor the sign changes observed for h_1. It does have a repetitive pattern of slow changes followed by rapid changes. The slow changes in h_2 coincide with the slow, linear changes in h_1, and the coupling of the two motions prevents high frequency oscillations from occurring in the acceleration and forcing function. The rapid changes in h_2 coincide with the rapid changes in h_1 and with the high frequency oscillations in the forcing function.

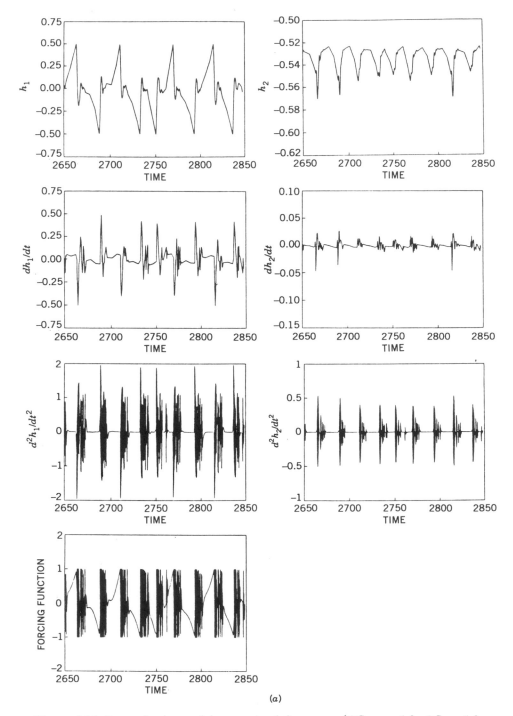

Figure 4.16. Large forcing and large natural frequency ($\beta C_D = -1.0$, $\beta C_L = 1.0$, $\mu = 0.0054$, and $k = 1.1k_S$): (a) Motions.

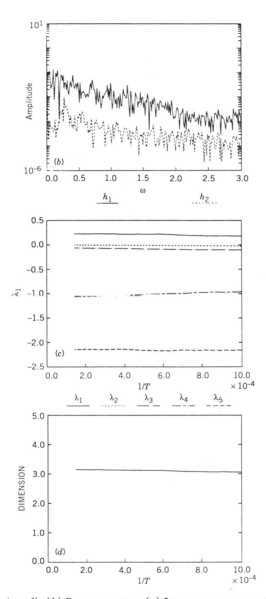

Figure 4.16. *(Continued).* (*b*) Power spectra; (*c*) Lyapunov exponents; (*d*) dimension.

For the large forcing cases, the Fourier transforms of the position are noisy and do not have any well defined peaks (Figure 4.16). This absence of peaks is caused by the continually changing period of the motion and by the sign changes.

The solutions for the medium forcing cases are chaotic and show no tendencies toward periodic or fixed point solutions. For $k > k_S$ the results are plotted in Figure 4.17. Although the peak velocity perpendicular to the free stream is approximately 2 orders of magnitude greater than the peak velocity parallel to the free stream, the motions in both directions are similar. The displacements, velocities, and accelerations have large variations in their amplitudes. They appear to die down to a constant value; then a sudden burst occurs and the amplitude increases dramatically. This bursting occurs when the high frequency oscillations in the aerodynamic forcing function die out, and the forcing function again can excite the system. This cycle of decay and excitation follows no regular pattern, and the Fourier transforms of the displacements in Figure 4.17 show small and very broad peaks at the Strouhal frequency of the system. Solutions started at large times $t = 10^6$ also show this cycle of decay and excitation. However, the peak amplitudes of the bursts are smaller, and the frequency of the oscillations in the burst region is higher. The amplitudes may become larger and the frequencies smaller as time increases, or this may represent a true change in the character of the solution, indicating that the system does again have a dependency on the time origin.

The short term solutions obtained for $k > k_S$ show an interesting transition to the chaotic behavior previously described. In Figure 4.17a, the motion initially appears periodic and the frequencies of both h_1 and h_2 are equal to the Strouhal frequency. There is a short transition period between this periodic motion and the chaotic motion that shows period doubling; the motion during this transition period is composed of terms with frequencies equal to the Strouhal frequency and to half the Strouhal frequency. This doubly periodic motion sustains itself for several hundred time units until the motion finally becomes chaotic.

These computed results do not agree with the experimental results reported in Bearman [23]. Although there are Reynolds numbers at which vortex shedding will produce chaotic motions, the experimental motions for these values of βC_D and βC_L are reported to be periodic. There are also no experimental indications that either the amplitude or the character of the motion changes as time increases. Furthermore, the experimental amplitudes are an order of magnitude greater than the computed amplitudes.

For the low forcing case with $k > k_S$ illustrated in Figure 4.18, the motions are predominantly periodic functions composed of multiple frequencies. The ratios of the amplitudes of these multiple frequency terms do vary with time; however, the long term motion shows no signs of becoming a one frequency motion or a nonperiodic motion. The peak amplitudes of wire velocity for the unmatched frequency case are generally small and not large

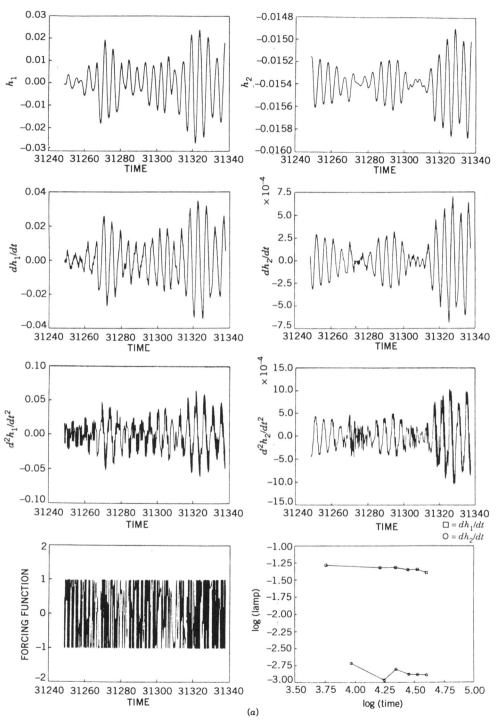

Figure 4.17. Medium forcing and large natural frequency ($\beta C_D = -0.0294$, $\beta C_L = 0.0147$, $\mu = 0.0054$, and $k = 1.1 k_S$): (*a*) Motions.

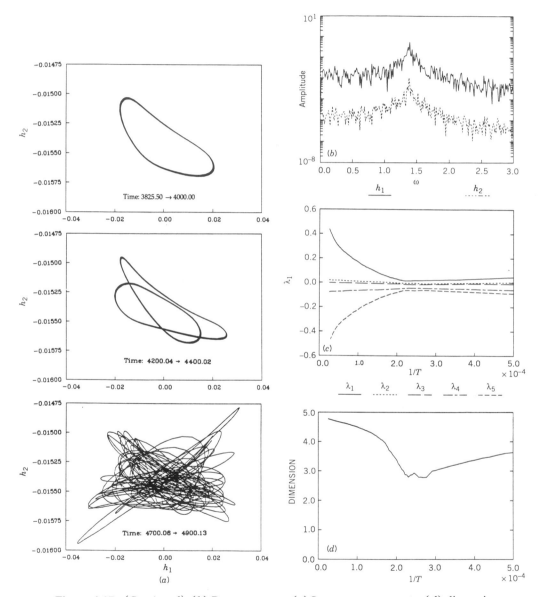

Figure 4.17. *(Continued).* (*b*) Power spectra; (*c*) Lyapunov exponents; (*d*) dimension.

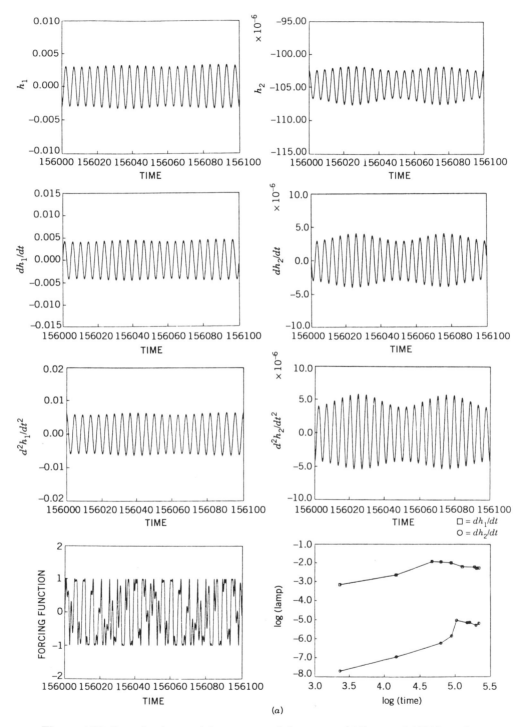

Figure 4.18. Low forcing and large natural frequency ($\beta C_D = -0.00020$, $\beta C_L = 0.00007$, $\mu = 0.00001$, and $k = 1.1k_S$): (a) Motions.

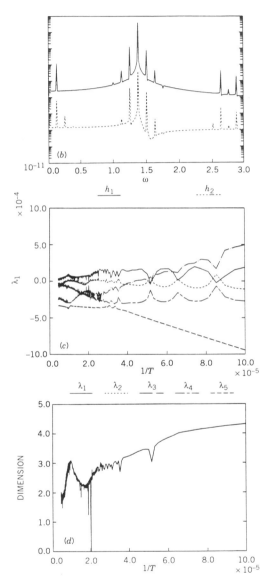

Figure 4.18. *(Continued)*. (*b*) Power spectra; (*c*) Lyapunov exponents; (*d*) dimension.

enough to introduce high frequency oscillations into the aerodynamic forcing function or the acceleration. The amplitudes do appear to increase slightly or remain more or less constant, and it is expected that as time increases, high frequency oscillations eventually will occur and the amplitude of the motion will decrease.

In experiments where the Strouhal frequency was not matched to a subharmonic of the natural frequency of the wire, the Fourier transforms in Van Atta and Gharib [18] did show peaks at the Strouhal frequency, at the closest subharmonic of the natural frequency, and at their sums and differences. These peaks are also observed in the computational results. However, in the experimental data, the Strouhal frequency peak was higher than that of the subharmonic of the natural frequency; in the computed data, the natural frequency peak is much higher than the Strouhal frequency peak.

For all three forcing amplitudes, the addition of a second degree of freedom to the system for the motion of a cylinder induced by vortex shedding does prevent the computed motion from assuming the long term motion of the 1-degree-of-freedom system, a motion that minimizes the aerodynamic forcing function and is periodic and damped. Depending upon the choice of the forcing amplitude, the 2-degrees-of-freedom system gives persistent and distinct periodic and chaotic motions; however, a dependency on the time origin of the system remains.

4.4. DIMENSIONALITY

With increased interest in nonlinear systems, the need to characterize their behavior has become very important. One sure indicator of chaos is the existence of a positive Lyapunov exponent; Lyapunov exponents measure the rate of expansion (positive) or contraction (negative) of individual components of a multidimensional dynamical system. For a dissipative dynamical system, the total space must be contracting, and the sum of the Lyapunov exponents therefore must be negative.

A method of characterizing a chaotic attractor is the dimensionality of the attracting subspace. Several methods are available for determining this dimension. An upper bound can be obtained using the Kaplan–Yorke conjecture with the Lyapunov exponents [26]. This method only provides an upper bound, and it is not easy to apply to experimental data where the Lyapunov exponents must be constructed. Several other methods for determining the dimension of a dynamical system that are readily applicable to experimental data also have come into use. These methods include the box-counting methods (e.g., Mandelbrodt [27]), the nearest-neighbor method of Pettis et al. [28], and the correlation method of Grassberger and Procaccia [10]. Although these methods for determining dimensions are theoretically valid only for an infinite number of data points in a bounded region, they have been effectively used for sets with finite numbers of data points.

However, there has been little done with regard to the limitations and accuracy of such methods under conditions for practical application.

In this section we will first examine the use of Lyapunov exponents and the Kaplan–Yorke conjecture on the 2-degrees-of-freedom oscillator described in the previous section. Second, we will examine the limitations and accuracy imposed by real world constraints on dimensions determined using the Grassberger–Procaccia algorithm.

4.4.1. Lyapunov Exponents and Dimension

Lyapunov exponents and the Kaplan–Yorke conjecture will be used to assist in the classification of the motions for the 2-degrees-of-freedom model of the oscillating cylinder. The Kaplan–Yorke conjecture provides an estimated upper bound to the dimension (Lyapunov dimension, D_L) of the system; this upper bound is given by

$$D_L = j + \frac{\sum_{i=1}^{j}\lambda_i}{|\lambda_{j+1}|} \qquad (4.9)$$

where j is the largest integer such that $\lambda_1 + \lambda_2 + \lambda_3 + \cdots + \lambda_j \geq 0$, and λ_i is the ith Lyapunov exponent.

For the large forcing case (Figure 4.16), there is one bounded, positive Lyapunov exponent, one Lyapunov exponent with a value of zero, and three bounded, negative Lyapunov exponents. The Kaplan–Yorke dimension is well behaved and smoothly approaches a value slightly greater than 3.

For the medium forcing case (Figure 4.17), there is a sudden change in the trends of the Lyapunov exponents and of the conjectured dimension. Initially, the motion is periodic or doubly periodic, and the magnitudes of all of the Lyapunov exponents decrease as time increases. The Kaplan–Yorke conjecture indicates that the dimension of the system is decreasing to some value less than 3. When the motion becomes chaotic and tries to fill the space, one of the Lyapunov exponents becomes positive and unbounded, one of the exponents becomes negative and unbounded, and the conjectured dimension approaches 5. The Lyapunov exponents and the Kaplan–Yorke dimension do reflect the change from a periodic to a chaotic solution.

For the low forcing case, the Lyapunov exponents shown in Figure 4.18 are small and noisy. However, these exponents do show definite trends. Two exponents, one positive and one negative, appear to asymptotically approach zero. The other three remain negative. This implies that the asymptotic solution lies on a two-torus and that the two frequencies observed in the solution are incommensurate. The Kaplan–Yorke dimension does seem to be approaching a value near 2; however, it is too early in the computation to obtain a valid asymptotic value for the dimension.

The calculated Lyapunov exponents and the Kaplan–Yorke dimension do reflect the differences between the types of motions obtained with this 2-degrees-of-freedom system. For computed motions, such as those for this system, the Lyapunov exponents are readily available. Unfortunately, as noted earlier, constructing the Lyapunov exponents and using them to characterize the dimension of an experimental system can prove difficult. There are other methods that work well with experimental data; however, the practical use of these methods with real data has not been explored extensively.

4.4.2. Grassberger and Procaccia Algorithm

In the study of real world constraints on dimension calculations, we will first examine the constraints associated with error-free data, such as finite number, distribution of data points, object scales, and lacunarity [27]. Second, the effects of measurement errors are considered. These two sets of issues will be illustrated using simple geometric objects and a cantor set. The third and final set of issues arises in the way data are collected in the study of dynamic systems, such as in Poincaré sections or partial phase space information. This final set of issues will be illustrated using a well studied dynamic system.

The correlation dimension ν is defined by the Grassberger–Procaccia algorithm as

$$C(r) = \lim_{\substack{N \to \infty \\ r \to 0}} \frac{1}{N^2} \sum_{\substack{i=1 \\ j=1}}^{N} \Theta\big(r - |\mathbf{x}_i - \mathbf{x}_j|\big) = Br^\nu$$

where N is the total number of points and Θ is the Heaviside function. The summation over j simply counts the number of data points x_j in a reference sphere of radius r centered on the point x_i. The summation over i is the summation over all possible reference spheres centered about all data points. In practice there is not an infinite number of data points, and the limits $N \to \infty$ and $r \to 0$ cannot be taken. Instead, the dimension is found using a range of rs, $r_2 < r < r_3$, where r_2 is proportional to the minimum spacing of the data and r_3 is related to the object size and the location of the reference points. Even with a finite but large number of data points, this algorithm can take considerable computer time to evaluate; hence, for this study it has been modified to use only a subset of the data points as reference points:

$$C'(r) = \frac{1}{N_{\text{ref}}} \sum_{i=1}^{N_{\text{ref}}} \frac{1}{N} \sum_{j=1}^{N} \Theta\big(r - |\mathbf{x}_i - \mathbf{x}_j|\big) = B'r^{\nu'} \qquad (4.10)$$

Here N_{ref} is the number of reference points and N is the number of data

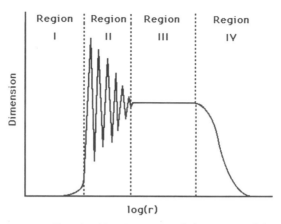

Figure 4.19. Sketch of local slope; log(C) versus log(r) curve.

points. Generally N_{ref} is much less than N. The dimension can be determined by finding the slope of the line $\log[C'(r)] = v' \log(r) + \log(B')$ or by plotting the local slope of $v'(r)$ versus $\log(r)$. In general, using a least squares fit to determine the slope of the line provides better results; the plots of the local slope identify regions where the least squares fit should be made.

In every plot of the local slope v' versus $\log(r)$ there are four distinct regions as shown in Figure 4.19. In region I, the radii of the reference spheres are smaller than the data spacing, hence no data points are added to the reference spheres as the radii are increased. Therefore, there is no information concerning the dimension of the object and local slope is zero. In region II, the local slope is erratic, a result of discrete data and noisy data. In region III, there is a plateau at the dimension of the system. In region IV, the reference spheres extend beyond the object size and the plateau rolls off. The extent of these regions can be determined theoretically, and minimal conditions for a plateau at the dimension of the object can be specified.

4.4.3. Data without Errors

Six factors that affect the length and quality of the plateau region are the number of data points, the number of reference points, the incremental increase in the radius of the reference sphere, the scales of the object, lacunarity, and the distribution of data points (equally spaced or random). The first five of these will be examined first with equally spaced data on simple objects, such as a straight line, a sawtooth line, a circle, and a triadic cantor set. The effect of the data distribution then will be demonstrated for randomly distributed data.

In Figure 4.20 local slopes are plotted for a straight line, $0 < x < 1$. The data points are equally spaced on the line. The reference points coincide with

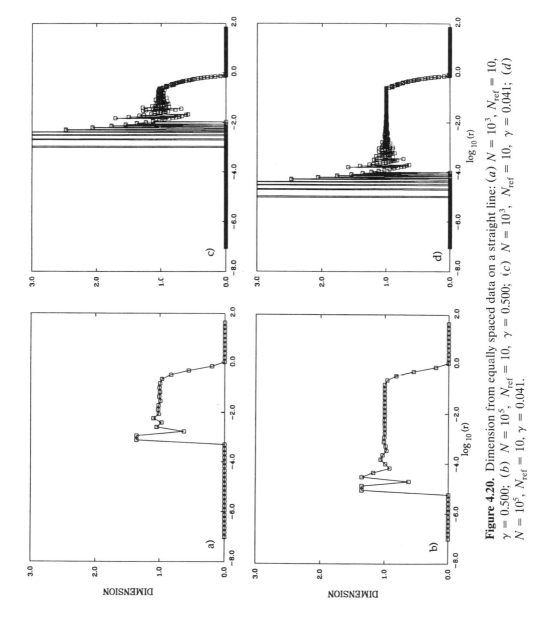

Figure 4.20. Dimension from equally spaced data on a straight line: (*a*) $N = 10^3$, $N_{\text{ref}} = 10$, $\gamma = 0.500$; (*b*) $N = 10^5$, $N_{\text{ref}} = 10$, $\gamma = 0.500$; (*c*) $N = 10^3$, $N_{\text{ref}} = 10$, $\gamma = 0.041$; (*d*) $N = 10^5$, $N_{\text{ref}} = 10$, $\gamma = 0.041$.

data points and are also equally spaced; however, they are confined to a region $0.25 < x < 0.75$.

Region I persists until there is an average of more than one point per reference sphere. Because the data are equally spaced, no data points other than the reference point lie inside each reference sphere as long as the radius of the reference sphere is less than the data spacing. However, when the radius of the reference sphere equals the data spacing, a data point is added to each reference sphere and region I ends abruptly. It can be readily shown that end of region I is given by

$$\log(r_1) = \frac{1}{D} \log(V) - \frac{1}{D} \log(N) \tag{4.11}$$

where r_1 is the radius of the reference sphere, D is the dimension of the system or object, V is the volume of the object, and N is the number of data points. This expression is in excellent agreement with the results shown in Figure 4.20; the end of region I moves to the left as the number of data points increases. Similar agreement between equation (4.11) and the end of region I were also obtained with a two-dimensional plane.

In region II the local slope oscillates wildly and does not immediately settle down to a constant value. The length of this region to the right of region I depends strongly on the rate of increase in the size of the reference spheres, on the dimension, and on the accuracy that one desires for the dimension. An expression for the length of region II is given as (see Appendix)

$$\log(r_2) - \log(r_1) = +\frac{1}{D} \log\left(\frac{2^{D-1}D!}{A}\right)$$
$$-\left\{ 2\log(1 + \gamma) + \frac{1}{D} \log[\varepsilon \ln(1 + \gamma)] \right\} \tag{4.12}$$

where γ is the rate of increase of the reference sphere radius, A is a geometrical constant, and ε is the acceptable level of error in the dimension. Note that the length of region II is independent of the number of data points; however, it is strongly dependent on the dimension of the object.

In Figure 4.20 the local slope is plotted for two choices of N and for $\gamma = 0.5$ and 0.041. As the number of data points increases, region II moves to the left but its length remains constant. However, when γ changes, the length of region II changes. Based on equation (4.12) the predicted length of region II for a line of unit length and a 1% error is 1.7 for $\gamma = 0.5$ and 3.1 for $\gamma = 0.041$. These results are in good agreement with the plotted results in Figure 4.20. Good agreement between equation (4.12) and the calculated end of region II was also obtained for a flat plane.

Region III is the plateau region at which the dimension of the object can be determined. It begins where region II ends and persists until the scales of the object begin to affect the local slope. Two factors determine where the plateau region ends. The first is object size. In Figure 4.20 the plot of local slope shows a decrease to zero as the radii of the reference spheres increase. There is some minimum radius where a portion of at least one of the reference spheres lies outside the object. As the radius is increased, only a portion of the points necessary to keep the local slope constant is added to that reference sphere, and a gradual drop off from the plateau region occurs. As the radius continues to increase, not only does the portion become smaller, but more reference spheres extend beyond the object, and the local slope continues to drop off. The radius at which the initial drop off occurs is simply the smallest distance from the origin of a reference sphere to the edge of the object. This is the beginning of region IV and can be stated as

$$r_3 = \kappa L \tag{4.13}$$

where κ (a number less than $\frac{1}{2}$) represents the distance to edge of the object in terms of the size of the object L, the smallest distance across the object.

For the line of unit length, the roll-off from the plateau region plotted in Figure 4.20 is smooth. It begins at $r_3 = 0.25$ ($\kappa = 0.25$) and continues until the line lies completely inside every sphere at $r = 0.75$. Region IV for other objects may be more complicated. There may be several rapid decreases in local slope as the radius is increased and additional reference spheres or portions of reference spheres extend past the object.

For equally spaced data on simple objects, the plots of local slope will be identical in regions I, II, and III for any number of reference points. The theoretical analysis indicates this and numerical examples (not shown) support this conclusion. It can change region IV, the roll-off region, because the choice of the number of reference points may alter their location relative to the edge of the object.

A second contributing factor to the onset of region IV is the presence of other scales such as folding or curvature of the object. For instance, in the case of a sawtooth line shown in Figure 4.21, there are only a finite number of folds, and the dimension of the object should be equal to the dimension of the line. The algorithm does give a value of 1 for a small range of radii. However, as the size of the reference spheres increases, they start picking up data on neighboring portions of the sawtooth line, and the local slope becomes greater than 1.

A circle can be used to illustrate how curvature can affect the dimension of even the simplest system. As the radii of the reference spheres approach the diameter of the circle, the number of data points added with increasing reference sphere radius increases rapidly and the local slope increases above the plateau dimension but remains bounded because of the finite nature of the data. This is illustrated in Figure 4.22 for $\gamma = 0.041$.

It is difficult to obtain specific analytic limits for the effects of these secondary object scales because they are object dependent. It is sufficient for

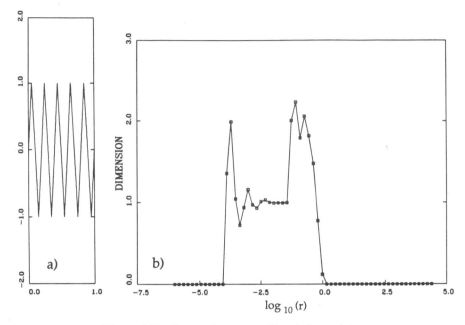

Figure 4.21. Sawtooth curve and local slope plot.

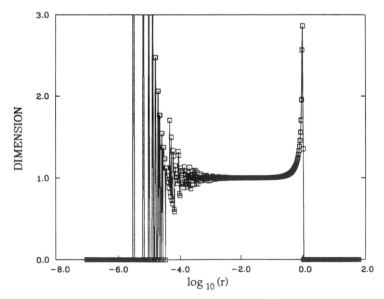

Figure 4.22. Local slope plots for a circle ($N = 10^6$, $N_{ref} = 10$, $\gamma = 0.041$).

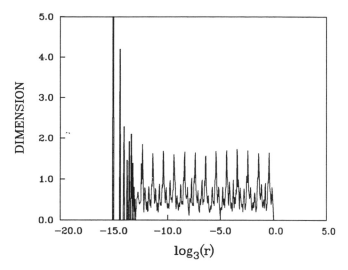

Figure 4.23. Local slope plot for a triadic Cantor set.

our purpose here to draw attention to the limitations imposed on the length of the plateau by these secondary scales.

For many systems, such as the line discussed previously, the plateau region is flat and the dimension can easily be determined from plots of the local slope. However, in cases where the object has fractal properties such as bandedness or what is generally referred to as lacunarity, a well defined plateau may not be observed. For instance, consider the local slope plotted for a triadic Cantor set in Figure 4.23. The data representing the Cantor set have been chosen as the end points of the closed line segments that make up the Cantor set on the 16th level. As the radii of the reference spheres are increased, empty regions are encountered where no data points are added to the reference spheres and sudden drops occur in the local slope plots. As the radii are further increased, a band of data points is encountered, data are added, and the local slope increases. Because the Cantor set is self-similar and the empty regions are of length $1/3^n$, the drops and increases in local slope should repeat themselves whenever the radii of the reference spheres increase by a factor of 3. This repetitive pattern is observed in Figure 4.23. The raggedness of the plateau region makes it impossible to determine the dimension from the local slope plots. However, a least squares fit of the slope of the curve $\log[C'(r)] = \nu' \log(r) + B$ does give the theoretical dimension of the triadic set ($= 0.6309$) to four significant digits.

All of the foregoing results and discussion were for equally spaced data. For randomly distributed data, the analysis previously given does not hold. Instead, statistical methods must be used to determine where the various regions begin. A line of unit length was used to study the effect of random distribution of data points. Although the data points were randomly dis-

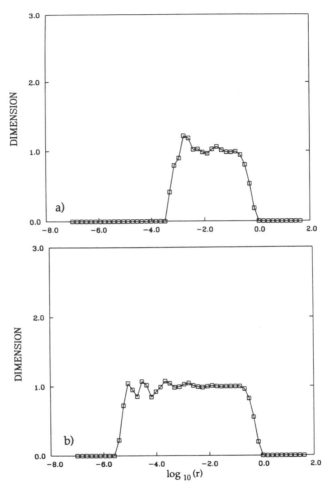

Figure 4.24. Dimension from randomly spaced data on a straight line: (*a*) $N = 10^3$, $N_{\text{ref}} = 10$, $\gamma = 0.500$; (*b*) $N = 10^5$, $N_{\text{ref}} = 10$, $\gamma = 0.500$; (*c*) $N = 10^3$, $N_{\text{ref}} = 100$, $\gamma = 0.500$; (*d*) $N = 10^5$, $N_{\text{ref}} = 100$, $\gamma = 0.500$.

tributed on the object, the reference points were not; they were equally spaced between $0.25 < x < 0.75$. The local slopes are plotted in Figure 4.24. Region I ends and region II begins when there is an average of more than one point per reference sphere. Region II can be defined to begin when there is a 50% chance that on the average there is one data point in each reference sphere. In general this occurs when

$$\log(r_1) = \frac{1}{D} \left[\log(V) - \log(N) - \log(A) \right]$$

For a line of unit length, this yields $\log(r_1) = -0.3 - \log(N)$. This agrees with the results shown in Figure 4.24.

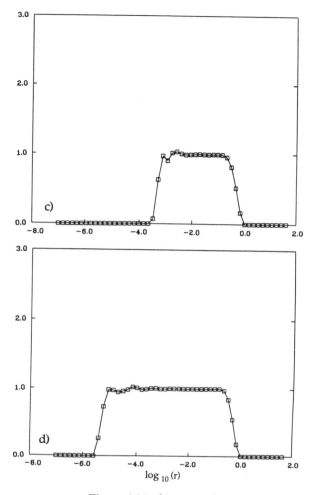

Figure 4.24. *(Continued)*

From statistical analysis, the criteria for the end of region II and the beginning of region III (see Appendix) is

$$\log(r_2) = -\frac{1}{D} \log\left(\frac{2 N_{\text{ref}} AN}{V}\right) - 2\left\{2\log(1 + \gamma) + \frac{1}{D}\log[\varepsilon \ln(1 + \gamma)]\right\}$$

$$(4.14)$$

where N_{ref} is the number of reference points. The length of region II is given by

$$\log(r_2) - \log(r_1) = -\frac{1}{D}\log(2 N_{\text{ref}})$$

$$- 2\left\{2\log(1 + \gamma) + \frac{1}{D}\log[\varepsilon \ln(1 + \gamma)]\right\}$$

As is the case for equally spaced data, the length of region II is independent of the number of data points. However, unlike the equally spaced data case, the number of reference points does have an effect on the length of region II. For a line of unit length, this expression becomes

$$\log(r_2) - \log(r_1) = -\log(2N_{ref}) - 2\{2\log(1 + \gamma) + \log[\varepsilon \ln(1 + \gamma)]\}$$

Because of the factor of 2 before the brackets, the length of region II can be longer for randomly distributed data than for equally spaced data and,

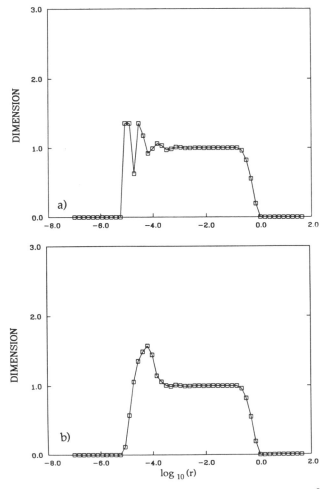

Figure 4.25. Dimensions determined from data with errors ($N = 10^5$, $N_{ref} = 10$, $\gamma = 0.500$): (*a*) Maximum error 10^{-5}; (*b*) maximum error 10^{-4}; (*c*) maximum error 10^{-3}; (*d*) maximum error 10^{-2}.

therefore, the length of the plateau region can be shorter. However, the number of reference points is important for randomly spaced data where it was not for equally spaced data. Increasing the number of reference points can decrease the length of region II and, hence, increase the length of the plateau region. For 10 reference points and a linear object, region II will be a little more than a decade longer for randomly spaced data than for equally spaced data. These results are well borne out by the plots shown in Figure 4.24.

One further point should be made concerning region II for the randomly distributed data. It does not have the rapid jumps observed for the equally

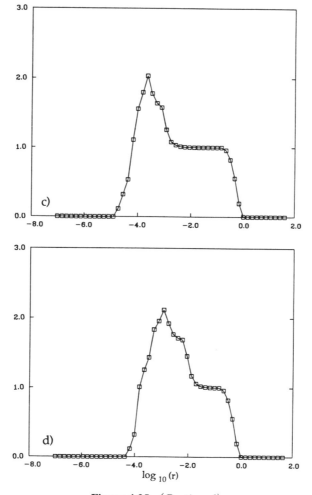

Figure 4.25. *(Continued)*

spaced data; instead, it is almost smooth. For equally spaced data, the average number of points per reference sphere can increase only by integer values, thus causing large jumps in plots of the local slope. For randomly distributed data, noninteger increases can occur, thus smoothing out the jumps.

Region IV, the roll-off from the plateau, occurs at the same point that it did for equally spaced data; it begins when the reference spheres extend beyond the boundary of the object or when other scales, such as folds and curvature, become important. This can be readily seen by comparing results in Figures 4.20 and 4.24.

4.4.4. Data with Errors

When measurement errors corrupt the data, the plots of the local slopes are affected. In Figure 4.25 the local slope is plotted for a line of unit length with equally spaced data in the x_1 direction (along the line) and with random errors of maximum size E in the x_2 direction. In this case, the reference points were chosen to coincide with data points that were away from the edge of the object. Region II begins when at least one additional point is added to one of the reference spheres. This occurs at the less negative of the following two expressions:

$$\log(r_1) = -\log(N) \quad \text{or}$$

$$\log(r_1) = \tfrac{1}{2}\big[\log(E) - \log(N) - \log(N_{\text{ref}}) - \log(\pi)\big]$$

For the plots in Figure 4.25, N and N_{ref} are 10^5 and 10, and region II should start at either -5.0 or $-3.2 + 0.5\log(E)$. This is in excellent agreement with the results shown in Figure 4.25.

The height of region II also depends on the size of the error. For errors less than or equal to the minimum data spacing, region II oscillates about the desired value of the dimension. For errors greater than the minimum data spacing, region II is definitely higher than the expected dimension. It should approach the value of the dimension of the object plus any additional dimensions contributed by the errors in the data, in the case shown an additional value of 1. For small errors this value is not approached, but as the errors become larger, the peak approaches a value of 2. The reason it cannot fully reach the value of 2 here is that it experiences roll-off when the radii of the reference spheres approach the size of the errors in the same manner that occurs when the radii approach the size of the object. Conditions can occur that allow two fully developed plateau regions to develop. This is shown in Figure 4.26. Everything is the same as for the case in the lower right of Figure 4.25 except the number of data points has been increased to 10^6. There is now a sufficient number of data points to establish a plateau at both the dimension of the object plus noise (2) and the object alone (1). For a small range of radii the apparent object is a plane. The roll-off from this plateau is very similar in shape to the roll-off from the plateau for the true

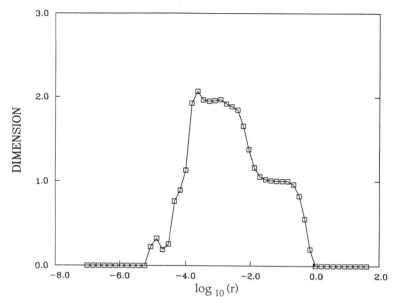

Figure 4.26. Dimension plateaus from errors and object ($N = 10^6$, $N_{\text{ref}} = 10$, $\gamma = 0.500$).

dimension of the line. All of the criteria for onset and roll-off of these regions follow from the same theoretical analysis described earlier if the proper scales are identified.

The influence of errors was also investigated for a self-similar fractal, the Cantor set. Here the points were constructed as described earlier and random errors with maximum size of $1/3^6$ were now introduced to these numbers. The results are shown in Figure 4.27. When the radii of the reference spheres are less than the error size, the object appears one dimensional; the error smears out the structure of the Cantor set. As the radius increases, the error becomes small compared both to the radii of the reference spheres and to the increase in the radii. The structure of the Cantor set now becomes apparent, the cyclical pattern is observed in the plateau region, and the dimension, on the average, approaches the value for the Cantor set. This, however, does not occur until the radii of the reference spheres are 10 times greater than the noise.

4.4.5. Process Dependent Effects

There is one last set of issues that needs to be considered when determining dimensions from real data sets. These issues occur in the study of dynamic systems where the object is now in phase space and we normally do not or cannot obtain all of the coordinates of the phase space or we choose instead to take a cross section, such as a Poincaré section, through the object. It is important to have confidence that when only a portion of the phase space is

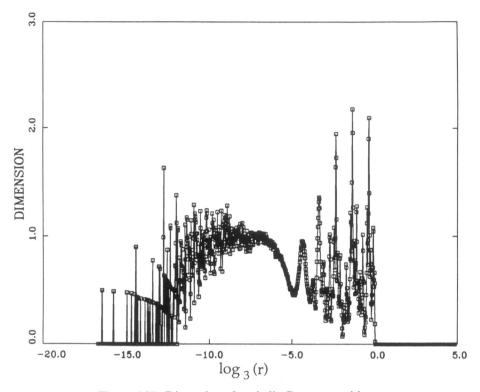

Figure 4.27. Dimension of a triadic Cantor set with error.

used, the dimension of the system is properly determined. We will illustrate these issues with a simple nonlinear dynamic system, a damped, two-well oscillator forced at some frequency. This system has been studied extensively [29] and is described by the differential equation

$$\ddot{x} + a\dot{x} - (1 - x^2)x/2 = f\cos(\omega t)$$

For the sake of this study $a = 0.10$, $f = 0.20$, and $\omega = 1.0$. For these values of the parameters, the object in phase space is a strange attractor with a dimension slightly larger than 2. Using data calculated with the preceding equation, the dimension of the system was found three different ways:

1. Using all three coordinates of the phase space directly.
2. Using eight different Pointcaré sections through the attractor at $\pi/4$ increments in the forcing phase.
3. Using a single component of the phase space information and employing the embedding theorem [30].

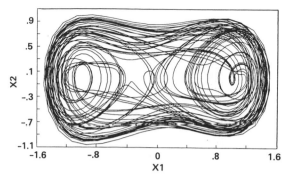

Figure 4.28. Projection of two-well potential trajectory onto the x_1-x_2 plane.

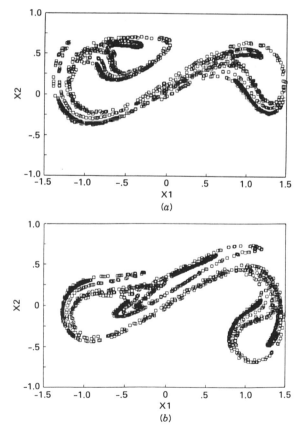

Figure 4.29. Poincaré sections of the two-well potential: (*a*) $\phi = \pi/4$; (*b*) $\phi = \pi/2$.

TABLE 4.1. Dimensions of Attractor from Poincaré Sections

ϕ (deg)	Dimension	ϕ (deg)	Dimension
45	2.364	225	2.356
90	2.341	270	2.315
135	2.280	315	2.317
180	2.291	360	2.321

These results are compared to the earlier results of reference [29] and to the upper bound on the dimension determined with the Kaplan–Yorke conjecture [26].

A plot of the projection of the attractor onto the $x_1 - x_2$ plane is shown in Figure 4.28. Using all coordinates, the correlation dimension was determined using equation (4.9) as 2.331. Data sets of 125,000 and 250,000 points were used with little difference in the resulting dimension. Two Poincaré

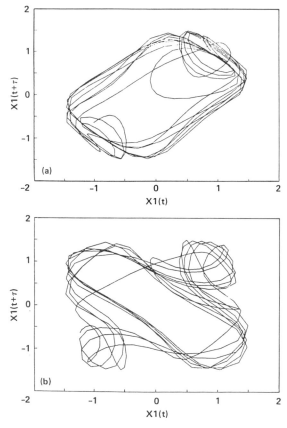

Figure 4.30. Embedded phase space trajectories: (a) $\tau = 1.7/2\pi$; (b) $\tau = 3.2/2\pi$.

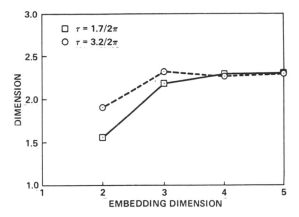

Figure 4.31. Object dimension versus embedding dimension.

sections for $\phi = \pi/4$ and $\pi/2$ are shown in Figure 4.29. Note the object is a highly folded structure with considerable banding. Dimensions were determined for eight different Poincaré sections. These are tabulated in Table 4.1. Note the determined dimension has been increased by 1 because the Poincaré section is a two-dimensional cut through an object in three-space. All of the Poincaré sections yielded dimensions in very close agreement with one another (2.321 ± 0.03). The number of points in each Poincaré section was varied between 15,635 and 31,250 with very little effect on the resulting dimension.

In determining the dimension using the embedding method, the x_1 coordinate was used. Discrete elements of $x_1(t_0), \ldots, x_1(t_n)$ were used to define a vector. Four additional vectors were constructed by taking this vector and shifting its time origin, $x_1(t + n\tau)$, where τ is some time shift and n is an integer from 1 to 4. Two examples of these are shown in Figure 4.30 for $n = 1$ and two different values of τ ($1.7/2\pi$ and $3.2/2\pi$). The dimensions for these embedded phase space objects are shown in Figure 4.31 as a function of n and for both values of τ. The asymptote of these curves is the

TABLE 4.2. Comparison of Dimensions

Determined From	Dimension
3-D phase space	2.331
Poincaré sections	2.321
Embedding	
Time delay = 170	2.292
Time delay = 320	2.304
Reference [29]	2.363
Kaplan–Yorke conjecture	2.489

proper dimension for the object. For the two values of τ chosen the dimensions agree very well (2.292 and 2.304, respectively).

In Table 4.2 the dimensions determined for this system using all of the different processes for examining the phase space are shown. The dimensions are in excellent agreement, indicating that the dimensions can be accurately determined even when all the coordinates of the phase space are not available. Also shown are the dimension determined in reference [29] and the dimension determined with the Kaplan–Yorke conjecture. The Kaplan–Yorke conjecture is supposed to represent an upper bound, and indeed it is slightly larger than the others.

4.5. CONCLUDING REMARKS

It is clear that chaos can play an important role in flight dynamics. Furthermore, it is clear that much of the complexity of the flight dynamic behavior has its roots in the aerodynamic characteristics, which can include strong nonlinearities, subcritical bifurcations, periodicity, and chaos even in the absence of vehicle dynamics. Hence, it is critical that modeling of the aerodynamic characteristics be given serious attention. Care must be taken with these models so that correct dynamic behavior is produced and not just transient phenomena. In addition when the fluid or flight dynamic behavior is chaotic, it is important to provide some means to characterize this behavior. The dimension of the attracting subspace is one such characteristic; however, when used, its accuracy needs to be stipulated.

All of the foregoing issues were addressed to some degree, but the full impact of chaotic behavior on the field of atmospheric flight dynamics has not been fully explored nor is it fully appreciated at this point. The proper modeling of complex aerodynamic characteristics is the key issue for successful studies in this field.

ACKNOWLEDGMENTS

Most of this work was completed while G. T. Chapman was a senior staff scientist at NASA Ames Research Center and L. A. Yates was a research associate with the National Research Council at NASA Ames Research Center; M. J. Szady was supported by NASA Ames Research Center. Further support for G. T. Chapman was provided by NASA Ames Consortium Agreement NCA 2-465 and for L. A. Yates by NASA Grant NCC 2-583.

APPENDIX

A.1. Uniformly Spaced Data

The length of region II for uniformly spaced data can be determined as follows. If the local slope, $\nu'(r)$, for the dimension calculations is found using central differences, then

$$v'(r_n) = \frac{\log C'(r_{n+1}) - \log C'(r_{n-1})}{\log(r_{n+1}) - \log(r_{n-1})}$$

Letting $C'(r_{n+1})$ and r_{n+1} be defined as

$$C'(r_{n+1}) = C'(r_{n-1}) + \Delta C'$$

$$r_{n+1} = (1 + \gamma)^2 r_{n-1}$$

gives

$$v'(r_n) = \frac{\log[1 + \Delta C'/C'(r_{n-1})]}{2\log(1 + \gamma)}$$

If the data distribution is continuous and the reference spheres lie completely within the object, the number of data points in each reference sphere is $C'' = \rho A r^D$. The number of data points added to each reference sphere as the radius is increased from r_{n-1} to r_{n+1} is

$$\Delta C'' = C''(r_{n+1}) - C''(r_{n-1}) = \rho A r_{n-1}^D \left[\left(\frac{r_{n+1}}{r_{n-1}} \right)^D - 1 \right]$$

where D is the dimension of the object, ρ is the point density, and A depends only on D. For objects with integer dimensions, $A = (\pi^{D/2})/\Gamma(D/2 + 1)$.

Because only finite amounts of discrete data exist, $\Delta C'$ makes integer jumps and does not change smoothly. The error $\delta C'$ in determining $\Delta C'(r_{n+1})$ is the difference between $\Delta C''$ and the actual number of points added to the sphere. An approximation to this error can be found by determining the number of points that lie on the edge of a reference sphere. Define the radius of the reference sphere as

$$r_i = \sqrt{M_1^2 + M_2^2 + \cdots + M_D^2} \, \Delta x = M_{\text{tot}} \Delta x$$

where Δx is the spacing between points. If a reference sphere is located at the point $(M_{0_1}, M_{0_2}, \dots, M_{0_D}) \Delta x$ and has a radius r_i, the points that lie on the edge of the sphere satisfy

$$\mathbf{x}_i - \mathbf{x}_j = \left(M_{0_1} - M_{j_1}, M_{0_2} - M_{j_2}, \dots, M_{0_D} - M_{j_D} \right) \Delta x$$

where

$$\sum_{k=1}^{D} \left(M_{0_k} - M_{j_k} \right)^2 = M_{\text{tot}}^2 \tag{A.1}$$

If the radius of the sphere is slightly less than r_i, none of these points will be included; if it is slightly larger than r_i, all of these points will be included. Therefore, an estimate of the maximum error in the number of data points included in each reference sphere can be approximated by

$$N_{\text{pts}} = \Sigma \frac{2^P D!}{\Pi N_{M_i}!}$$

where the sum is over the unique sequences of $|M_{0_i} - M_{j_i}|$s that satisfy equation (A.1), N_{M_i} is the number of terms in the sequence that have the same value, and P is the number of terms in the sequence that are nonzero. Approximating the sum by its largest term gives $\delta C' \approx 2^D D!$. Using this estimate for the maximum error in the total number of points, the local slope becomes

$$\nu' + \delta\nu' = \frac{\log[1 + (\Delta C'' + \delta C')/C'(r_{n-1})]}{2\log(1 + \gamma)}$$

$$= \frac{\log\left\{1 + \left[(r_{n+1}/r_{n-1})^D - 1\right] \pm 2^D D!/(\rho A r_{n-1}^D)\right\}}{2\log(1 + \gamma)}$$

$$= \frac{\log\left[(1 + \gamma)^{2D} \pm 2^D D!V/(NA r_{n-1}^D)\right]}{2\log(1 + \gamma)}$$

where N is the number of data points and V is the volume of the object. If the number of points at the edge of the reference sphere is small compared to the number of points inside the reference sphere, then

$$\nu' + \delta\nu' = D \pm \frac{D!V}{2(1 + \gamma)^{2D} \ln(1 + \gamma) AN} \left(\frac{2}{r_{n-1}}\right)^D$$

This can be rewritten to give the minimum radius for a desired error ε in the dimension:

$$r_{n-1} = \frac{1}{(1 + \gamma)^2} \left(\frac{V 2^{D-1} D!}{AN\varepsilon \ln(1 + \gamma)}\right)^{1/D}$$

or

$$\log(r) = \frac{1}{D} \log\left[2^{D-1} D! \frac{V}{AN}\right] - \left\{2\log(1 + \gamma) + \frac{1}{D}\log[\varepsilon \ln(1 + \gamma)]\right\}$$

A.2. Randomly Distributed Data

Region I ends and region II begins when there is an average of more than one point per reference sphere. If there are N_{ref} reference points and the spheres associated with these points do not overlap, the probability that any given data point (other than data points associated with the centers of the reference spheres) will lie in one of the reference spheres is $Ar^D N_{ref}/V$ where Ar^D is the volume of a reference sphere and V is the volume of the object. The probability that it will not lie inside any of the spheres is $1 - Ar^D N_{ref}/V$. If there are N data points, the probability that a total of M points will lie inside all the reference spheres is

$$
P(M, r) = \frac{N!}{M!(N - M)!} \left(\frac{Ar^D N_{ref}}{V} \right)^M \left(1 - \frac{Ar^D N_{ref}}{V} \right)^{N-M}
$$

$$
\approx \frac{\exp\left[-(M - NAr^D N_{ref}/V)^2 / (2NAr^D N_{ref}(1 - Ar^D N_{ref}/V)) \right]}{\sqrt{2\pi N(Ar^D N_{ref}/V)(1 - Ar^D N_{ref}/V)}}
$$

Noting that $M_{av} = Ar^D N/V$ is the number of points per reference sphere for continuous data and assuming that $N_{ref} Ar^D \ll V$ gives

$$
P(M, r) \approx \frac{1}{\sqrt{2\pi M_{av} N_{ref}}} \exp\left[\frac{-(M - N_{ref} M_{av})^2}{2 N_{ref} M_{av}} \right]
$$

The most probable number of points in all the reference spheres is simply $N_{ref} M_{av}$, which is as expected. Because this distribution is symmetric about $N_{ref} M_{av}$, there is a 50% chance that more than $N_{ref} M_{av}$ data points will be inside the reference spheres and a 50% chance that fewer points will be inside the reference spheres. If the centers of the reference spheres do not coincide with data points, then region II can be defined to begin when there is a 50% chance that more than an average of one data point is in each reference sphere, or when $M_{av} \geq 1$. In general, this occurs when

$$
\log(r) = -\frac{1}{D} \log\left(\frac{AN}{V} \right)
$$

If the center of each reference sphere is a data point, then region II begins when there is at least a 50% probability that a data point in addition to those at the reference sphere centers lies within one of the reference spheres, or when $N_{ref} M_{av} \geq 1$. This occurs when

$$
\log(r) = -\frac{1}{D} \log\left(\frac{N_{ref} AN}{V} \right)
$$

From our previous discussion for equally spaced data, region II ends and region III begins when

$$\frac{\log\left[(1 + \delta C')/((1 + \gamma)^{2D}C')\right]}{2\log(1 + \gamma)} < \varepsilon$$

where ε is the specified error in the dimension, $\delta C'$ is the error in the number of points added to the reference sphere as it increases from a radius of r_{n-1} to r_{n+1}, and C' is the number of points initially in the reference sphere. The probability that an average of $\Delta C'$ data points are added to each reference sphere when the radius is increased from r_{n-1} to r_{n+1} is

$$P(\Delta C', r_n) = \int_0^\infty P(M, r_{n-1}) P(\Delta C' + M, r_{n+1}) \, dM$$

$$\approx \frac{\exp\left\{ -\frac{N_{\text{ref}}\left[\Delta C' - M_{\text{av}}(r_{n+1}) + M_{\text{av}}(r_{n-1})\right]^2}{2\left[M_{\text{av}}(r_{n-1}) + M_{\text{av}}(r_{n+1})\right]} \right\}}{\sqrt{2\pi N_{\text{ref}}\left[M_{\text{av}}(r_{n-1}) + M_{\text{av}}(r_{n+1})\right]}}$$

The most probable value of $\Delta C'$ is the value obtained for continuous data, $M_{\text{av}}(r_{n+1}) - M_{\text{av}}(r_{n-1})$, and the dispersion is $\sigma^2 = [M_{\text{av}}(r_{n-1}) + M_{\text{av}}(r_{n+1})]/N_{\text{ref}} \approx 2M_{\text{av}}(r_n)/N_{\text{ref}}$. Because there is more than a 50% probability that $\Delta C'$ will lie between $M_{\text{av}}(r_{n+1}) - M_{\text{av}}(r_{n-1}) - \sigma$ and $M_{\text{av}}(r_{n+1}) - M_{\text{av}}(r_{n-1}) + \sigma$, the maximum error $\delta C'$ can be approximated by σ, and region II should end when

$$\varepsilon \ln(1 + \gamma) = \frac{\sigma}{2(1 + \gamma)^{2D}C'} = \frac{1}{(1 + \gamma)^{2D}\sqrt{2N_{\text{ref}}M_{\text{av}}}}$$

or when

$$\log(r) = \frac{1}{D}\log\left[\frac{V}{2ANN_{\text{ref}}}\right] - 2\left\{2\log(1 + \gamma) + \frac{1}{D}\log[\varepsilon \ln(1 + \gamma)]\right\}$$

NOMENCLATURE

A	Geometrical constant
B	Constant
C	Number of data points contained in reference spheres
C_D	Drag coefficient
C_L	Lift coefficient

D	Dimension
D_L	Lyapunov dimension
d	Diameter
f	Forcing amplitude
h, h_1, h_2	Displacement
k	Spring constant
k_S	Strouhal frequency
L	Characteristic length
M	Total number of points in reference spheres
M_{av}	Average number of data points in reference spheres
M_i	Integer giving location of point for equally spaced data
m	Mass
N	Number of data points
N_{ref}	Number of reference spheres
q	Angular velocity
r	Radius of reference sphere
t	Time
U	Resultant velocity
U_0	Free-stream velocity
u, v	Velocity components
V	Volume of object
$\mathbf{x}_i$	Location of a point
x, x_i	Cartesian coordinates
α	Angle of attack
β	Forcing parameter
γ	Increase in the radius of the reference spheres
E	Maximum error in data
ε	Maximum allowed error in dimension calculations
Θ	Heaviside function
κ	Scalar factor indicating onset of roll-off
λ_i	Lyapunov exponents
μ	Mechanical damping coefficient
ν	Correlation dimension
ρ	Density
τ, ξ	Time parameters
ω	Frequency

REFERENCES

1. G. H. Bryan, *Stability in Aviation*. MacMillan, London, 1911.
2. B. Etkin, *Dynamics of Flight—Stability and Control*. Wiley, New York, 1959.
3. M. Tobak, On the use of the indicial function concept in the analysis of unsteady motions of wings and wing-tail combinations. NACA Report 1188, 1954.

4. M. Tobak and W. E. Pearson, A study of nonlinear longitudinal dynamic stability. NASA Report R-209, September 1964.

5. M. Tobak and L. B. Schiff, Aerodynamic mathematical modeling—Basic concepts, AGARD Lecture Series No. 114 on Dynamic Stability Parameters, Lecture No. 1. AGARD, Neuilly sur Seine, France, March 1981.

6. G. T. Chapman and M. Tobak, Nonlinear problems in flight dynamics. In *Proceedings of the Berkeley–Ames Conference on Nonlinear Problems in Control and Fluid Dynamics*. Math Sci Press, Brookline, MA, 1985.

7. M. Tobak, G. T. Chapman, and L. B. Schiff, Mathematical modeling of the aerodynamic characteristics in flight dynamics. In *Proceedings of the Berkeley–Ames Conference on Nonlinear Problems in Control and Fluid Dynamics*. Math Sci Press, Brookline, MA, 1985.

8. M. Tobak, G. T. Chapman, and A. Unal, Modeling aerodynamic discontinuities and the onset of chaos in flight dynamical systems. *Ann. Telecommun.* **42**(5-6) 300–314 (1987).

9. G. T. Chapman and M. Tobak, Bifurcations in unsteady aerodynamics—Implications for testing. Proceedings of Workshop II on Unsteady Separated Flow, Paper No. FJSRL-TR-88-0004, USAF Systems Command, Colorado Springs, CO, pp. 7–23, September 1988.

10. P. Grassberger and I. Procaccia, Characterization of strange attractors. *Phys. Rev. Lett.* **50**(5) 346 (1983).

11. M. Fiechter, Über Wirbelsysteme an schlanken Rotationskörpern und ihren Einfluss auf die aerodynamischen Beiwerte. Report 10/66, Deutsch-Französisches Forschungs-Institut Saint-Louis, December 1966.

12. G. N. Malcolm and M. H. Clarkson, Wind tunnel testing with a rotary-balance apparatus to simulate aircraft spin motions. In *Proceedings of the AIAA 9th Aerodynamics Testing Conference, June 1976*, pp. 143–146. AIAA, Washington, DC, 1976.

13. K. W. McAlister and L. W. Carr, Water-tunnel experiments on an oscillating airfoil at Re = 12,000. NASA Report TM-78446, March 1978.

14. K. W. McAlister, L. W. Carr, and W. J. McCroskey, Dynamic stall experiments on the NACA 0012 Airfoil. NASA Report TP-1100, January 1978.

15. A. S. Arena, Jr., and R. C. Nelson, The effects of asymmetric vortex wake characteristics on a slender delta wing undergoing wing rock motion. *AIAA Atmospheric Flight Mechanics Conference, August 1989*, Paper No. 89-3348. AIAA, Washington, DC, 1989.

16. S. Thompson, S. Batill, and R. Nelson, The separated flow field on a slender delta wing undergoing transient pitching motion. *AIAA 27th Aerospace Sciences Meeting, January 1989*, Paper No. 89-0194. AIAA, Washington, DC, 1989.

17. C. Hwang and W. S. Pie, Investigation of steady and fluctuating pressures associated with the transonic buffeting and wing rock of a one-seventh scale model of the F-5A aircraft. NASA Report CR-3061, 1978.

18. C. W. Van Atta and M. Gharib, Ordered and chaotic vortex streets behind circular cylinders at low Reynolds numbers. *J. Fluid Mech.* **174** 113–133 (1987).

19. C. C. Feng, The measurement of vortex-induced effects in flow past stationary and oscillating circular and D-section cylinders. Master's thesis, University of British Columbia, Vancouver, 1968.

20. J. G. Leishman and K. Q. Nguyen, State-space representation of unsteady airfoil behavior. *AIAA J.* **28**(5) 836–844 (1990).

21. W. J. Chyu, and L. B. Schiff, Nonlinear aerodynamic flap modeling of flap oscillations in transonic flow: A numerical validation. *AIAA J.* **21**(1) 106–113 (1983).

22. D. V. Truong and M. Tobak, Aerodynamic mathematical modeling based on indicial response approach: Derivation from Navier–Stokes equations. Part I. Time-invariant equilibrium state. NASA Report TM-102856, October 1990.

23. P. W. Bearman, Vortex shedding from oscillating bluff bodies. *Ann. Rev. Fluid Mech.* **16** 300–314 (1984).

24. A. C. Lazer and P. J. McKenna, Large-amplitude periodic oscillations in suspension bridges: Some new connections in nonlinear analysis. *SIAM Rev.* **32**(4) 537–578 (1990).

25. L. A. Yates, A. Unal, M. Szady, and G. T. Chapman, One-degree-of-freedom motion induced by modeled vortex shedding. NASA Report TM-101038, July 1989.

26. J. Kaplan and J. Yorke, In *Functional Differential Equations and Approximation of Fixed Points* (H-O. Preiten and H-O. Walther, eds.) p. 228. Springer-Verlag, Berlin, 1979.

27. B. Mandelbrot, *The Fractal Geometry of Nature*. W. H. Freeman, San Francisco, 1984.

28. K. W. Pettis, T. Bailey, N. K. Jain, and R. C. Dubes, An intrinsic dimensionality estimator from near-neighbor information. *IEEE Trans. Pattern Anal. Machine Intelligence* **ASMI-1**(1) 25 (1979).

29. F. C. Moon and G. X. Li, The fractal dimension of two-well potential strange attractor. *Physica* **17D** 99 (1985).

30. N. H. Packard, J. P. Crutchfield, J. D. Farmer, and R. S. Shaw, Geometry from a time series. *Phys. Rev. Lett.* **45** 712 (1980).

CHAPTER 5

NEW APPLICATIONS OF CHAOS IN CHEMICAL ENGINEERING: INTUITION VERSUS PREDICTION

J. M. OTTINO
Department of Chemical Engineering
Northwestern University
Evanston, IL 60208-3100

The theoretical foundations of chaos are on firm footing and chaotic behavior has been studied by analytical, computational, and experimental means. Manifestations of chaos involve a wide range of systems: mechanical, electrical, and optical devices, hydrodynamic processes of various length scales, transport processes and chemical reactions, neurophysiological processes, ecological and urban systems, and even the realms of the very large and very small, such as celestial mechanics and quantum processes. However, more often than not the findings have been a posteriori: explaining ongoing complex behavior and demonstrating that roots of the complexity can be traced back to a deterministic explanation.

On the other hand, actual applications of the concepts to engineering contexts, that is, the development and design of *new* processes and products are more rare and the field is open to new ideas. In this presentation I concentrate on uses of chaos and demonstrate how some of the theoretical concepts can lead to new applications of interest in chemical engineering. For example, during the past few years there have been considerable improvements in the basic understanding of stirring and mixing—due in large measure to the introduction of concepts from chaos theory—and some of the basic developments are ready for implementation in practical situations. The

Applied Chaos, Edited by Jong Hyun Kim and John Stringer.
ISBN 0-471-54453-1   © 1992 John Wiley & Sons, Inc.

most obvious possibilities are the use of basic concepts for the design of new mixing devices, such as those encountered in the polymer industry and biomedical applications, and enhancing a variety of transport-limited processes: speeding up the rate of heat and mass transfer near walls, improving the cleaning of surfaces with microcavities, accelerating the rate of dissolution of solids suspended in fluids, and tailoring flows to accelerate and control the course of diffusion-controlled chemical reactions. Other applications can be found in *flow structuring*: flow manipulation to produce tailored dispersions and aggregates. This list, however, is far from exhaustive and more applications will undoubtedly emerge. It should be stressed that some of the ideas presented here might provide a rational foundation to time-tested methodologies arrived at in an empirical or intuitive way.

5.1. SETTING

During the past few years there has been considerable interest in dynamical chaos. It is now widely accepted that simple systems are capable of generating complex behavior, and there is ample evidence that the occurrence of chaos is widespread among physical, chemical, and biological systems. In fact, in the context of dynamical systems, absence of chaos seems to be the exception rather than the rule. Moreover, the theoretical foundations of the subject are on firm footing and demonstrations of chaos have been firmly established by analytical, computational, and experimental means.

However, the applications of the mathematical apparatus to physical, chemical, and biological systems have been largely a posteriori, that is, explaining existing complex behavior and demonstrating that the complexity stems from a deterministic cause leading to chaos. Thus, for example, current theory has been able to provide new insights into the phase-locked behavior of biological systems [44], the propagation of diseases [45], the irregularities in the orbits of planets and asteroids [26, 27], as well as the presence of complex oscillations in chemical reactions [34–37]. Chaos might also have something to say regarding general aspects of order–disorder transitions arising in systems far from thermodynamic equilibrium [39] as well as the unstable behavior exhibited by large networks of computers [81]. In a more classical engineering sense, it is also likely that already existing theory might be able to provide rational explanations to problems of vibrations in heat exchangers, oscillations in electrochemical reactions and conventional chemical reactors, adaptive control problems, and several problems in fluid mechanics involving mixing and transitional flows. Therefore, it appears that engineering analysts would do well to become familiar with some of the ideas and terminology of chaos theory.

TABLE 5.1

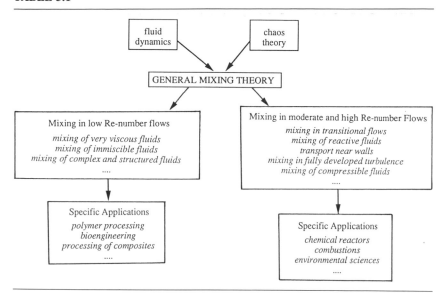

Nevertheless, it is rather naive to expect that direct applications of existing theory will lead to immediate practical results. What is more likely is that chaos theory might contribute in some essential way to the basic understanding of a given topic. To give an example I am familiar with, chaos has already served to clarify several important issues in the mixing of fluids [46–48]. In this particular case a combination of chaos, fluid mechanics, and transport phenomena, along with a heavy dose of kinematics, has produced a general framework [49] that now can be used in a variety of practical situations (Table 5.1).

Unfortunately, there seems to be relatively little accumulated work around the idea "knowing that chaos is possible, is there any class of problems that can benefit from it?" In this presentation I concentrate precisely on this issue. The aim is to demonstrate how some of the theoretical concepts can lead to the discovery of new applications of interest in engineering. The amount of review on dynamical chaos will be kept to a minimum because many of these issues are well covered in the literature and, in particular, by several chapters in this book. The ideas are presented in the context of two questions: Is chaos inherently bad? Can chaos be used as an innovative concept for inspiring and/or improving the design or operation of systems of interest in engineering? Mostly due to limitations of the author, the illustrations are restricted to fluid mechanics and transport processes.

5.2. A BRIEF REVIEW OF DYNAMICAL CHAOS

A dynamical system is a prescription for the future [1–4]. Typically this prescription is of the form of a set of ODEs or a map

$$\frac{d\mathbf{x}}{dt} = \mathbf{f}(\mathbf{x}, p), \qquad \mathbf{x}_{n+1} = \mathbf{g}(\mathbf{x}_n, p)$$

where the components of the vector $\mathbf{x} = (x_1, \ldots, x_n)$ might have either a transparent physical meaning or not, according to the problem in question (for example, in chaotic advection $\mathbf{x}$ denotes the actual physical space [49, 50]). The space spanned by $\mathbf{x}$ is called the phase space of the system and p, or a set of ps, are parameters such as the Rayleigh, Reynolds, or Strouhal numbers, the flow rate in a continuous reactor, and so on. According to the form of $\mathbf{f}(\mathbf{x})$, we can speak of two kinds of dynamical systems and, consequently, two kinds of chaos. The first kind of chaos corresponds to systems that contract volume in phase space—the so-called *dissipative systems*; the other kind corresponds to those systems that conserve volume in phase space, and of those the most important subclass is given by the so-called *Hamiltonian systems* (a system can be volume-preserving and not be Hamiltonian; however, if it is Hamiltonian, it is volume-preserving). The prototypical example of a dissipative system is the forced pendulum with friction [1]; the prototypical Hamiltonian system is a forced pendulum without friction [2] or a collection of particles interacting with each other by conservative forces [4, 5]. For example, if one were trying to describe what molecules placed in a box do, the system representing the behavior of the molecules would be a Hamiltonian system.

Dissipative systems are associated with one-dimensional maps (such as the logistic equation [10, 11]), volume-contracting systems of ordinary differential equations (such as in the Lorenz equations [12]), and strange attractors characterized by fractal dimensions [7]. If the model is continuous, a dissipative system must consist of at least three (autonomous) ordinary differential equations in order to exhibit chaos (as in the Lorenz model). On the other hand, if the model is represented by a mapping $\mathbf{x}_{n+1} = \mathbf{g}(\mathbf{x}_n, p)$, it can display chaos in one dimension, that is, with $\mathbf{x}_n$ being real (as in the logistic equation). By contrast, a volume-preserving mapping must be at least two-dimensional to be chaotic.

Hamiltonian systems have been studied for many years. In fact, as has been noted repeatedly, the first ever hints of chaos ever were noted by Poincaré in connection with studies in celestial mechanics at the end of the 19th century (by contrast, the emphasis on dissipative systems is relatively recent and can be traced to work of Lorenz [12], Feigenbaum [11], and Ruelle and Takens in the 1960s and 1970s [13]). As opposed to dissipative systems, Hamiltonian systems have no stable steady states, the phase space does not contract, and there are no attractors, strange or otherwise. A few aspects of the theory of chaos in Hamiltonian systems have been studied in considerable detail, especially for systems with a small number of degrees of freedom [5]. For example, when the perturbations from integrability are small —roughly speaking, when the forcing is small—the problem is relatively clear and several theorems, such as the Kolmogorov–Arnol'd–Moser (KAM) theorem [16, 17] describe the behavior of the system. However, much less is known when the perturbations are large.

Dissipative and Hamiltonian systems have their own ways of "going chaotic." However, both types of system have a few things in common. One of the connections is a stretching-and-folding mechanism in phase space. Stretching and folding is the fingerprint of chaos; the cleanest of the stretching and folding transformations go by the names of the "baker's transformation" [9] and the "Smale horseshoe" map [14, 15]. The discovery of a baker's or horseshoe map is sufficient to label a system as chaotic.

Probably over 90% of the current applications of chaos involve dissipative systems (the composition of this book is an excellent example). Consequently,

TABLE 5.2

from equations to chaos	from chaos(?) to characterization
physical situation	time signal
model assumptions	time series analysis
PDE's ODE's	time delay embedding dimension
Poincaré sections reduction in dimensions	attractor?
chaos?	dimension(s) Liapunov exponents
characterization of chaos	possible model?
example: thermal convection leading to Lorenz equation	example: analysis of Taylor-Couette flow

one of the main questions when facing a system (or signal) suspected of being chaotic is to determine whether it possesses a strange attractor. The techniques can be based on the measurement of one or more component of the vector **x**. This entails selecting a time-delay and an embedding dimension (see Chapter 15). Subsequently, the "amount of chaos" in the projection of the attractor can be characterized by determining its dimensions, by measuring one or more Lyapunov exponents, and so on. Some physical systems, such as the Taylor–Couette flow [23], have been analyzed from this viewpoint in considerable detail (see Table 5.2). Naturally, there are instances when the analysis starts with the equations themselves (for example, the Navier–Stokes equations). However, in many cases the equations are unmanageable and they must be transformed in a way that is suitable for analysis. The most famous example belonging to this class is the reduction of the Rayleigh–Bénard flow problem to the Lorenz equations [12] (see Table 5.2).

5.3. BENEFITING FROM CHAOS: POSSIBLE USES IN CHEMICAL ENGINEERING

Current chaos theory has done quite well. It has provided explanations for a variety of phenomena once deemed as too complex to admit simple mechanistic descriptions. At the same time it has set bounds on the expectations on the kinds of things that can and cannot be predicted; for example, a precise description of the weather might be forever out of reach. However, we should also recognize that many of the uses of chaos, no matter how elegant and intellectually satisfying they might be, are just explanations after the fact. Something has been observed, either in the laboratory or in nature, and chaos theory provides a clean explanation for the complexity of the results. In some cases we might even be able to use theory to anticipate how things will change if parameters are changed. These explanations can be based entirely on analysis of the output, a model of the process, or both (see Table 5.2).

Chaos manifests itself under rather ordinary circumstances, and simple laboratory demonstrations are common. Two such examples are the so-called double pendulum [24] and the dripping faucet [8]. Another interesting example in the context of fluid mechanics and dissipative systems is the so-called salt-water oscillator [70]. This example involves a beaker of salty water, containing a small orifice or thin tube in its bottom, placed in a larger receptacle containing pure water. The interface between the two fluids is unstable and in general the liquid from the beaker will flow into the clear liquid in time-periodic spurts until the system equilibrates itself. The time period might be of the order of seconds to minutes and the entire process might last from a few hours to a few days, according to experimental conditions (it appears that the oscillations are due to density differences and not to molecular diffusion). This result has been known since 1970, when it was published in a geophysics journal [71]. However, what happens if two

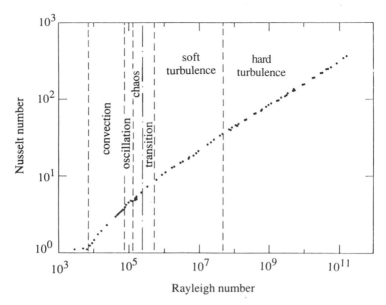

Figure 5.1. Heat transfer rate in Rayleigh–Bénard flow. (Adapted from reference [66].)

orifices are made in close proximity? In this case we have two coupled nonlinear oscillators and experiments show that chaos is possible [70]. This example, as well as many others, suggests several questions and possible viewpoints. Does the example illustrate in a clear way any basic concept of chaos? Is the example just an intellectual curiosity or is there more to it? That is, knowing that this happens what kinds of heretofore unexplained processes can the phenomenon exemplified by the experiment actually explain? (for example, how relevant is the example to geophysical processes?) From a practical viewpoint, however, the question is a different one: Is there any way to *benefit* from this result in any practical way?

This brings us to an important point. It is rather obvious that the fact that something is chaotic might be very important or completely unimportant in a given physical situation according to the application and the questions one might ask. For example, the Rayleigh–Bénard flow produces a series of complex bifurcations leading to chaotic behavior [65, 66]. However, from the point of view of heat transfer between the two plates, these details are relatively unimportant [66]. In fact, soon after the onset of convection, the variation of Nusselt number versus Rayleigh number is fairly insensitive to inner motion of the fluid; there are no changes of a nature similar to those showed by the friction factor in the laminar–turbulent transition in pipe flow (see Figure 5.1).

On the other hand, if the question is whether chaotic advection, which occurs after the onset of time-periodic flow, is important in the dispersion of a tracer in the direction parallel to the walls, the answer is that chaotic advection is terribly important ([67, 68]; the rate of transport can be increased by several orders of magnitude over the steady flow). However, whether or not some amount of chaos might be beneficial in engineering applications remains to be seen. In fact, it is fair to say that, consciously or not, many things are designed to stay away from chaos. For example, chaos might be present in fiber spinning (variations in fiber diameters), coating operations (variations in coating thickness), rheometry (chaotic signal response when shear rate exceeds some critical value in a shear rheometer), and in various types of control operations. In such instances, chaos is unwanted but a careful analysis of the responses might reveal interesting aspects of the system dynamics.

Can chaos actually be used to produce beneficial consequences? In the following sections we will argue that the answer to this question is yes. The applications are developed mostly in the context of fluid mechanics and transport. An obvious application where chaos is beneficial is fluid mixing, and theoretical concepts already in place can be used in an engineering context. A brief review of chaotic advection [46] is presented in Sections 5.4 and 5.5, and a few examples are presented in Sections 5.6 and 5.7. However, the possibility of generating efficient stirring and mixing suggests other possibilities that go beyond the examples presented in Sections 5.6–5.9, and it is probably convenient to discuss a few aspects here. One important issue is the *enhancement of transport processes*, such as in mass- and heat-transfer operations involving transport near rigid walls. Use of these ideas can be found in systems used in devices in the biomedical industry (e.g., blood oxygenation [77]), and studies along these lines are being pursued at Princeton University by Professor G. Karniadakis, at Lehigh University by Professor K. Stephanoff, and at Notre Dame by Professor H.-C. Chang. Related issues are the mixing and dispersion of solid particles suspended in a fluid and the enhancement of mass transfer between the suspended particles and a surrounding fluid; such a situation arises in the dissolution of solids. A somewhat related issue in dispersion of solid particles within a fluid is the coupling between inertial effects and mixing [62, 78]. Given a chaotic flow, there seems to be an optimal amount of inertia that maximizes the mixing. Calculations reveal that if the particles are too light (no inertia), they might be unable to cross loosely connected chaotic regions (separated by KAM surfaces [49] in two dimensions); on the other hand, if the particles are too heavy (too much inertia), they are largely unaffected by the underlying chaotic flow, and the mixing is poor. Another broad area, currently investigated by our group, involves the tailoring of flows leading to flow structuring, that is, formation of carefully controlled dispersions of two-phase fluids and aggregates of particles suspended in the fluid [61]. Things remain at a basic level but the possibilities for applications are evident.

Aggregation and breakup, as well as diffusion and processes involving fast reactions, can be viewed as a population of microstructural elements whose behavior is dictated by a macroscopic (chaotic) flow ([61]; [49], Chapter 9). In the case of drop breakup, for example, the relevant parameters are the capillary number Ca $= G\mu_e r_0/\sigma$, which measures the relative importance of viscous and interfacial tension forces (G is the local shear rate experienced by different parts of the extended droplet of filament), the viscosity ratio $p = \mu_i/\mu_e$, and the size of the drop r_0 (the subscripts i and e represent the droplet and suspending fluid, respectively).

If the viscous force is strong enough to overcome the interfacial tension, droplets placed in chaotic regions deform into long filaments. The filament is then subject to a complex deformation history including stretching, compression, and folding, resulting in drops of widely distributed sizes. Experiments reveal that the primary mode of breakup is capillary wave instabilities; models can be based on the evolution of an infinitely extended filament—with disturbances of all wavelengths—driven by a local flow rate. Experiments show as well that the equilibrium size distributions corresponding to high-viscosity-ratio drops ($p > 1$) are more nonuniform than those corresponding to low viscosity ratios ($p < 1$) and that, in general, the mean drop size decreases as the viscosity ratio p increases. Drops with $p < 1$ stretch passively but extend relatively little before they break, resulting in the formation of large droplets; these droplets undergo subsequent stretching, folding, and breakup. Under identical conditions, drops with $p > 1$ stretch substantially $O(10^3–10^4)$] before they break, producing very small fragments; these small fragments rarely break again. However, when properly scaled, drop size distributions corresponding to different operating conditions collapse into two scaling solutions with a possible crossover mechanism depending upon viscosity ratio and shear rate. It is likely that extensions of these studies will lead to the design of motions capable of producing controlled drop size distributions.

Another instance of flow structuring is aggregation [59]. Studies in this area remain at the theoretical level, but ongoing computer simulations suggest some interesting possibilities. Flow-induced aggregation produces fractal aggregates reminiscent of those formed by diffusion-limited aggregation and suggests possibilities for the fabrication of fractal clusters with various degrees of anisotropy [61]. In this case, the flow field can be tailored to produce clusters with flow-dependent fractal dimensions.

It should be clear, however, that the uses of chaos in chemical engineering are not reduced to the invention of new processes. In fact, chaos might have been used, implicitly, in many applications and it is undoubtedly *already* present in many applications (for example, the frequency of bubble detachment in a distillation plate, regimes in fluidized beds, interfacial waves in two-phase flows, and many others). Another interesting concept is that chaos provides built-in flexibility and makes changes in outputs relatively easy to implement by small changes in operating conditions [40]. Thus, if a system is chaotic, it might be possible to control the process and to select a specific time-periodic cycle in such a way as to produce a desired output. It thus appears that knowledge of chaos might allow engineers to look at old processes with new eyes.

5.4. CHAOTIC ADVECTION IN TWO-DIMENSIONAL FLOWS

During the past few years it has been established, by analysis, computation, and experiments, that two-dimensional time-periodic flows can display chaotic behavior (for a review, see reference [49]). This finding was based on an analogy with classical theory: The equations governing the particle path in a two-dimensional fluid flow are identical to those of a Hamiltonian system [46, 49]. However, couching a description of mixing in terms of standard Hamiltonian theory and concepts such as "integrability" and "nonintegrability" seems a bit too restrictive. As discussed earlier a substantial part of Hamiltonian theory applies to systems with small perturbations from integrability, but in many cases in mixing flows there is, in fact, no integrable picture to speak of. An alternative, also based on classical concepts in chaos, is to try to devise flows capable of producing horseshoe maps. In fact, this is the way we invented the cavity flow described next. Nevertheless, it is also clear that more theoretical developments seem to be required as well.

For a fluid flow to produce a horseshoe map, it must be capable of stretching and folding a region of fluid and returning it—stretched and folded—to its initial location. A necessary, but not sufficient, condition is that streamline portraits at two successive times show crossing of streamlines. In continuous or duct flows (see Section 5.5), this requirement translates into transverse streamline crossing at different axial locations. These effects can be accomplished in a variety of ways; one possibility, the one most studied to date, is to have time-dependent boundary conditions; another, less-studied possibility is to have a time-dependent change of geometry.

One of the two-dimensional flows studied in detail during the past few years has been the cavity flow [51]. The basic flow consists of a rectangular region capable of producing a two-dimensional velocity field (Figure 5.2). Two opposing walls can be moved in a steady or a time-dependent manner, and objects, such as baffles, can be attached to the moving walls to produce time-dependent changes of geometry.

A dynamical systems approach has produced a fairly complete picture of mixing in the cavity flow and, in general, in two-dimensional time-periodic flows. The key to understanding resides in establishing the location and the character of periodic points in the flow. *Periodic points* are points that return to their initial locations after an exact number of periods of the flow; they can be classified as hyperbolic or elliptic according to the deformation of the fluid in the neighborhood of the point itself (see Figure 5.3). Near an elliptic point there is a twist; near a hyperbolic point there is stretching and contraction [49]. Elliptic points are surrounded by *islands* that do not exchange matter

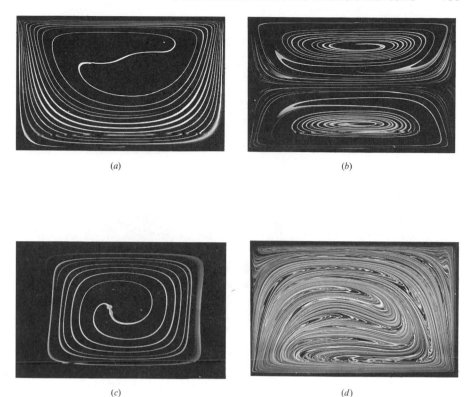

(a)   (b)

(c)   (d)

Figure 5.2. Comparison of mixing in steady and chaotic flows: The initial condition is a line of tracer placed vertically along the cavity. The flows in (a)–(c) are steady whereas the flow in (d) corresponds to discontinuous time-periodic flow. All flows consume roughly the same energy; it is apparent that the time-periodic flow producing chaotic advection is much more efficient in terms of stretching and dispersion. (From reference [51], reproduced with permission.)

with the rest of the fluid. Experimentally, if the point is elliptic, it appears as a hole (or not-dye-filled region) unless the dye was located in the island itself at the very beginning of the experiment, in which case the dye is unable to escape from it. Conversely, hyperbolic points have associated stable and unstable manifolds that might intersect transversely forming homoclinic and/or heteroclinic trajectories. These intersections are responsible for the chaotic behavior and the large-scale mixing depends critically upon the interaction of manifolds belonging to different periodic points (see reference [49]). This body of basic knowledge, briefly sketched here, can be used in engineering applications; examples are the improvement of mixing in screw extruders and static mixers [79, 80] discussed in Sections 5.7 and 5.8 (for a

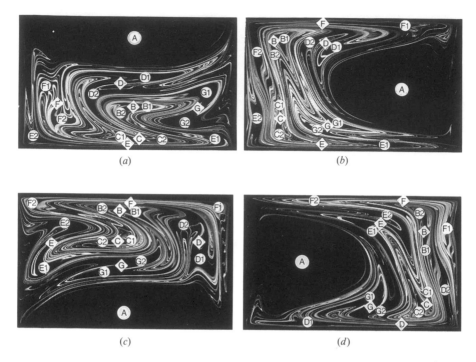

Figure 5.3. Motion of islands and periodic points in a time-periodic cavity flow (from reference [51]): An idea of the complexity of the motion can be obtained by following the motion of periodic points. The elliptic periodic points are shown by circles; the hyperbolic points by diamonds. The tracer exhibits the characteristic stretching in chaotic flows whereas the islands move in closed orbits and return to their original locations after a number of periods; the placement of points becomes symmetric when the system is viewed at specific times. (From reference [51], reproduced with permission.)

more complete treatment of chaotic mixing and applications, the reader should consult references [49], [50], and [61]).

5.5. CHAOTIC ADVECTION IN SPATIALLY PERIODIC FLOWS

Another class of widely studied chaotic flows is the spatially periodic duct flows. One such flow is the partitioned-pipe mixer (PPM) [54] (see Figure 5.4). This system consists of a pipe partitioned into a sequence of semicircular ducts by rectangular plates placed orthogonally to each other; it can be regarded as an idealized version of a commercial mixer called the Kenics static mixer [79].

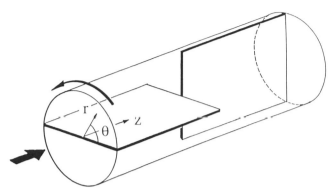

Figure 5.4. The partitioned pipe mixer. (From reference [54], reproduced with permission.)

A Kenics mixer consists of a pipe within which are alternating helical elements of left-handed and right-handed pitch. The trailing edge of one element forms a right angle with the leading edge of the next element. Ideally, such arrangement should be able to subdivide two streams, A and B, at a rate 2^n, where n is the number of elements (Figure 5.5). The mixer is able to shuffle the material over the cross section by stretching and cutting; in particular, this action is able to remove the material near the walls and place it in the center of the mixer, thereby resetting temperature or concentration gradients and enhancing the rate of heat and mass transfer between the bulk of the fluid and the walls of the pipe. However, a real mixer does not work as well as this idealized picture and one would like to have the possibility of doing studies on more manageable systems. The PPM is one such system.

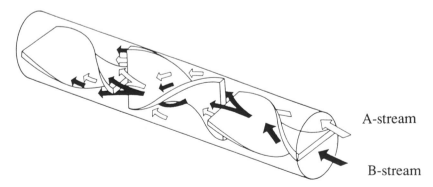

A-stream

B-stream

Figure 5.5. Schematic representation of helical elements and stream splitting in the Kenics static mixer. (Adapted from reference [79].)

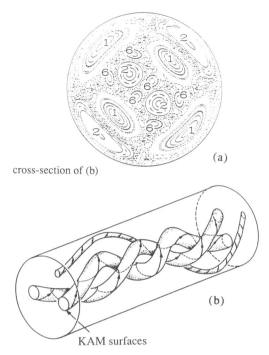

cross-section of (b)

(a)

(b)

KAM surfaces

Figure 5.6. Mixing in the PPM: (*a*) cross-sectional mixing structure (Poincaré section) showing periodicity of KAM tubes; in general, the smaller the island the higher the period; (*b*) three-dimensional view of the KAM tubes (adapted from reference [54]).

In the PPM, if one neglects developing flows, a fluid particle jumps from stream surface to stream surface in between adjacent elements. Maps $x_{n+1} = g(x_n, \beta)$, where β represents the operating condition [54], are then generated by recording every intersection of a trajectory with the surfaces of section in a very long (ideally, infinitely long) mixer and then projecting all the intersections onto a plane parallel to the surfaces (a so-called Poincaré section, see Figure 5.6*a*).

These kinds of studies show that the axial flow has a major effect on the Poincaré sections and thus on the cross-sectional mixing and dispersion. For example, the Poincaré sections corresponding to plug axial flow (perfect slip at the wall) are quite different from those corresponding to Poiseuille flow [54]. Another parameter that has a considerable effect on the Poincaré section, and thus on the mixing, is the sense of rotation in the adjacent elements. The Poincaré section for the counter-rotating case, which corresponds (roughly) to the configuration in the Kenics mixer, indicates that the flow is chaotic over most of the cross section and seems to mix better than the co-rotating case. As long as the system is spatially periodic there are KAM tubes (the three-dimensional version of islands surrounded by KAM

injection

Figure 5.7. Mixing in the partitioned pipe mixer: This particular experiment shows the simultaneous injection of two streaklines; one of them undergoes considerable stretching and folding while the other hardly suffers any stretching at all. (H. A. Kusch, University of Massachusetts, 1991.) See Plate 1 in insert.

surfaces) that prevent effective cross-sectional mixing (Figure 5.6*b*). Such tubes are also evident in experiments (Figure 5.7). However, the influence of KAM tubes can be minimized by symmetry manipulation (Section 5.7).

5.6. IMPROVING MIXING IN A SINGLE-SCREW EXTRUDER

Single-screw extruders are used widely in the plastics industry as mixers and reactors (for an introduction see references [79] and [80]). The simplest type of extruder consists of a screw that rotates inside a close-fitting cylindrical barrel (Figure 5.8*a*). The fluid between the threads of the screw is pushed forward due to the drag motion produced by the rotation of the screw. Virtually all studies on the fluid mechanics of single-screw extruders are based on the parallel plate approximation, or unwound channel representation, of the screw channel (Figure 5.8*b*), which is an excellent approximation if the aspect ratio of the channel, H/W, is small. The relative motion of the screw and the barrel appears as a plate or lid that moves diagonally. The fluid motion (or flow) in the channel can be decomposed into two compo-

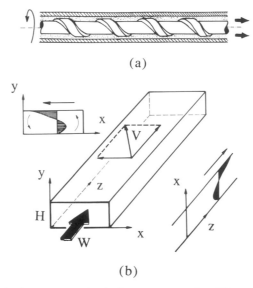

(a)

(b)

Figure 5.8. (*a*) Typical geometry of a single-screw extruder; (*b*) unwound, or flat-plate, representation of an extruder channel; the flow in the *x*–*y* direction corresponds to the cavity flow of Figure 5.2*a*.

nents: a cross flow in the *x*–*y* plane and an axial flow in the *z* direction. The axial flow is responsible for the pumping; however, the most important contribution to mixing in the screw channel is due to the cross *x*–*y* flow. However, because the cross flow is steady, the pathlines are confined to concentric stream surfaces and the mixing is poor (see reference [80]; the cross flow corresponds to Figure 5.2*a*; the helical paths are similar to those of Figure 5.13*a*).

We see in Section 5.5 that time-periodic forcing of two-dimensional flows can produce chaotic mixing. Unfortunately, there does not seem to be a clean way of translating this concept into the extruder (for example, a time-dependent motion of the screw, besides being impractical, does not generate chaos). There is, however, another possibility. The cavity flow, which resembles the *x*–*y* flow in the extruder channel, can be used to test time-dependent changes of *geometry*. One such flow can be produced by attaching a small baffle to one of the moving walls as indicated in Figure 5.9. Experiments carried out in our laboratory show that the flow, in certain operating conditions, is chaotic (unpublished experiments by C. W. Leong, University of Massachusetts, 1989).

Even though a time-periodic change in geometry is still difficult to implement in the extruder channel, it is possible, however, to achieve the same effect by translating the time-periodic change of geometry into a *spatially periodic* change of geometry in the unwound channel representation. This

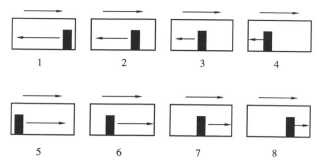

Figure 5.9. Time-dependent changes of geometry in the cavity flow apparatus: The lower wall, which has a small baffle attached to it, moves in from right to left and left to right in a time-periodic manner.

concept is illustrated in Figure 5.10. In this particular case we arrive at a "snakelike baffle" within the primary extruder channel. Obviously, this is only one of the many ways of achieving a spatially periodic change in geometry leading to a chaotic flow, and there are several other design possibilities using similar ideas. In fact, variations of this concept seem to have been arrived at, empirically, in the design of screw channels by manufacturers of extruders. A theoretical foundation, however, should allow for optimization of the design. In the next section we turn to some of these issues.

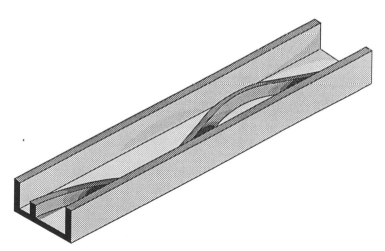

Figure 5.10. Unwound channel representation of a time-periodic change of cavity geometry: The *time-dependent* change of the cavity geometry translates into a *spatially periodic* change in the geometry of the cross section of the channel. This results in a snakelike baffle within a parallel channel, corresponding to the black rectangle of Figure 5.9 moving within the cavity of Figure 5.8*b*.

5.7. IMPROVING MIXING IN STATIC MIXERS

KAM islands—or tubes in spatially periodic systems—are the most important obstruction to mixing and an important question is how to minimize their effects. We know that good mixing can be accomplished by using a time-periodic sequence of actions. However, we know as well that, according to the operating conditions, large islands might form. The question, then, is how to destroy or prevent island formation in an efficient manner. One possibility, investigated recently by our group, is to exploit the underlying symmetries present in time-periodic and spatially periodic chaotic systems. A time or spatial sequence is then designed in such a way that possible islands are "shifted" into locations of already well mixed material. This idea has been applied to mixing in time-periodic cavity flows but the extension to spatially periodic flows (such as static mixers and the extruder with the twisted channel of Section 5.5) is trivial. A periodic sequence HVHVHV... of horizontal (H) and vertical (V) plates in the PPM results, in general, in the formation of KAM tubes. In general, something entirely similar happens in any spatially periodic system—in particular, in the extruder with the twisted channel of Figure 5.9. A basic knowledge of symmetries [56] allows the design of sequences of Vs and Hs such that the formation of the tubes is prevented [57]. The translation of this idea to the extruder channel entails building a secondary channel with a suitably designed non-periodic pitch.

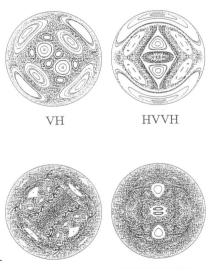

VH   HVVH

VHHVHVVH   HVVHVHHV
VHHVHVVH

Figure 5.11. Improvement in the cross-sectional mixing of the PPM by systematic shifting of symmetries [57].

5.8. THREE-DIMENSIONAL FLOWS SUITABLE FOR MIXING VERY VISCOUS FLUIDS AND DELICATE FLUIDS

Another class of continuous chaotic mixers is that of time-periodic duct flows. One example belonging to this class is the eccentric helical annular mixer (EHAM). The cross section of this system corresponds to the flow between two eccentric cylinders (which has been studied extensively in our laboratory [53]), and the axial flow is a pressure-driven Poiseuille flow. This flow produces chaotic mixing, and recent experiments show a rich and complex behavior.

One of the practical advantages of both the PPM and the EHAM to mix viscous fluids (i.e., low-Reynolds-number flows) is that the axial and cross flows are independent; this allows for control of the mixing before the material exits the system. Obviously, this is not the case in flows relying entirely on inertial effects [55]. However, inertial effects (such as transitions to Taylor–Couette flow and others) can be used to tailor the mixing in the systems.

The most revealing visualization, both computationally and experimentally, is provided by streaklines (dye injection). According to the location of injection, the streaklines can undergo complex trajectories or just "shoot through" the mixer, undergoing relatively little stretching. It is significant that, even though the axial flow always moves forward, the streaklines can "go backward," because they can wander into regions of low axial velocity, whereas other parts of the streakline can bulge forward (Figure 5.12). Intermittency is also possible; because the regular regions (islands) move

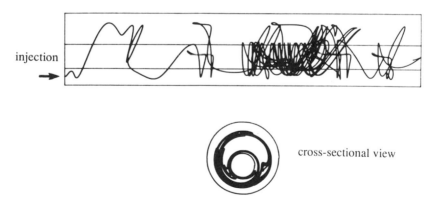

injection

cross-sectional view

Figure 5.12. Mixing in the eccentric helical annual mixing flow (EHAM): The cross section of this system corresponds to a combination of a journal-bearing flow and the axial flow, pressure-driven Poiseuille flow. The figure shows (top) a lateral view and (bottom) an end view of a streakline injected into the flow (H. A. Kusch, University of Massachusetts, 1991).

injection ↓   injection ↓

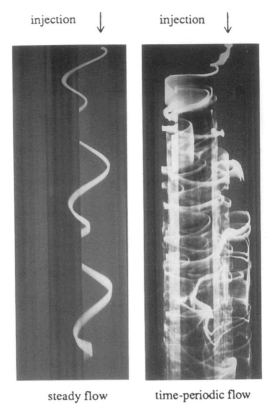

steady flow   time-periodic flow

Figure 5.13. Experimental study of mixing in the EHAM: (*a*) corresponds to steady rotation of the cylinders and leads to a regular helix; (*b*) corresponds to the same axial flow rate, but involves time-modulation of both cylinders. It is rather obvious that a lot of stretching and folding can be "packed up" in the mixer before the dye leaves the system (from H. A. Kusch, University of Massachusetts, 1991). See Plate 2 in insert.

throughout space, the streakline can find itself in a regular domain for some amount of time, then be trapped in a chaotic region, then escape and undergo relaminarization, and so on. Figure 5.13 shows the advantages of time-periodic operation: Figure 5.13*a* corresponds to steady rotation of the cylinders and leads to a regular helix; Figure 5.13*b*, corresponding to the same axial flow rate, operates under time-modulation of both cylinders. It is rather obvious that a lot of stretching and folding can be accumulated in the mixer before the dye stream leaves the system.

Mixers based on chaotic advection can be designed in such a way that a variety of design conditions are satisfied. For example, the mixing can be tailored to increase the speed of mass-transfer processes and to accelerate the rate of diffusion-limited chemical reactions. If the fluids are "delicate"

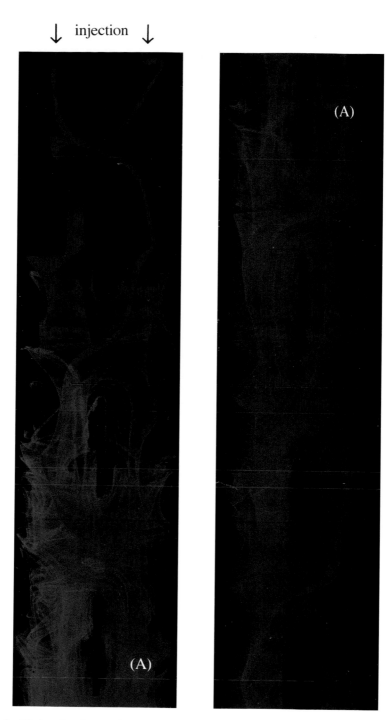

Plate 1. Mixing in the partitioned pipe mixer. This particular experiment shows the simultaneous injection of two streaklines; one of them undergoes considerable stretching and folding while the other hardly suffers any stretching at all (from H.A. Kusch, University of Massachusetts, 1991).

injection ↓   injection ↓

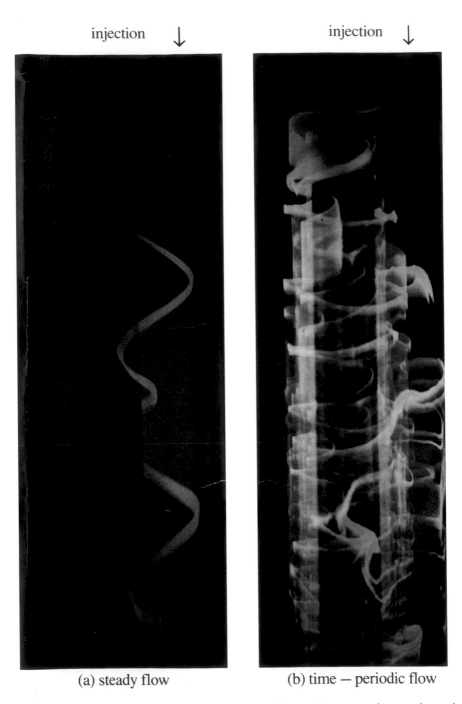

(a) steady flow   (b) time − periodic flow

Plate 2. Experimental study of mixing in the EHAM. Figure (a) corresponds to steady rotation of the cylinders and leads to a regular helix; the system in figure (b), while corresponding to the same axial flow rate, involves time-modulation of both cylinders. It is rather obvious that a lot of stretching and folding can be "packed up" in the mixer before the dye leaves the system (from H.A. Kusch, University of Massachusetts, 1991).

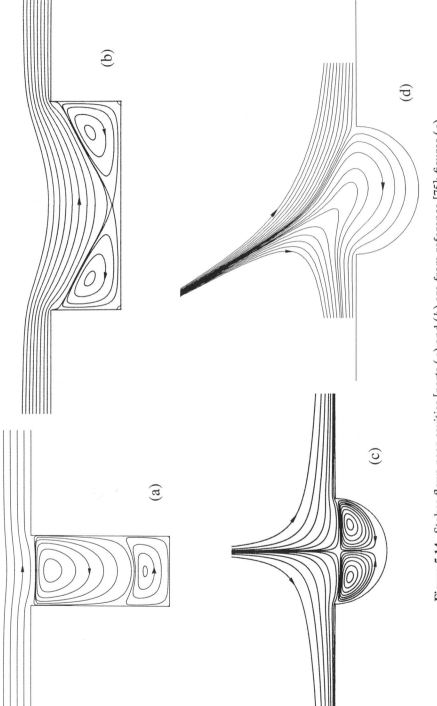

Figure 5.14. Stokes flows near cavities [parts (*a*) and (*b*) are from reference [75]; figures (*c*) and (*d*) are from reference [76]; reproduced with permission].

(such as biofluids), it is possible to design mild shear-rate histories to avoid breakage and degradation. In many instances mixing requires narrow residence time distributions; this also can be optimized by modifying the forcing in the flow.

5.9. REMOVAL OF IMPURITIES USING CHAOTIC-ENHANCED CONVECTION AND CHEMICAL REACTION

Another possible industrial application of chaotic advection is the cleaning of surfaces such as those found in semiconductor applications [72–74]. Typically the surfaces are cleaned by soaking, flushing, or wiping; however, these kinds of treatments might be insufficient to remove all traces of impurities. The surface might contain micron-sized cavities and crevices, and removal of impurities by liquids might be rather inefficient because the process is dominated primarily by molecular diffusion. The removal process can be aided somewhat by reaction as well as by convection, for example, by flowing or impinging a fluid over the cavity (the typical Reynolds numbers are of the order 1–100). However, the efficiency of the removal might be seriously hindered by the presence of steady flows with closed streamlines (see Figure 14a–c; [75, 76]). According to the geometry of the cavity and pressure gradients in the flow, it might be possible to impinge jets in such a way that there are no closed streamlines within the cavity (Figure 14d; [84]). However, in other cases this might be impossible to achieve and other methods to speed up the process are called for.

The efficiency of the process can be substantially increased if the flow is made chaotic. This can be accomplished by, for example, changing the angle of the impinging jets of fluids in a time-periodic manner (recall that the streamlines have to cross and a purely tangential variation of speed is not able to do the job). If the cavities are deeper than they are wide (Figure 14a), there is another issue that should be taken into account. Low-Reynolds-number flows can produce a succession of eddies in long cavities and corners (Figure 14a and 14b); typically the strength of the eddies changes by a factor of 1000 as one gets closer to the bottom of the cavity. The issue of generating chaotic motions that go deep into the cavity needs to be explored. We are carrying out studies in this area.

5.10. CONCLUSIONS

There seems to be little doubt that concepts based on chaotic dynamics will be used in the context of chemical engineering applications. Mixing is a clear example where controlled chaos is beneficial. The application of these ideas might be widespread: mixing is common to many processes and applications within chemical engineering. We presented examples based on mixing design, transport enhancement, and flow structuring, but many other possibilities are open. A possibility not discussed here concerns the interplay of mixing with chemical reactions. It should be clear, however, that the uses of chaos should not be confined to the invention of new processes. The knowledge that there is such a thing as chaos provides a new way of looking at things and rationalizing ongoing processes and providing a foundation for empirical designs and rules-of-thumb. Undoubtedly, chaos is *already* present in applications (for example, the frequency of bubble detachment in a distillation plates, regimes in fluidized beds, interfacial waves in two-phase flows, and many others) and, in such cases, a knowledge of chaos might aid in the interpretation of experimental information. Even in cases where chaos is undesirable: for example, fiber spinning, coating operation, rheometry, and others, a careful analysis of the responses might reveal interesting aspects of the system dynamics. A knowledge of chaos might allow engineers to look at old processes with new eyes and open some of the engineering arts, such as mixing, to rigorous analysis.

ACKNOWLEDGMENTS

I would like to thank several of my former students, C. W. Leong, P. D. Swanson, J. G. Franjione, H. A. Kusch, and M. Tjahjadi, for their assistance in the preparation of this article. Our research in mixing of fluids is supported by the Department of Energy, Division of Basic Energy Sciences, the Air Force Office of Scientific Research, and the National Science Foundation.

REFERENCES

Introduction to Dynamics and Chaos: Basic References

An accessible introduction to most of the terminology of dynamics and chaos, mostly in the context of the forced pendulum is, given by:

1. G. K. Baker and J. P. Gollub, *Chaotic Dynamics: An Introduction.* New York: Cambridge University Press, New York, 1989.

General reference for analytical techniques are:

*2. J. Guckenheimer and P. Holmes, *Nonlinear Oscillations, Dynamical Systems, and Bifurcations of Vector Fields.* Springer-Verlag, New York, 1983.
3. S. Wiggins, *Global Bifurcations and Chaos—Analytical Methods.* Springer-Verlag, New York, 1988.

*References marked with an asterisk (*) have fairly comprehensive bibliographies.

A general reference, devoted mostly, but not exclusively, to Hamiltonian systems is:

*4. A. J. Lichtenberg and M. A. Lieberman, *Regular and Stochastic Motion*. Springer-Verlag, New York, 1983.

A broader coverage of Hamiltonian systems is given in:

5. R. Z. Sagdev, D. A. Usikov, and G. M. Zaslavsky, *Nonlinear Physics: From the Pendulum to Turbulence and Chaos*. Harwood Academic Publishers (Gordon and Breach), New York, 1988.

An accessible introduction to chaos in dissipative and Hamiltonian systems is given in:

*6. M. F. Doherty and J. M. Ottino, Chaos in deterministic systems: Strange attractors, turbulence, and applications in chemical engineering. *Chem. Eng. Sci.* **43** 139–183 (1988).

A good and short review of "almost everything" in chaotic (dissipative) systems is given in:

*7. C. Grebogi, E. Ott, and J. E. Yorke, Chaos, strange attractors, and fractal basin boundaries in nonlinear dynamics. *Science* **238** 632–638 (1987).

A very accessible review of the most important concepts in chaos is given in:

8. J. P. Crutchfield, J. D. Farmer, N. H. Packard, and R. S. Shaw, Chaos. *Scientific American* **255** 46–57 (1986).

A good survey of one-dimensional maps is given in:

*9. H. G. Schuster, *Deterministic Chaos: An Introduction*, 2nd ed. Physik-Verlag, Weinheim, 1984.

The original references on one-dimensional maps are:

*10. R. M. May, Models with very complicated dynamics. *Nature* **261** 459–467 (1976).

11. M. J. Feigenbaum, Universal behavior in non-linear systems. *Los Alamos Science* **1** 4–27 (1980).

The original treatment of the Lorenz system is given in:

12. E. N. Lorenz, Deterministic nonperiodic flow. *J. Atmos. Sci.* **20** 130–141 (1963).

Once the earliest connections between chaos and turbulence is given in:

13. D. Ruelle and F. Takens, On the nature of turbulence. *Commun. Math. Phys.* **20** 167–192.

The Smale horseshoe map is discussed in:

14. J. Moser, *Stable and Random Motion in Dynamical Systems*. Princeton University Press, Princeton, NJ, 1973.

The original presentation of the Smale horseshoe map as well as a mathematician's view of dynamical systems is given in:

15. S. Smale, Differentiable dynamical systems. *Bull. Amer. Math. Soc.* **73** 747–817 (1967).

The original references on the Kolmogorov–Arnol'd–Moser theorem are:

16. A. N. Kolmogorov, On conservation of conditionally periodic motions under small perturbations of the Hamiltonian. *Dokl. Akad. Nauk SSSR* **98** 527–530 (1954).

17. A. N. Kolmogorov, The general theory of dynamical systems and classical mechanics. *Proceedings of the 1954 Congress in Mathematics.* North-Holland, Amsterdam, pp. 315–333. Translated as Appendix (pp. 741–757) in *Foundations of Mechanics* (R. H. Abraham and J. E. Marsden, eds). Benjamin/Cummings, Reading, MA, 1978.

18. V. I. Arnol'd, Small denominators and problems of stability in classical and celestial mechanics. *Russian Math. Surveys* **18**(6) 85–191 (1963).

The most quoted paper regarding embedding theorems is:

19. F. Takens, in *Dynamical Systems and Turbulence. Lecture Notes in Mathematics* **898** (D. A. Rand and L. S. Young, eds.), p. 366. Springer-Verlag, Berlin, 1981.

Various definitions of dimensions are summarized in:

20. J. D. Farmer, E. Ott, and J. A. Yorke, The dimension of chaotic attractors. *Physica D* **7** 153–180 (1986).

A discussion of delay coordinates can be found in:

21. N. H. Packard, J. P. Crutchfield, J. D. Farmer, and R. S. Shaw, Geometry from a time series. *Phys. Rev. Lett.* **45** 712–716 (1980).

Examples of Chaos

A very useful general discussion of chaos in various physical systems appears in:

*22. H. L. Swinney, Observations of complex dynamics and chaos. In *Fundamental Problems in Statistical Mechanics VI* (E. G. D. Cohen, ed.), pp. 253–289. Elsevier Science Publishers, Amsterdam, 1985. [The Appendix of this paper was published in *Physica D* **7** 3–15 (1983).]

A complete case study for the characterization of a chaotic system is:

23. A. Brandstater and H. L. Swinney, Strange attractors in turbulent Couette–Taylor flow. *Phys. Rev. A* **35** 2207–2220 (1987).

Chaos in mechanical systems:

A specific example is presented in:

24. P. H. Richter and H.-J. Sholz, Chaos in classical mechanics: The double pendulum. In *Stochastic Phenomena and Chaotic Behaviour in Complex Systems* (P. Schuster, ed.), pp. 86–97. Springer-Verlag, Berlin, 1984.

A more general presentation is given in:

25. F. C. Moon, *Chaotic Vibrations: An Introduction for Applied Scientists and Engineers.* Wiley, New York, 1987.

Chaos in celestial mechanics:

26. J. Wisdom, Chaotic behaviour in the solar system. In M. V. Berry, I. C. Percival, and N. O. Weiss, eds., Dynamical chaos. *Proc. Roy. Soc. Lond. Ser. A* **413** 109–129 (1987). Reprinted for the Royal Society and the British Academy by Cambridge University Press, 1987, 1988, 1989.

27. L. M. Polvani, J. Wisdom, E. deJong, and A. P. Ingersoll, Simple dynamical models of Neptune's great dark spot. *Science* **249** 1393–1398.

Chaos in quantum systems:

28. M. V. Berry, The Bakerian lecture, 1987. Quantum chaology. In M. V. Berry, I. C. Percival, and N. O. Weiss, eds. Dynamical chaos. *Proc. Roy. Soc. Lond. Ser. A* **413** 189–198 (1987). Reprinted for the Royal Society and the British Academy by Cambridge University Press, 1987, 1988, 1989.

A book devoted to the connection between chaos in classical and quantum mechanics is:

29. M. C. Gutzwiller, *Chaos in Classical and Quantum Mechanics.* Springer-Verlag, New York, 1990.

Chaos in simple electrical circuits is discussed in:

30. P. S. Lindsay, Period doubling and chaotic behavior in a driven anharmonic oscillator. *Phys. Rev. Lett.* **47** 1349–1352 (1981).

31. J. Testa, J. Perez, and C. Jeffries, Evidence for universal chaotic behavior of a nonlinear driven oscillator. *Phys. Rev. Lett.* **48** 714–717 (1982).

32. R. W. Rollins and E. R. Hunt, Intermittent transient chaos at interior crises in the diode resonator. *Phys. Rev. A* **29** 3327–3334 (1984).

33. C. Jeffries and J. Perez, Observation of a Pommeau–Manneville intermittent route to chaos in a nonlinear oscillator. *Phys. Rev. A* **26** 2117–2122 (1982).

A nice introduction to oscillating chemical reactions is given in:

34. I. R. Epstein, K. Kustin, P. De Kepper, and M. Orbán, Oscillating chemical reactions. *Scientific American* **248**(3) 112–123 (1983).

Another introductory treatment is given in:

35. I. R. Epstein, Complex dynamical behavior in "simple" chemical systems. *J. Phys. Chem.* **88** 187–198 (1983).

Chaos in chemical reactors is discussed in:

36. J. L. Hudson, J. C. Mankin, and O. E. Rössler, Chaos in continuous stirred chemical reactors. In *Stochastic Phenomena and Chaotic Behavior in Complex Systems* (P. Schuster, ed.), pp. 98–105. Springer-Verlag, Berlin, 1984.

The role of chaos in chemistry is discussed in:

37. C. Vidal and H. Lemarchand, *La Reaction Creatrice: Dynamique des Systemes Chimiques*. Hermann, Paris, 1980.

An up-to-date compilation of chaos in chemical reactions is given in a recent issue of J. Phys. Chem. in honor of Richard M. Noyes:

38. *J. Phys. Chem.* **93**(7) (1989).

The relationship between chaos and thermodynamics systems far from equilibrium is discussed in:

39. A. V. Gaponov-Grekhov and M. I. Rabinovich, Disorder, dynamical systems, and chaos. *Physics Today* **43**(7) 30–38 (1990).

The possibility of controlling chaos is addressed in:

40. E. Ott, C. Grebogi, and J. E. Yorke, Controlling chaos. *Phys. Rev. Lett.* **64** 1196–1199 (1990).

The relationship between chaos and turbulence is explored in:

*41. O. E. Landford, The strange attractor theory of turbulence. *Ann. Rev. Fluid Mech.* **14** 347–364 (1982).

*42. J. Guckenheimer, Strange attractors in fluids: Another view. *Ann. Rev. Fluid Mech.* **18** 15–31 (1986).

Chaos in optical systems is discussed in:

43. N. B. Abraham and U. Hübner, Homoclinic and heteroclinic chaos in a single-mode laser. *Phys. Rev. Lett.* **61** 1587–1590 (1988).

Applications of chaos to physiological systems are discussed in:

44. L. Glass and M. C. Mackey, *From Clocks to Chaos: The Rhythms of Life.* Princeton University Press, Princeton, NJ, 1988.

The applications of chaos to propagation of diseases is discussed in:

45. L. F. Olsen and W. M. Schaffer, Chaos versus noisy periodicity: Alternative hypotheses for childhood epidemics. *Science* **249** 499–504 (1990).

Applications to Fluid Mechanics and Transport Processes

The relationship between chaos and mixing of fluids is discussed in:

46. H. Aref, Stirring by chaotic advection. *J. Fluid Mech.* **143** 1–21 (1984).
47. J. M. Ottino, C. W. Leong, H. Rising, and P. D. Swanson, Morphological structures produced by mixing in chaotic flows. *Nature* **333** 419–425 (1988).
48. J. M. Ottino, The mixing of fluids. *Scientific American* **260** 56–67 (1989).

A general presentation of mixing, including background in dynamical systems, is given in:

49. J. M. Ottino, *The Kinematics of Mixing: Stretching, Chaos, and Transport.* Cambridge University Press, Cambridge, 1989. (Reprinted 1990.)

A presentation of mixing and chaotic advection, including the connection between chaotic advection and turbulence, is given in:

50. J. M. Ottino, Mixing, chaotic advection, and turbulence. *Ann. Rev. Fluid Mech.* **22** 207–254 (1990).

A detailed study to date on chaos in cavity flows is given in:

51. C. W. Leong and J. M. Ottino, Experiments on mixing due to chaotic advection in a cavity. *J. Fluid Mech.* **209** 463–499 (1989).

A more extensive presentation is given in:

52. C. W. Leong, Chaotic mixing of viscous fluids in time-periodic cavity flows. Ph.D. thesis, University of Massachusetts, Amherst, 1990.

A rather complete study comparing experiments and computations in the journal bearing flow appears in:

53. P. D. Swanson and J. M. Ottino, A comparative computational and experimental study of chaotic mixing of viscous fluids. *J. Fluid Mech.* **213** 227–249 (1990).

The "partitioned pipe" mixer is discussed in:

54. D. V. Khakhar, J. G. Franjione, and J. M. Ottino, A case study of chaotic mixing in deterministic flows: The partitioned pipe mixer. *Chem. Eng. Sci.* **42** 2909–2926 (1987).

The "twisted pipe" mixer is discussed in:

55. S. W. Jones, O. M. Thomas, and H. Aref, Chaotic advection by laminar flow in a twisted pipe. *J. Fluid Mech.* **209** 335–357 (1989).

The use of symmetries in mixing is addressed in:

56. J. G. Franjione, C. W. Leong, and J. M. Ottino, Symmetries within chaos: A route to effective mixing. *Phys. Fluids A* **1** 1772–1783 (1989).

The use of symmetries in continuous flows is addressed in:

57. J. G. Franjione, Development and applications of techniques for the analysis of chaotic mixing in two- and three-dimensional flows. Ph.D. thesis, University of Massachusetts, Amherst, 1991.

A brief introduction to mixing of viscoelastic fluids is given in:

58. C. W. Leong and J. M. Ottino, Increase in regularity by polymer addition during chaotic mixing in two-dimensional flows. *Phys. Rev. Lett.* **64** 874–877 (1990).

Coagulation in chaotic flows is discussed in:

59. F. J. Muzzio and J. M. Ottino, Coagulation in chaotic flows. *Phys. Rev. A* **38** 2516–2524 (1988).

Mixing of immiscible fluids is discussed in:

60. M. Tjahjadi and J. M. Ottino, Stretching and breakup of droplets in chaotic flows. *J. Fluid Mech.* **232** 191–219 (1991).

A general presentation of the behavior of microstructures in chaotic flows is given in:

61. J. M. Ottino, Unity and diversity in mixing: Stretching, diffusion, breakup, and aggregation in chaotic flows. *Phys. Fluids A* **5** 1417–1430 (1991).

The generation of chaotic mixing due to inertial effects in oscillatory flows is discussed in:

62. T. Howes, M. R. Mackley, and E. P. L. Roberts, The simulation of chaotic mixing and dispersion for periodic flows in baffled channels. *Chem. Eng. Sci.* **46** 1669–1677 (1991).

The relation between chaos and particle dispersion is addressed in:

63. L.-P. Wang, T. D. Burton, and D. E. Stock, Chaotic dynamics of heavy particle dispersion: Fractal dimension versus dispersion coefficients. *Phys. Fluids A* **2** 1305–1308 (1990).

A computational example dealing with chaos in three-dimensional flows in the context of "clean-air" rooms is given in:

64. S. Lichter, A. Dagan, W. B. Underhill, and H. Ayanle, Mixing in a closed room by the action of two fans. *Trans. ASME* **57** 762–768 (1990).

The Rayleigh–Bénard flow is discussed in:

65. F. H. Busse, Transition to turbulence in Rayleigh–Bénard convection. In *Hydrodynamic Instabilities and the Transition to Turbulence*, 2nd ed. (H. L. Swinney and J. P. Gollub, eds.), *Topics in Applied Physics*, vol. 45, pp. 97–137. Springer-Verlag, New York, 1985.

A short presentation is given in:

66. A. Libschaber, From chaos to turbulence in Bénard convection. In M. V. Berry, I. C. Percival, and N. O. Weiss, eds. Dynamical chaos. *Proc. Roy. Soc. Lond. Ser. A* **413** 63–69 (1987). Reprinted for the Royal Society and the British Academy by Cambridge University Press, 1987, 1988, 1989.

Transport in the Rayleigh–Bénard flow is discussed in:

67. T. H. Solomon and J. P. Gollub, Chaotic particle transport in time-dependent Rayleigh–Bénard convection. *Phys. Rev. A* **38** 6280–6286 (1988).
68. R. Camassa and S. Wiggins, Chaotic advection in Rayleigh–Bénard flow. *Phys. Rev. A* **43** 774–797 (1990).

The Taylor–Couette flow as well as the Rayleigh–Bénard flow are discussed in:

69. F. H. Busse, J. P. Gollub, S. A. Marslowe, and H. L. Swinney. In *Hydrodynamic Instabilities and the Transition to Turbulence*, 2nd ed. (H. L. Swinney and J. P. Gollub, eds.), *Topics in Applied Physics*, vol. 45, Springer-Verlag, New York, 1985.

Experiments using the salt-water oscillator leading to chaotic behavior are discussed in:

70. K. Yoshikawa, N. Oyama, M. Shoji, and S. Nakata, Use of a saline oscillator as a simple non-linear dynamical system: Rhythms, bifurcation and entrainment. *Amer. J. Phys.* **59** 137–141 (1991).

Additional References

The original treatment of oscillations due to density differences is given in:

71. S. Martin, A hydrodynamic curiosity; the salt oscillator. *Geophys. Fluid Dynam.* **1** 143–160 (1970).

Cleaning of microcavities is discussed in the following papers (the first two present modeling, the third one, experimental information)

72. R. Chilikuri and S. Middleman, Circulation, diffusion and reaction within a liquid trapped in a cavity. *Chem. Eng. Comm.* **22** 127–138 (1983).

73. R. Chilikuri and S. Middleman, Cleaning of a rough rigid surface: Removal of a dissolved contaminant by convention-enhanced diffusion and chemical reaction. *J. Electrochem. Soc.* **131** 1169–1173 (1984).

74. S. Tighe and S. Middleman, An experimental study of convection-aided removal of a contaminant from a cavity in a surface. *Chem. Eng. Comm.* **33** 149–157 (1985).

Flow fields in microcavities are discussed in:

75. J. J. L. Higdon, Stokes flow in arbitrary two-dimensional domains: Shear flow over ridges and cavities. *J. Fluid Mech.* **159** 195–226 (1985).

76. J. J. L. Higdon, Effect of pressure gradients on Stokes flows over cavities. *Phys. Fluids A* **2** 112–114 (1990).

The use of time-periodic flows in blood oxygenation is discussed in:

77. B. J. Bellhouse, F. H. Bellhouse, C. M. Curl, T. I. MacMillan, A. J. Gunning, E. H. Spratt, S. B. MacMurray, and J. M. Nelems, A high efficiency membrane oxygenator and pulsatile pumping system and its application to animal trials. *Trans. Amer. Soc. Artif. Int. Org.* **19** 77–79 (1973).

The generation of effective mixing and dispersion due to oscillatory flow is discussed in:

78. I. J. Sobey, Dispersion caused by separation during oscillatory flow through a furrowed channel. *Chem. Eng. Sci.* **40** 2129–2134 (1985).

For a description of mixing in static mixers and extruders, see:

79. S. Middleman, *Fundamentals of Polymer Processing*. Wiley, New York, 1977.

Mixing in the flow field of a single-screw extruder is described in:

80. R. Chella and J. M. Ottino, The fluid mechanics of mixing in a single screw extruder. *Ind. Eng. Chem. Fundam.* **24** 170–180 (1985).

81. B. A. Huberman, paper presented at the Workshop on Chaos, San Francisco, December 4–7, 1990 (at the Electric Power Research Institute).

Further developments on the algebra of symmetries and symmetry manipulation in continuous flows presented in Section 5.7 appear in:

82. J. G. Franjione and J. M. Ottino, Symmetry concepts for the geometric analysis of mixing flows, *Phil. Trans. Roy. Soc. Lond.* **338** 301–323 (1992).

An update on the material presented in Section 5.8 and in-depth investigation of the flows presented in Figures 5.7 and 5.13 are given in:

83. H. A. Kusch and J. M. Ottino, Experiments on mixing in continuous chaotic flows, *J. Fluid Mech.* **236** 319–348 (1992).

Illustrations of the concepts presented in Section 5.9 and flow manipulation of velocity fields such as those shown in Figure 5.14 are given in:

84. S. Jana and J. M. Ottino, Chaos-enhanced transport in cellular flows, *Phil. Trans. Roy. Soc. Lond.* **338** 519–532 (1992).

CHAPTER 6

CHAOTIC MIXING FOR HEAT TRANSFER ENHANCEMENT

H.-CHIA CHANG
Department of Chemical Engineering
University of Notre Dame
Notre Dame, IN 46556

MIHIR SEN
Department of Aerospace and Mechanical Engineering
University of Notre Dame
Notre Dame, IN 46556

Chaos is a phenomenon that has been both theoretically and experimentally demonstrated to exist in many physical systems. One of its manifestations is in the chaotic mixing of fluids. In this chapter we analyze the possibility of using it to enhance heat transfer. Analytical and numerical methods are used in the model problem of flow between eccentric cylinders, to determine the mechanism of heat transfer as well as to estimate its increase for regular and chaotic mixing. One significant conclusion is that heat transfer enhancement can be reached with little accompanying increase in power input—an important consideration for chaotic mixing to be applied to practical heat transfer processes.

6.1. INTRODUCTION

The vast majority of work on chaos is on temporal chaos, that is, chaos in the time-evolution of a small number of variables. However, it is also possible to have similar behavior for spatial displacements of fluid particles. Chaotic mixing, which is the result, has bestowed on laminar flows many of the good mixing characteristics that turbulent flows have long been known to enjoy.

Applied Chaos, Edited by Jong Hyun Kim and John Stringer.
ISBN 0-471-54453-1   © 1992 John Wiley & Sons, Inc.

That swirling flow generates better mixing was established by Aref [1]. Similar concepts were proposed by Holmes [8], Chien, Rising, and Ottino [5], and Chaiken et al. [3]. See reference [16] for a review and reference [15] for a book-length exposition on mixing. Terms such as *chaotic mixing*, *chaotic advection*, and *Lagrangian turbulence* have been suggested for this phenomenon, in which an initial blob or mass of fluid is eventually dispersed over a large area. There are many numerically computed examples of this process in the literature in addition to the ones just listed. Khakhar, Rising, and Ottino [12] consider some model two-dimensional flows, Chaiken et al. [4] look at the Stokes flow between two eccentric cylinders rotating alternately, and Jones and Aref [9] analyze the pulsed source–sink system.

In a three-dimensional flow, furthermore, it is not even necessary to have time-dependent boundary conditions. Dombre et al. [7] use analytical and numerical means to study chaotic particle trajectories produced by a steady three-dimensional flow, the so-called ABC flow. Changes in the internal secondary flow due to duct curvature can also be effectively used for chaotic mixing [13]. Similar work by Jones, Thomas, and Aref [10] indicates that a sequence of pipe bends, each one in a plane different from the previous one, is able to produce chaotic mixing due to the swirling secondary flow field in the pipe. The tube axis acts as the time coordinate and the time-varying forcing function is represented by the spatial variation of the swirls. More recently, Bajer and Moffat [2] show that steady nonaxisymmetric streamlines within a spherical bubble can induce chaotic particle paths.

A few designs have actually been suggested for using chaotic mixing in practical devices. One is the partitioned pipe mixer of Khakhar, Franjione, and Ottino [11], in which a central partition and rotation of the pipe as a whole are used to produce the mixing of material. Another is the so-called static or Kenic mixer [19]. More work needs to be done along these lines to incorporate these concepts into actual designs.

6.2. HEAT TRANSFER ENHANCEMENT BY MIXING

In this chapter we investigate the possibility of augmenting heat transfer by chaotic mixing. Although this has been suggested in the past, a rigorous proof of its feasibility has not been offered in the literature. The most elemental mechanism of heat transfer in a fluid is that of thermal conduction in a motionless fluid where the thermal energy is transported by motion at a molecular level. Even if the fluid flows in such a way as to maintain parallel streamlines, the transverse heat transfer is still by conduction. Convection is the term generally reserved for heat transfer in a moving medium wherein the heat transfer is not only by thermal conduction but also is due to the physical advection of the fluid particles in the direction of the heat flux. The geometry of the streamlines determines the advective component of the heat flux. It must be remembered that, at a wall, heat transfer is due to conduction alone because the fluid there does not move with respect to the wall itself. Good mixing in the fluid flattens the temperature profile in the mixed regions and steepens it at the walls, resulting in a larger heat flux throughout the flow. For the same applied temperature, the convective heat flux is greater for increased mixing.

Regular mixing such as that due to vortices, recirculation regions, or cellular convection increases heat transfer by increasing the advection of fluid [17, 18, 14]. One characteristic of this kind of mixing is that particle paths are nonchaotic, that is, the trajectories of neighboring particles remain reasonably close to each other. If, however, conditions are such that particle paths are chaotic, there occurs another increase in heat transfer derived from the seemingly random motion of the fluid elements. A similar increase is obtained once again for turbulent flows in which turbulent mixing is responsible for high heat fluxes.

It must be noted that any kind of mixing is also accompanied by transverse momentum transfer, leading to higher wall shear stresses and larger mechanical power requirements to maintain the flow. The question remains whether the advantage gained in increased convection is larger than the price paid in power input. We investigate this engineering trade-off here with an analytical study of small-amplitude temporal perturbation of a two-dimensional flow field and a numerical analysis of large-amplitude perturbation. In addition to offering physical insight into the basic mixing mechanisms, which are sometimes obscured by brute-force numerical integration, small-amplitude analysis is often of practical importance if large perturbations are either impossible or undesirable.

6.3. ECCENTRIC ANNULUS MODEL SYSTEM

In order to understand the basic mechanism behind chaotic enhancement of heat transfer, we begin with the simplest system that has all the essential characteristics of more complicated flows. With this in mind we study the flow in the two-dimensional space between eccentric cylinders. It has been numerically and experimentally established that this flow field can give rise to regular mixing without perturbation and to chaotic mixing with perturbation of the speed of rotation of the cylinders [3, 1, 15]. Another reason for choosing this flow is that, under suitable approximations, the velocity field can be easily determined. Thus the fluid dynamics itself is not a complicating factor. The purpose of our analysis is to study the effect that this flow has on convective heat transfer across the annulus.

Even though the fluid dynamics of the chosen system is fairly straightforward, estimates of the heat transfer are more difficult to obtain. The reason is that it depends to a large extent on the "degree of mixing" that exists for given parameter values. For our purpose we prefer to analyze the ratio of heat fluxes with and without mixing. This factor depends not only on the fluid mechanics but also on the molecular diffusivity of the fluid, in effect the Peclet number, Pe. We have thus an advection–diffusion system where the advective part may be regular of chaotic. For certain parameter values, mixing in the fluid within the annular space is regular, whereas for others it is chaotic. Thus quantitative comparisons can be made between the heat transfer rates in the two cases, as well as the power requirements to maintain rotation. One can thus weigh the benefits of chaotic mixing versus the price paid in terms of additional mechanical energy expended.

There are several nondimensional parameters that control the fluid mechanics and heat transfer in the annulus. In the absence of time-periodic perturbation, that is, when both the cylinders are rotating at constant angular velocities, there are three dimensionless parameters that characterize the steady flow: the dimensionless eccentricity ε, the speed ratio δ, and the clearance ratio λ. For time-periodic perturbation there are two additional parameters, the frequency ω and amplitude of perturbation α. The relative importance of thermal conduction versus advection in the transport equations is expressed through Pe, which is also a governing factor in the results.

6.4. FLOW FIELD AND HEAT TRANSFER

For simplicity we assume two-dimensional Stokes flow in the gap between the cylinders. The fluid is assumed to be Newtonian with constant properties. We assume a small gap width and use the lubrication approximation. In the limit

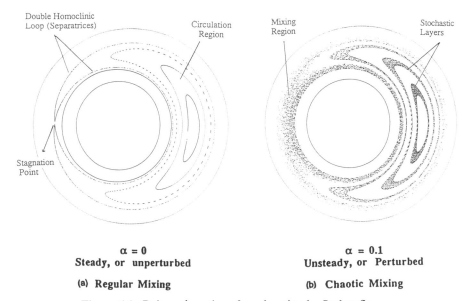

$\alpha = 0$	$\alpha = 0.1$
Steady, or unperturbed	**Unsteady, or Perturbed**
(a) Regular Mixing	**(b) Chaotic Mixing**

Figure 6.1. Poincaré sections for advection by Stokes flow.

of small gap width, characterized by the clearance ratio $\lambda(= c/R_2) \ll 1$ (Figure 6.1), we use local Cartesian coordinates in the gap to obtain the equations of motion in the lubrication approximation. In these coordinates, the azimuthal direction is represented by x, and the inward normal from the outer cylinder is represented by y. The inner boundary is approximated by a leading-order expansion in the clearance ratio. The properly nondimensionalized governing equations under the present assumptions are

continuity
$$\frac{\partial u}{\partial x} + \frac{\partial v}{\partial y} = 0 \tag{6.1a}$$

x momentum
$$0 = -\frac{\partial p}{\partial x} + \frac{\partial^2 u}{\partial y^2} \tag{6.1b}$$

y momentum
$$0 = -\frac{\partial p}{\partial y} \tag{6.1c}$$

The boundary conditions

$$u(x,0) = -\delta(1 + \alpha \sin \omega t) \tag{6.1d}$$

$$u(x, 1 + \varepsilon \cos x) = 1 \tag{6.1e}$$

$$v(x,0) = 0 \tag{6.1f}$$

$$v(x, 1 + \varepsilon \cos x) = \frac{dh}{dx} \tag{6.1g}$$

represent time-periodic forcing of frequency ω and amplitude α at one wall. All quantities have been appropriately nondimensionalized, and the gap width is approximated by $h = 1 + \varepsilon \cos x$. These equations are solved for the velocity and pressure, and the resulting expressions for velocities are utilized to write the equations of motion in the Hamiltonian formulation as

$$\frac{dx}{dt} = -\frac{\partial \psi}{\partial y} \tag{6.2a}$$

$$\frac{dy}{dt} = \frac{\partial \psi}{\partial x} \tag{6.2b}$$

where the stream function ψ, which serves as the Hamiltonian, has a time-independent and a time-dependent portion. Thus

$$\psi = \psi_0(x, y; \varepsilon, \delta) + \alpha \psi_1(x, y, t; \varepsilon, \delta) \tag{6.3}$$

where

$$\psi_0 = \delta y - \varepsilon y^3 (1 - \delta) \cos x$$

$$-\frac{y^2}{2} \left[\frac{1 + \delta}{1 + \varepsilon \cos x} - 3\varepsilon (1 - \delta) \cos x (1 + \varepsilon \cos x) \right]$$

$$\psi_1 = \delta \sin \omega t \left[\varepsilon y^3 \cos x - \frac{y^2}{2(1 + \varepsilon \cos x)} - \frac{3}{2} \varepsilon y^2 \cos x (1 + \varepsilon \cos x) + y \right]$$

From the time-independent stream function ψ_0, we obtain the flow field of Figure 6.1 ($\alpha = 0$). A separation bubble is clearly evident for all eccentric cylinders ($\varepsilon \neq 0$). The stream functions that bound the bubble can be estimated from a simple analysis of the flow field within the separation bubble; closed streamlines surround an elliptic fixed point (stagnation point). The bounding streamlines of the bubble are separatrices made up of homoclinic loops that connect to a hyperbolic fixed point. The circuit time around the bubble approaches infinity. Without temporal perturbation and in the limit of small diffusivity (Peclet number), the closed streamlines within the bubble produce a mixing mechanism that enhances the heat transfer across the gap. The bottleneck of the heat transfer process, however, lies in the slow diffusion across the homoclinic loops. This is evident from our numerical solution in Figure 6.2 of the energy equation with the flow field given previously. A flattened temperature profile is seen within the circulation bubble which enhances the local heat flux across the gap. However, it is also evident from Figure 6.2 that the bottleneck process of the radial heat transfer

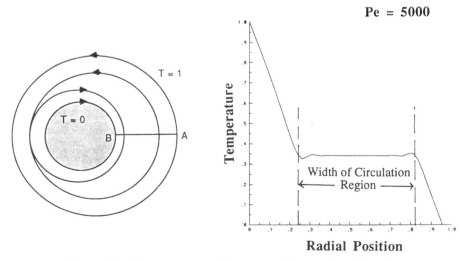

Figure 6.2. Temperature profile across widest section of annulus.

mechanism lies in the bounding homoclinic loops. The slow diffusion across these boundaries stipulates that the major mechanism outside the bubble is still conduction. We shall show that a time-periodic perturbation of the flow field breaks up the separatrices and allows chaotic exchange of fluid elements inside and outside the separation bubble.

The observation that there is a flattening of the temperature profile within the bubble allows us to analyze heat transfer enhancement by the time-dependent flow field. A perturbation analysis of the stream function in (6.3) about the concentric case $\varepsilon = 0$ allows one to estimate the position and width of the separatrix. Although a precise estimate of the heat transfer can be obtained, we shall only report a qualitative analysis of the effects of ε and speed ratio δ. To leading order, the width of the circulation region is

$$l \sim 4\sqrt{\varepsilon}\,\frac{\delta}{(1+\delta)^2}(1 + \cos x_s) \tag{6.4}$$

where x_s is the azimuthal location of the hyperbolic stagnation point. It is then evident that, if other factors are kept constant, a maximum enhancement occurs at $\delta = 1$. If one assumes a flat temperature profile within the circulation region, the local flux at a particular azimuthal position can also be estimated by a leading order $(1/\text{Pe} \to 0)$ perturbation theory. The overall enhancement can then be obtained by integrating over the entire annulus. Both the prediction of a maximum at $\delta = 1$, and the approach of the numerical solution to our zero-order theory as Pe increases, are evident in Figure 6.3.

With the introduction of a small-amplitude periodic perturbation on the rotation of the outer cylinder, stochastic layers are seen to form within the separation bubble, as in Figure 6.1b. The stable and unstable manifolds of the double homoclinic loop are seen to break up and become finite-width

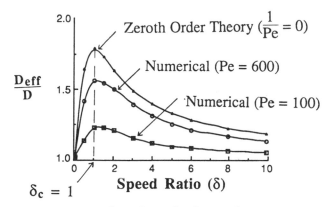

Figure 6.3. Effect of speed ratio on enhancement.

layers, leading to an increase in the overall bubble size. The chaotic particle motion is confined within stochastic layers, or resonant bands, that are bounded by invariant KAM tori that serve as impenetrable loops in the flow domain and correspond to those periodic orbits that survive the perturbation. Within the stochastic layers, particle motion displays extreme sensitivity to initial location and is therefore chaotic. The stochastic layer in the vicinity of the separatrix arises out of the homoclinic tangles. Because the concentration field is already well mixed within the bubble even without external perturbation, the homoclinic tangle is the only mechanism that extends the mixing process to the region outside the bubble and hence represents the major contribution of chaotic mixing to heat transfer enhancement. We hence develop an analytical theory to analyze the homoclinic tangle and to estimate its contribution to heat transfer enhancement.

6.5. WIDTH OF MIXING LAYER NEAR THE SEPARATRIX

For small amplitudes of periodic forcing, that is, $\alpha \ll 1$, the dynamics of stochastic particle motion in the vicinity of the separatrix is best examined by constructing a Poincaré surface of section at the section of largest clearance (Figure 6.2). Because the perturbation stream function ψ_1 is periodic in both x and t, its variation is bounded and small due to $\alpha \ll 1$. Thus, following Chirikov [6], we compute only the aperiodic changes in ψ ($\Delta\psi_0$) due to the perturbation. These may accumulate over many oscillations of the periodic forcing, giving rise to instability and chaotic particle paths. For small amplitude of periodic forcing, $\alpha \ll 1$, we reduce the dynamics of particle motion in the vicinity of the separatrix to a separatrix mapping,

$$\psi_0^{n+1} = \psi_0^n - \alpha M(\tau_0^n) \tag{6.5a}$$

$$\tau_0^{n+1} = \tau_0^n + \omega T(\psi_0^{n+1}) \tag{6.5b}$$

where M is the Melnikov function and τ_0 is the phase of M. $T(\psi_0)$ is the circuit time for motion along a streamline whose stream function is ψ_0. The mapping gives changes in ψ_0 and thus in the time-periodic flow that a fluid element close to the separatrix experiences a series of short kicks or jumps. When changes in ψ_0 over successive iterations are small, the separatrix mapping is further reduced to the well-known *standard mapping*,

$$I^{n+1} = I^n + K \sin \theta^n \tag{6.6a}$$

$$\theta^{n+1} = \theta^n + I^{n+1} \tag{6.6b}$$

where I and θ represent action-angle coordinates. Action I is related to ψ_0, and θ is related to the phase τ_0. The reduction essentially involves a linearization of the mapping in ψ_0 about resonant orbits close to the homoclinic orbit Γ. The stochasticity parameter of the mapping, K, is given by

$$K = T(\psi_0)\alpha\omega A_\Gamma(\varepsilon, \delta, \omega) \tag{6.7}$$

and A_Γ is the positive maximum of the Melnikov function corresponding to Γ. (Because particle paths are chaotic for nonzero α, stable and unstable manifolds of the homoclinic orbit intersect and thus M changes sign for some $\tau_0 \in [0, 2\pi]$.) Because T is simply the rate of change of the circuit time with variations in the stream function ψ_0, K is not only a function of the parameters but is also a function of position, characterized by the stream function ψ_0. The dependence of K on position in the phase space, which is also the flow domain, arises out of the fact that resonant orbits are everywhere dense in the phase space and the linearization is about an arbitrary resonant orbit close to the homoclinic orbit Γ. The stability border of the stochasticity parameter K has been studied extensively in the literature [6]. It is well known that for $|K| > K_c$, where K_c is the stability border of K and is approximately 1, all stable domains are completely destroyed and variations in I become stochastic or random. Thus K_c gives us an estimate of the stochastic layer width.

The most interesting variation is seen with the frequency of periodic forcing, ω. The dependence of $1/K_c$, which is a measure of the layer thickness is shown in Figure 6.4. A maximum at a critical frequency implies maximum width of the stochastic layer at a frequency of periodic forcing

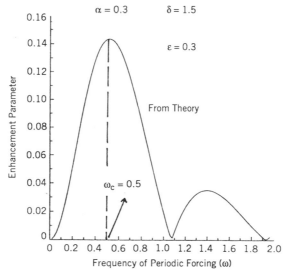

Figure 6.4. Analytical estimate of heat transfer enhancement due to variation of forcing frequency.

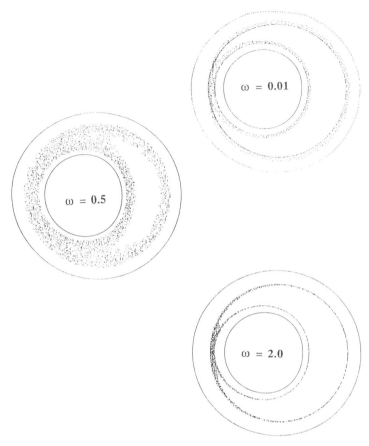

Figure 6.5. Poincaré sections for different ω, with $\varepsilon = 0.3$, $\delta = 1.5$, $\alpha = 0.3$.

corresponding to ω_c. Therefore, the analysis predicts an optimal frequency of periodic forcing. Numerically obtained Poincaré snapshots of a single particle near the separatrix being convected by the perturbed flow field with different frequencies of periodic forcing confirms the analytical prediction (Figure 6.5): The width of the stochastic layer is infinitesimally small at very low ω, is maximum at the optimal frequency of periodic forcing, ω_c, and reduces sharply as ω becomes large.

For $K \gg K_c$, when we are far beyond the stability border, the standard mapping is stochastic in the action I and a diffusion in I must occur. The diffusion will develop only in I because the motion (in I–θ space) is unbounded in this direction; the standard mapping is periodic in θ, $\theta \in [0, 2\pi]$, and for $K \gg K_c$ the entire interval is passed in only a few iterations. In the approximation of limiting stochasticity ($K \gg K_c$) the map can be assumed to be ergodic in θ, and the variations in I correspond to a diffusion-like process. Thus, using the ergodicity of the standard mapping, we estimate a mean diffusivity within the stochastic layer in the vicinity of the separatrix, which is normalized with respect to the molecular diffusion D,

$$\frac{D_{\text{eff}}}{D} = \frac{\text{Pe } \alpha^2 A^2}{4\tau} \tag{6.8}$$

where τ is the characteristic "kick time" of the separatrix map, or the mean time interval between successive iterations of the mapping.

Heat transfer is related not merely to the thickness of the stochastic layer but also to the local diffusivity within it. When combined with analytical estimates of the effective diffusivity within the circulation region, it gives a quantitative estimate of the heat transfer enhancement. Theoretical predictions of the convective enhancement as a function of the perturbation parameters are compared against numerical results in Figures 6.6 and 6.7, where both have been evaluated at large Pe, which is the only interesting limit in the context of heat transfer enhancement. In Figure 6.6 the enhancement with respect to the frequency of periodic forcing, ω, shows a peak at ω_c, which is the optimal frequency of periodic forcing shown in Figure 6.4. Also, the numerical solutions generated for the enhancement as a function of the amplitude of periodic forcing, α, show a monotonic increase as one would expect because the fluid agitation increases with α. At the low α limit, that is, at small amplitudes of the periodic forcing, which is essentially where the theory is valid, theoretical predictions match closely the numerical result. It is important to mention the significance of the dashed line: We only have a

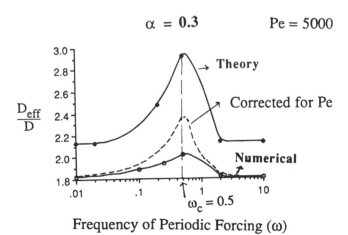

Figure 6.6. Relation between theory and numerical results.

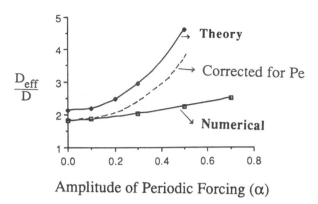

Figure 6.7. Effect of perturbation amplitude on enhancement.

zero-order theory in the $1/Pe$ expansion for the effective diffusivity within the circulation region at high Pe, and the dashed line is a correction of the theoretical result for finite Pe. This is carried out by simply using the numerically computed value for the time-independent case instead of our zero-order analytical prediction in Figure 6.3. The contribution of the time-independent flow field is, however, derived from the analytical theory. As is evident, the corrected prediction is in excellent agreement with the numerical result except near ω_c, where the width of the mixing layer is too large for our local theory to be accurate. A detailed analysis for the higher-order terms in $1/Pe$, without resorting to any numerical result, is underway and will be reported subsequently.

6.6. SUMMARY AND CONCLUSION

The mechanism by which regular and chaotic mixing enhances heat transfer has been rigorously demonstrated for a model system. It is noteworthy that the analysis shows that an optimal forcing frequency ω_c exists for maximization of chaotic enhancement. Although this prediction comes from a small-amplitude analytical theory, it is also observed from numerical simulations of large-amplitude perturbations. Because the power input into the system increases with ω, an optimum ω_c has the important practical implication that one can enhance heat transfer without a corresponding expenditure in power input to compensate for the inevitable momentum loss. The resolution of this important engineering trade-off is an important step toward the practical application of chaotic mixing.

ACKNOWLEDGMENTS

The authors acknowledge the support of the Gas Research Institute under Contract No. 5090-260-1971. They also thank S. Ghosh and N. Acharya for carrying out most of the work described here.

REFERENCES

1. H. Aref, Stirring by chaotic advection. *J. Fluid Mech.* **143** 1 (1984).
2. K. Bajer and H. K. Moffatt, On a class of steady confined Stokes flows with chaotic streamlines. *J. Fluid Mech.* **212** 337 (1990).
3. J. Chaiken, R. Chevray, M. Tabor, and Q. M. Tan, Experimental study of Lagrangian turbulence in a Stokes flow. *Proc. Roy. Soc. London Ser. A* **408** 165 (1986).
4. J. Chaiken, C. K. Chu, M. Tabor, and Q. M. Tan, Lagrangian turbulence and spatial complexity in a Stokes flow. *Phys. Fluids* **30** 687 (1987).
5. W.-L. Chien, H. Rising, and J. M. Ottino, Laminar mixing and chaotic mixing in several cavity flows. *J. Fluid Mech.* **170** 355 (1986).
6. B. V. Chirikov, A universal instability of many-dimensional oscillator systems. *Phys. Rep.* **52** 263 (1979).
7. T. Dombre, U. Frisch, J. M. Greene, M. Hénon, A. Mehr, and A. M. Soward, Chaotic streamlines and Lagrangian turbulence: The ABC flows. *J. Fluid Mech.* **167** 353 (1986).
8. P. Holmes, Some remarks on chaotic particle paths in time-periodic, three-dimensional swirling flows. *Contemp. Math.* **28** 393, 404 (1984).
9. S. W. Jones and H. Aref, Chaotic advection in pulsed source–sink systems. *Phys. Fluids* **31** 469 (1988).
10. S. W. Jones, O. M. Thomas, and H. Aref, Chaotic advection by laminar flow in a twisted pipe. *J. Fluid Mech.* **209** 335 (1989).
11. D. V. Khakhar, J. G. Franjione, and J. M. Ottino, A case study of chaotic mixing in deterministic flows: The partitioned pipe mixer. *Chem. Eng. Sci.* **42** 2909 (1987).
12. D. V. Khakhar, H. Rising, and J. M. Ottino, Analysis of chaotic mixing in two model systems. *J. Fluid Mech.* **172** 419 (1986).
13. C. S. Lee, J. J. Ou, and S. H. Chen, Quantification of mixing from the Eulerian perspective: Flow through a curved tube. *Chem. Eng. Sci.* **42** 2484 (1987).
14. P. McCarty, and W. Horsthemke, Effective diffusion coefficient for steady two-dimensional convective flow. *Phys. Rev. A* **37**(6) 2112 (1988).
15. J. M. Ottino, *The Kinematics of Mixing: Stretching, Chaos, and Transport*. Cambridge University Press, 1989.
16. J. M. Ottino, Mixing, chaotic advection, and turbulence. *Ann. Rev. Fluid Mech.* **22** 207 (1990).

17. M. N. Rosenbluth, H. L. Berk, I. Doxas, and W. Horton, Effective diffusion in laminar convective flows. *Phys. Fluids* **30**(9) 2636 (1987).

18. B. I. Shraiman, Diffusive transport in a Rayleigh–Bénard convection cell. *Phys. Rev. A* **36**(1) 261 (1987).

19. E. Villermaux, and J. P. Hulin, Chaos Lagrangien et mélange de fluides visqueux. *Eur. J. Phys.* **11** 179 (1990).

CHAPTER 7

CONTROLLING THE DYNAMICS OF CHAOTIC CONVECTIVE FLOWS*

J. J. DORNING and W. J. DECKER
Department of Nuclear Engineering
University of Virginia
Charlottesville, VA 22903-2442

JAMES PAUL HOLLOWAY
Department of Electrical and Computer Science Engineering
University of Michigan
Ann Arbor, MI 48100

Motivated by a desire to modify the dynamics of the flow in convective layers and closed convection loops, which commonly are present in engineering systems associated with the technology of power generation and heat removal, we explore some of the dynamics of parametrically driven Lorenz equations. We find that, through simple sinusoidal parametric driving, systems that are chaotic when autonomous can be made to behave periodically and can be led back to simple chaotic behavior via various period-doubling paths. We also find that autonomous chaotic systems can be made periodic with a frequency entrained by the driver frequency and a phase that depends sensitively upon initial conditions because of the presence of fractal basin boundaries. Autonomous chaotic systems, after being driven to periodicity, also can be led back to simple chaotic behavior through intermittency.

7.1. INTRODUCTION

Numerous components of electrical power generation systems and distribution systems include heat-generating devices that are cooled by the passage of a fluid over them via natural, forced, or mixed convection. Usually, in the

*This research was sponsored by the Electric Power Research Institute.

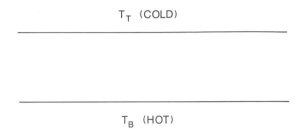

Figure 7.1. A horizontal thermal convection layer subjected to a high temperature T_B at the bottom boundary and a low temperature T_T at the top boundary.

case of power failures to the components that drive the forced convection such as pumps, fans, or compressors, the heat-generating devices depend upon natural convection alone for their cooling and often therefore for their integrity and safety. Frequently, a simple geometric model comprised of a horizontal layer of fluid exposed to a hot lower surface and a cold upper surface (Figure 7.1) provides a relevant representation of the dynamics of the natural convective cooling of a heat-generating component whose top supplies heat to the lower boundary of the fluid layer. In many cases of technological interest, this lower surface is not retained at a constant temperature, but rather varies in some time-dependent way (such as periodic, etc.). Hence, in order to ensure the efficient operation (in some cases, even the safety) of those components in electrical power generating and distribution systems that are dependent upon convective cooling, it is important to introduce and study the basic nonlinear dynamics of a generic model for the fluid motion above such devices and thereby the resulting heat transfer from them.

More important, even when the heat generating component supplies heat at a constant rate, it may be desirable to control the dynamical motion of the cooling fluid to enhance net heat transfer, avoid hot spots on the component surface, and so on, by adding a time-dependent externally operated control —such as an additional modulated heat source—either at the top of the heat generating component (lower boundary of the fluid) or at the cold surface (upper boundary of the fluid) that connects the system to the heat sink. The latter seems more practical both because it does not directly raise the already high temperature at the hot surface on the component and because the upper boundary also typically is more physically accessible in real engineering systems.

The convective fluid motion in such a fluid layer is well modeled by the Navier–Stokes equations for mass and momentum conservation, with the buoyancy terms represented in the Boussinesq approximation and the system augmented by the energy conservation equation for the fluid temperature. When these equations are combined with the time-varying upper- or lower-surface temperature, they are somewhat more complicated than those usually studied in the context of the classic Rayleigh–Bénard problem in which both surface temperatures are constant in time. Nevertheless, it is straightforward to arrive analytically at the space- and time-dependent no-flow temperature profile

$$T(z,t) = T_B + (T_T - T_B)z + T_1 \, \mathrm{Re}\left\{ \frac{\sinh[i\omega(1-z)/\kappa L^2]\exp[i\omega t]}{\sinh[i\omega/\kappa L^2]} \right\}$$

where T_1 is the amplitude of the sinusoidal temperature variation added to the lower temperature T_T at the top boundary. Although this solution is more complicated than its simple steady-state analog that arises in the classic Rayleigh–Bénard problem, it can be subtracted from the temperature field in the Navier–Stokes–Boussinesq equations to arrive at a new set of nonlinear fluids equations for the density, the velocity components, and the deviation of the temperature field from the space- and time-dependent no-flow temperature field. Then, following Lorenz [1] and making Fourier expansions of the stream function and temperature field and truncating them at the same low order as Lorenz did, it is straightforward to arrive at nonautonomous parametrically forced Lorenz equations

$$\dot{x}(t) = -\sigma x(t) + \sigma y(t),$$
$$\dot{y}(t) = [\rho_0 + \rho_1 \cos(\omega t)]x(t) - y(t) - x(t)z(t),$$
$$\dot{z}(t) = x(t)y(t) - bz(t)$$

where $x(t)$, $y(t)$, and $z(t)$ are, respectively, the time-dependent Fourier coefficients of the first term in the expansion of the stream function and the first two terms in the expansion of the temperature field. These parametrically forced Lorenz equations provide a starting point for a "first-cut" engineering analysis of the nonlinear dynamic response of a layer of gas or

T_T (COLD)

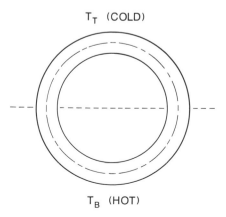

T_B (HOT)

Figure 7.2. A schematic diagram of a circular thermosyphon, with a circular cross section, subjected to a high temperature T_B on its bottom half and a low temperature T_T on its top half.

Figure 7.3. The well-known Lorenz "butterfly" chaotic attractor, in x–z projection, for $\rho_0 = 26.5$ and $\rho_1 = 0.0$.

liquid above a heat-generating component with sinusoidal temperature output in some electrical power generating or distribution system, or of the system response to a time-dependent boundary control at the top boundary of the coolant. Hence, it seems prudent to carry out dynamical studies of these equations before embarking on large-scale space–time numerical solutions to the full Navier–Stokes–Boussinesq equations necessary for a more complete engineering analysis.

Also common to engineering systems, especially those related to power generating systems and to heat-generating systems in general, are closed

Figure 7.4. The "shaken butterfly" chaotic attractor, in x–z projection, for $\rho_0 = 26.5$, $\rho_1 = 5.0$, and $\omega = 1.0$.

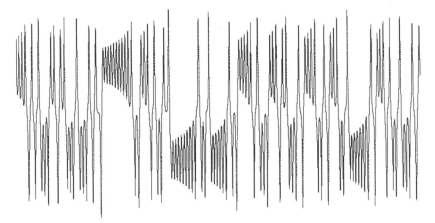

Figure 7.5. The time series for $x(t)$ in the Lorenz attractor for $\rho_0 = 26.5$, $\rho_1 = 0.0$.

convection loops that rely on natural thermal convection to transfer heat from some high-temperature heat source to a low-temperature heat sink somewhere above the source. In fact, some proposed designs of advanced nuclear power reactors depend upon such closed convection loops—also upon various open convection loops for the anticipated heat removal that will assure their safety and integrity during emergency shutdowns. A simple generic model for such a closed convection loop is the classic circular thermosyphon—a circular tube bent into a circle to form a large-aspect-ratio torus standing vertically—heated at the bottom and cooled at the top (see

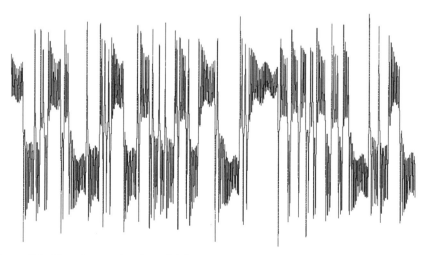

Figure 7.6. The time series for $x(t)$ in the "shaken" Lorenz attractor for $\rho_0 = 26.5$, $\rho_1 = 5.0$, and $\omega = 1.0$.

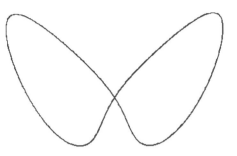

Figure 7.7. The periodic orbit, in x–z projection, to which the chaotic Lorenz attractor for $\rho_0 = 26.0$ is forced using a simple sinusoidal control with $\omega = 9.0$ and $\rho_1 = 2.5$ (less than 10% of the mean value of the time-dependent Rayleigh number).

Figure 7.2). It is well known that such a thin thermosyphon is better modeled by the Lorenz equations than is a Bénard convection layer of the type discussed previously. Thus, the Lorenz equations also are of interest as a simple generic model of a single-phase closed convection loop in some complex engineering system. Moreover, if such a loop is heated periodically at the bottom or cooled periodically at the top to control the motion of the fluid and the heat transfer within it, then the simple thermosyphon model of the convective loop is well described by the parametrically forced Lorenz equations introduced previously.

Motivated by the fact that the Lorenz equations provide a simple model for the dynamics of convection layers and for the dynamics of closed convection loops—both of which are extremely common in practical engineering systems—computer studies have been performed on the feasibility of

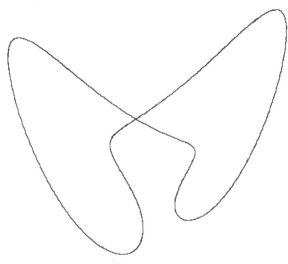

Figure 7.8. The simple periodic orbit, in x–z projection, to which the chaotic Lorenz attractor for $\rho_0 = 28.0$ is forced using a simple sinusoidal control with $\rho_1 = 11.0$ and $\omega = 8.0$.

modifying the chaotic behavior of these equations by simple parametric forcing through a sinusoidal Rayleigh number, with the objective of gaining insight into the possibility of boundary control of the chaotic dynamics of convection systems through sinusoidal or otherwise modulated surface temperatures.

7.2. MODIFYING CHAOTIC BEHAVIOR TO PERIODIC BEHAVIOR

A familiar Lorenz butterfly attractor calculated for $\rho_0 = 26.5$ is shown in Figure 7.3. This value of ρ_0 is well above the value at which the first homoclinic explosion occurs, and it is above the value at which the two fixed points become unstable through a subcritical Hopf bifurcation [2]. If one shakes an attractor, no doubt one ought to expect to see a shaken attractor! Figure 7.4 shows the attractor that results from parametrically "shaking" the system for the same value of ρ_0 but with the amplitude of the sinusoidal forcing function ρ_1 equal to 5.0 and the frequency ω equal to 1.0. The time series for $x(t)$ for the autonomous and nonautonomous systems are shown in Figures 7.5 and 7.6, respectively. The modulation of the amplitude of $x(t)$ during the more quiescent periods of the trajectory is clear from comparison of Figure 7.6 with Figure 7.5.

Notwithstanding the appearance of an occasional "shaken attractor," it is extremely easy to find parameter regimes where the chaotic autonomous system can be made periodic by simple parametric forcing. Figure 7.7 shows

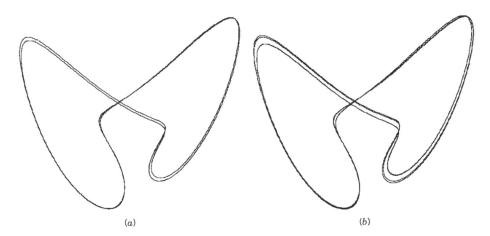

(a)   (b)

Figure 7.9. The period-doubling sequence back to chaos as ρ_1, the amplitude of the simple sinusoidal control, is increased from 11.0 to 11.1655 with $\rho_0 = 28.0$ and $\omega = 8.0$: (a) the period-2 orbit for $\rho_1 = 11.11$; (b) the period-4 orbit for $\rho_1 = 11.16$.

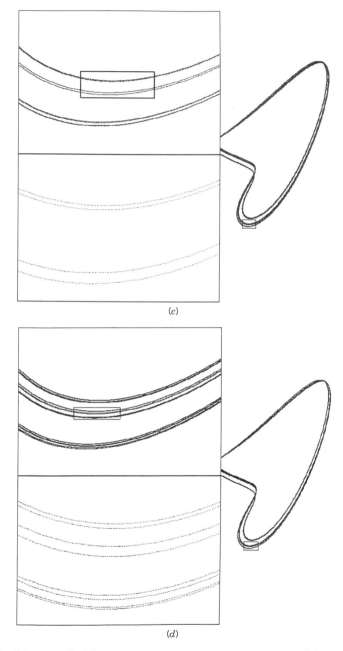

Figure 7.9. *(Continued).* (*c*) The period-16 orbit for $\rho_1 = 11.165$; (*d*) the period-64 orbit for $\rho_1 = 11.16542$.

an example: The chaotic attractor for the autonomous system with a Rayleigh number of 26.0 is converted to a simple periodic orbit when the system is forced at a frequency $\omega = 9.0$ with an amplitude ρ_1 as low as 2.5—less than 10% of the mean value of the time-dependent Rayleigh number. Thus our first objective was easily achieved: a system that behaves chaotically when autonomous was modified to behave periodically, without introducing an excessively large forcing function so that the cold-side temperature was not significantly increased.

7.3. BACK TO SIMPLE CHAOS

Another example of a chaotic Lorenz attractor made periodic by adding parametric forcing to the autonomous system is given in Figure 7.8, which shows a simple periodic orbit for ρ_0 equal to 28.0, $\omega = 8.0$, and $\rho_1 = 11.0$. As the amplitude ρ_1 is decreased to zero, naturally, the system returns to the chaotic Lorenz butterfly. However, as ρ_1 is increased from 11.0, the chaotic system that has been made periodic by parametric forcing is led through a period-doubling sequence to chaos arrived at by a value of ρ_1 equal to 11.1655. Figure 7.9a–d shows the period-2, -4, -16, and -64 orbits, and Figure 7.10 shows the chaotic attractor arrived at via the period-doubling sequence generated by increasing ρ_1.

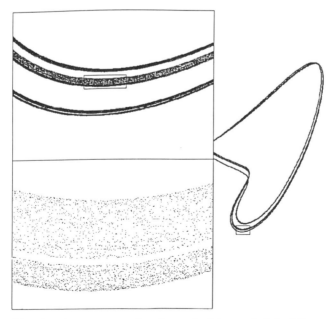

Figure 7.10. The chaotic attractor, after the end of the period-doubling sequence, for $\rho_0 = 28.0$, $\rho_1 = 11.1655$, and $\omega = 8.0$.

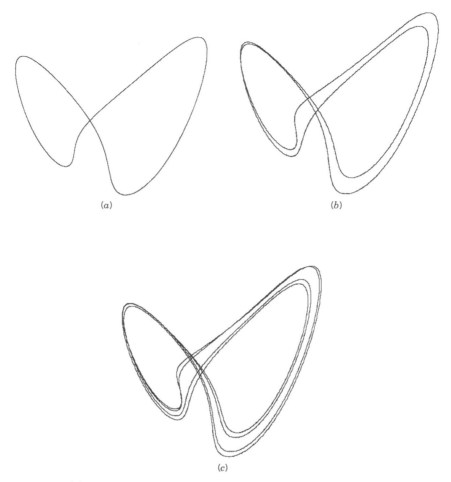

Figure 7.11. (*a*) The simple periodic orbit, in *x*–*z* projection, to which the chaotic Lorenz attractor for $\rho_0 = 28.0$ is forced using a simple periodic control with $\rho_1 = 5.0$ and $\omega = 8.0$; (*b*)–(*e*) the period-doubling sequence back to chaos as ω, the frequency of the simple sinusoidal control, is decreased from 8.0 to 7.622: (*b*) the period-2 orbit for $\omega = 7.7$; (*c*) the period-4 orbit for $\omega = 7.65$.

Analogous period-doubling sequences are easily found by varying the frequency of the driver while its amplitude is held fixed. Figure 7.11*a* shows the simple periodic orbit achieved by a system with an average Rayleigh number of 28.0—for which the autonomous system is chaotic—by driving it with an amplitude $\rho_1 = 5.0$ and a frequency of $\omega = 8.0$. Figure 7.11*b*–*e* shows the period-2, -4, -16, and -64 orbits, and Figure 7.11*f* shows the chaotic orbit arrived at when the frequency is decreased to 7.622. Thus, it is clear that some chaotic autonomous Lorenz systems can be modified so that they exhibit very simple periodic behavior and that they also can be led back

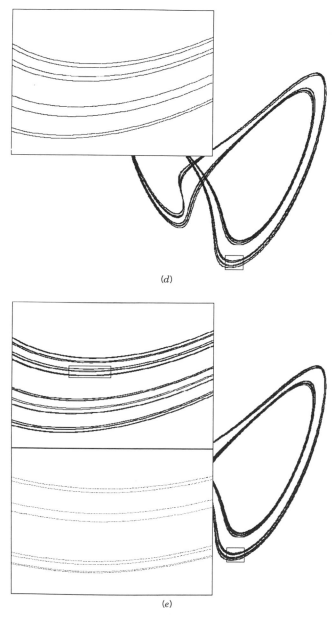

(d)

(e)

Figure 7.11. *(Continued).* (*d*) The period-16 orbit for $\omega = 7.623$; (*e*) the period-64 orbit for $\omega = 7.62225$.

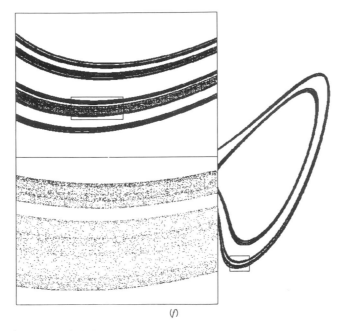

(f)

Figure 7.11. *(Continued).* (f) The chaotic attractor, after the end of the sequence, for $\omega = 7.622$.

to chaos through simple period-doubling sequences. This in turn suggests that a fairly significant amount of boundary control of natural thermal convection may, in fact, be possible.

7.4. FREQUENCY ENTRAINMENT AND SENSITIVE DEPENDENCE OF PHASE

When a system that originally is chaotic as an autonomous system at a Rayleigh number of 28.0 is forced with an amplitude of 11.4 and a frequency of 15.0, the chaotic attractor collapses to the simple periodic orbit shown from two vantage points in Figure 7.12*a* and 7.12*b*. The parametric forcing, in addition to modifying the behavior of this system from chaotic to periodic, also entrains its frequency; the time–T map of the system leads to a three-dimensional Poincaré section on which the attractor appears as two points—a discrete period-2 orbit, indicating that the system frequency is entrained at half the driver frequency. Naturally, a time–$2T$ map leads to two separate fixed point attractors on the three-dimensional Poincaré section. This being the case, it is interesting then to explore the basins of

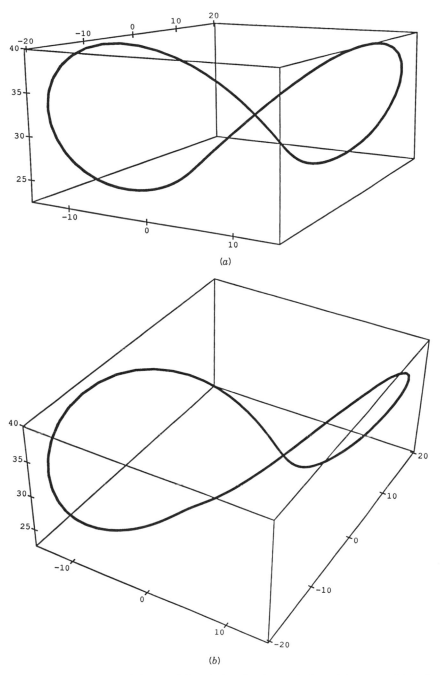

(a)

(b)

Figure 7.12. Two views, (a) and (b), of the simple periodic orbit onto which the chaotic autonomous system for $\rho_0 = 28.0$ is forced by the sinusoidal control with $\rho_1 = 11.4$ and $\omega = 15.0$.

Figure 7.13. The basins of attraction of the two fixed points of the time–$2T$ map on the three-dimensional Poincaré section for $\rho_0 = 28.0$, $\rho_1 = 11.4$, and $\omega = 15.0$. The sets of initial conditions (black and white) attracted to the two fixed points of the map are separated by a fractal basin boundary.

attraction of these two fixed points. As shown in Figure 7.13, these two basins are separated by a fractal boundary. Under the action of the time–$2T$ map the initial conditions indicated by black points in Figure 7.13 go to one fixed point of the map, whereas those indicated by white points go to the other under the action of the map. Because these two fixed points correspond to two points that are 180° out of phase on the periodic attractor of the parametrically forced system, it follows that the driven system displays sensitive dependence of phase upon initial conditions. Although two arbitrarily close initial conditions in the parametrically driven system both will follow trajectories to the periodic attractor, they can evolve onto it 180° out of phase with each other. Thus, although this originally chaotic system is made to behave periodically and its phase even is entrained, there is sensitive dependence upon initial conditions associated with the fractal basin boundary that separates the initial conditions that are entrained at the two different phases.

When the amplitude of the forcing function is decreased to 11.346, while keeping the frequency at 15.0 and the average Rayleigh number at 28.0, this system undergoes a transition from the stable periodic orbit shown in Figure 7.12a and 7.12b to type-I intermittency. Figure 7.14a shows the stable periodic orbit at $\rho_1 = 11.42$, and Figure 7.14b shows the trajectory for $\rho_1 = 11.346$ as it "falls off" the "ghost" of that previously stable orbit (shown

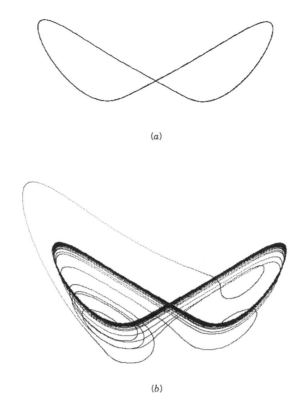

(a)

(b)

Figure 7.14. (a) The stable periodic orbit for $\rho_0 = 28.0$, $\rho_1 = 11.42$, and $\omega = 15.0$; (b) the trajectory, after ρ_1 is decreased to 11.346, as it "falls off" the "ghost" of the previously stable orbit after a long laminar period in the neighborhood of that orbit.

in Figure 7.14a) and embarks on a rapid chaotic transient following a long laminar period in the neighborhood where the periodic orbit previously had existed. The corresponding intermittent time series for $x(t)$ comprising laminar regions separated by chaotic bursts is shown in Figure 7.15a and 7.15b; the other two phase variables behave similarly. The attractor for the time–T map is shown in Figure 7.16. Finally, an estimate was made of ρ_1^*, the value of ρ_1 at which the inverse tangent bifurcation and the transition from the stable periodic orbit to type-I intermittency occurs. This was done by measuring the lengths of the laminar regions for a sequence of values of ρ_1 first in coarse steps of 0.0001 for the interval (11.3450, 11.3464) and then in fine steps of 0.00001 for the interval (11.34638, 11.34645). The maximum lengths L of the laminar regions were plotted against ρ_1 on log–log scales and a least squares fit was made of $L = \alpha(\rho_1 - \rho_1^*)^{-1/2}$, the leading-order

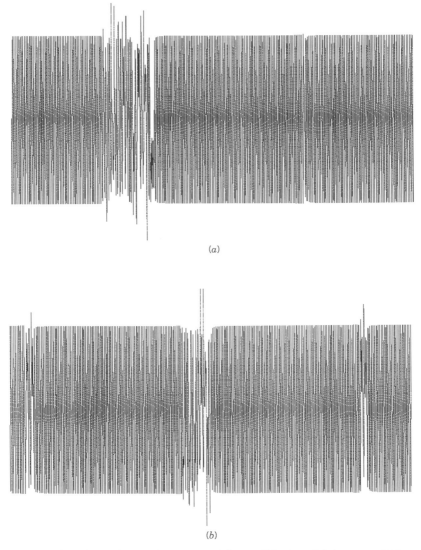

(a)

(b)

Figure 7.15. The intermittent time series, (*a*) and (*b*), for $x(t)$ for $\rho_0 = 28.0$, $\rho_1 =$ 11.346, and $\omega = 15.0$, showing laminar periods separated by intermittent bursts.

term in the expansion of L in terms of $(\rho_1 - \rho_1^*)$ for the inverse tangent bifurcation [3]. The resulting estimates for ρ_1^* were 11.346429 ± 0.0000003 from the coarse steps and $11.34645476 \pm 0.000000001$ from the fine steps, where the error estimates are based on the variances in the least squares fit projected onto the ρ_1 axis. Simulations were run using a fourth-order Runge–Kutta solver with time steps of 0.001, 0.0005, and 0.0001 to ensure

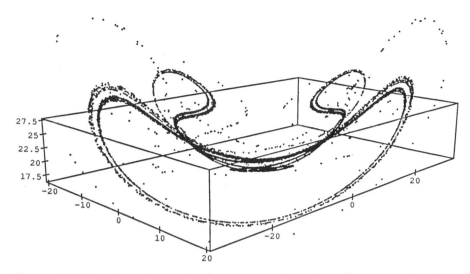

Figure 7.16. The intermittent chaotic attractor of the time$-T$ map on the three-dimensional Poincaré section for $\rho_0 = 28.0$, $\rho_1 = 11.346$, and $\omega = 15.0$.

that the bifurcation value was not sensitive to time steps of the order of 0.001, which was the value used to obtain the preceding estimates.

7.5. THE REVERSE: MAKING THE PERIODIC AUTONOMOUS SYSTEM BEHAVE CHAOTICALLY

It is well known that the Lorenz system behaves periodically for very large Rayleigh number [2]. Thus, the possibility of starting from an autonomous system that behaves periodically and parametrically driving it to make it behave chaotically, that is, doing the reverse of what was described previously, also was explored. Not surprisingly, as the Lorenz system with the Rayleigh number of 350 was driven parametrically it first took on the behavior of a limit cycle evolving with a finite winding number on the surface of the torus. As the amplitude of the driver was increased, the behavior became chaotic.

7.6. SOME ENGINEERING IMPLICATIONS

What engineering implications for practical technological systems are to be drawn from these computer experiments? Separate from the question of how well horizontal fluid convection layers and circular thermosyphons with small circular cross sections represent convection pools and closed convection loops in real engineering system is the question of how well the Lorenz equations represent the Navier–Stokes–Boussinesq equations for convective layers—surely not very well—and how well the Lorenz equations represent the simple circular thermosyphons. The answers to these questions are not completely encouraging. Even the autonomous Lorenz equations are known to exhibit some dynamical behavior that is not found experimentally in the systems that they are meant to model. Nevertheless, keeping this caveat in mind it may be useful to use the Lorenz equations as first-cut models for preliminary studies related to engineering systems. Certainly, the time and expense saved is strong motivation for preliminary studies using them, instead of the full partial differential equations that describe the related convective flows. Moreover, in many parameter regimes the systems that the Lorenz equations are meant to model exhibit complex dynamics similar to those of the Lorenz equations. Hence, the idea of exploring the feasibility of controlling the behavior of these convection systems by first studying the possibility of modifying the behavior of the Lorenz system through parametric excitation seems useful. The preliminary results reported here, in connection with both driving chaotic systems to periodicity and entraining the frequency of the resulting periodic systems, suggest that further studies along these lines would be worthwhile—both studies using simple models such as parametrically forced Lorenz equations and studies using a hierarchy of more complex and more realistic models. It was very easy to find parameter values of the single forcing function studied thus far that suppressed chaos and gave rise to periodicity and could then even be used to lead systems back to other types of chaotic behavior through period-doubling paths and inverse tangent bifurcations. This suggests that modification of the dynamics of convective cooling systems in a systematic way is feasible and that true boundary control of such systems may be possible.

REFERENCES

1. E. N. Lorenz, Deterministic nonperiodic flow. *J. Atmos. Sci.* **20** 130–141 (1963).
2. C. Sparrow, *The Lorenz Equations: Bifurcations, Chaos and Strange Attractors.* Springer-Verlag, New York, 1982.
3. Y. Pomeau and P. Manneville, Intermittent transition of turbulence in dissipative dynamical systems. *Comm. Math. Phys.* **74** 189 (1980).

CHAPTER 8

CHAOTIC TRANSIENTS AND FRACTAL STRUCTURES GOVERNING COUPLED SWING DYNAMICS*

Y. UEDA and T. ENOMOTO
Department of Electrical Engineering
Kyoto University
Kyoto 606
Japan

H. B. STEWART
Mathematical Sciences Group
Department of Applied Science
Brookhaven National Laboratory
Upton, NY 11973

Numerical simulations are used to study coupled swing equations modeling the dynamics of two electric generators connected to an infinite bus by a simple transmission network. In particular, the effect of varying parameters corresponding to the input power supplied to each generator is studied. In addition to stable steady operating conditions, which should correspond to synchronized, normal operation, the coupled swing model has other stable states of large-amplitude oscillations that, if realized, would represent non-synchronized motions: The phase space boundary separating their basins of attraction is fractal, corresponding to chaotic transient motions. These fractal structures in phase space and the associated fractal structures in parameter space will be of primary concern to engineers in predicting system behavior.

*This work was supported by the Applied Mathematical Sciences program of the U.S. Department of Energy under Contract No. DE-AC02-76CH00016.

Applied Chaos, Edited by Jong Hyun Kim and John Stringer.
ISBN 0-471-54453-1 © 1992 John Wiley & Sons, Inc.

8.1. INTRODUCTION

The stability of electric power systems has a long history of research; for background, see references [1]–[9]. The question of transient stability has been studied primarily by examining solutions of systems of differential equations based on the swing equation, or driven pendulum. Almost all research has relied on application of an energy integral or Lyapunov function to determine stability in a neighborhood of a desired solution; see, for example, references [10]–[12]. Although these methods may give sufficient conditions for practical operation, this approach by its nature does not address the question of the global structure in phase space of the basin of attraction of the desired stable operating condition, and it seems possible that an additional margin of safety may exist that cannot be identified by existing approaches. In any case, it is the basin of attraction that represents the proper focus of concern for the engineer, and this is the problem we shall study here. We note that our focus on basin boundary structure rather than stability near a desired solution closely parallels the new approach to ship stability criteria proposed by Thompson, Rainey, and Soliman [13].

One difficulty with the study of basin structures is their complexity, involving fractal basin boundaries [14]. The geometric theory of dynamic systems, including invariant manifolds, offers conceptual tools that may be helpful in elucidating these fractal structures. Here we make no final judgment on the practical utility of invariant manifold analysis, but simply present it with the aim of clarifying what we feel is the true and proper concern of engineering stability analysis, namely, the structure of basins of attraction and their boundaries. The model treated here uses the simplest expression for the generator, although in applications a more accurate model of the generator should be used. The geometric theory of dynamical systems is very general and could, in principle, easily accommodate more realistic models.

We consider a simple model of the dynamics of two electric generators connected to an infinite bus by a simple network consisting of two transmission lines. The configuration of this network is shown in Figure 8.1. Generator 1 is driven by mechanical input power P_{m1} and delivers its electrical

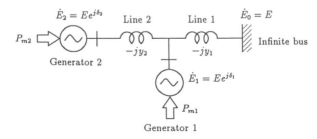

Figure 8.1. Two machines operating onto an infinite bus system.

output to transmission line 1 with transfer admittance y_1, while generator 2 with input power P_{m2} delivers its output to transmission line 2 with transfer admittance y_2, which is connected through line 1 to the same load, that is, infinite bus.

The equations used here to describe the dynamics of this system are

$$\frac{d\delta_1}{dt} = \omega_1$$

$$\frac{d\omega_1}{dt} = \frac{1}{m_1}\{-d_1\omega_1 + p_1 - \sin(\delta_1 - \delta_2) - k\sin\delta_1\}$$

$$\frac{d\delta_2}{dt} = \omega_2$$

$$\frac{d\omega_2}{dt} = \frac{1}{m_2}\{-d_2\omega_2 + p_2 - \sin(\delta_2 - \delta_1)\} \tag{8.1}$$

Here δ_1 and δ_2 are angular positions of the generator rotors and ω_1 and ω_2 are the angular velocities; these four state variables of the system are measured relative to the rotating reference of the infinite bus voltage. Both terminal voltages of the generators are regulated to maintain the same voltage as the infinite bus. The quantities p_1 and p_2 are the input powers normalized by the quantity $y_2 E^2$. The d_i are damping coefficients and the m_i are inertia constants, both likewise normalized by $y_2 E^2$, and $k = y_1/y_2$ is equal to the ratio of the admittances of the two transmission lines. These equations are derived under the assumption of small deviations in the angular frequencies from true 60-Hz oscillation, but the angular positions make large excursions.

Following the geometric method of Poincaré, we take the four state variables to be the coordinates of a four-dimensional phase space; an initial value is a point in this space, and solutions of the differential equations over time will trace out trajectories, or orbits, in this phase space. Because we cannot visualize this phase space directly, we shall consider two-dimensional subspaces such as (δ_1, δ_2), for example, and either look at the intersection of trajectories with the subspace or the projection of trajectories onto the subspace. Note that topologically the phase space is equivalent to the Cartesian product of a torus $(\delta_1, \delta_2) \in S^1 \times S^1 = T^2$ with a plane (ω_1, ω_2) $\in R^2$; for the torus we take the usual projection of T^2 unwrapped to the square $[0, 2\pi) \times [0, 2\pi)$.

The remaining quantities in the system (8.1) are parameters, held constant while a given trajectory is solved by integration. In a real power network, the d_i, m_i, and k are fixed characteristics of the network, whereas the input powers p_i are under the control of the engineer. It is therefore natural that p_1 and p_2 should enjoy a special role; we call them control parameters, or controls. In this study we fix

$$m_1 = m_2 = 0.1$$
$$d_1 = d_2 = 0.005$$
$$k = 1$$

throughout, and we consider the effect on solutions of choosing different values of p_1 and p_2. Following the comprehensive geometric viewpoint of Thom [15] and Abraham [16], we consider a control–phase space that is the Cartesian product of the control plane $(p_1, p_2) \in R^2$ with the phase space. Of particular interest in this study is the occurrence of fractal structure in the control plane, which according to the comprehensive view is a consequence of coherent structure in control–phase space linking the fractal structure in control space with fractal structure in phase space. Here we shall be content with visualizing the phenomenon in the (p_1, p_2) plane and, separately, in subspaces of phase space, with the hope that the reader will keep in mind the existence of a more comprehensive view; see, for example, Abraham and Shaw [17] for a visual introduction to the comprehensive view.

8.2. REGULAR BASIC MOTIONS

The first step in constructing a geometric phase portrait of the system (8.1) is to find the solutions corresponding to regular motions, that is, the equilibrium points corresponding to steady motions, and the closed trajectories corresponding to periodic motions. In either case the regular motion may be stable or unstable. The stable motions are attractors, motions that can be sustained in a real system. The unstable motions are not typically observed in real systems, but they are nevertheless important in understanding the geometric structure of phase space and the overall dynamical behavior. Of particular interest are the unstable motions that lie in the boundaries separating basins of attraction of the various stable motions.

The equilibrium points of (8.1) are found by setting all time derivatives to zero, leading to the conditions

$$\omega_1 = 0$$
$$\omega_2 = 0$$
$$k \sin \delta_1 + \sin(\delta_1 - \delta_2) = p_1$$
$$\sin(\delta_2 - \delta_1) = p_2 \qquad (8.2)$$

which are to be solved for the state-variable coordinates of the equilibrium points. We consider the region of control space

$$|p_1 + p_2| \le k = 1$$
$$|p_2| \le 1 \tag{8.3}$$

where there are four fixed-point solutions of (8.2). Clearly, $\omega_1 = \omega_2 = 0$ in all cases; the δ_i coordinates are

$$\delta_1^{(1)} = \text{Sin}^{-1} \frac{p_1 + p_2}{k} \qquad\qquad \delta_2^{(1)} = \text{Sin}^{-1} p_2 + \text{Sin}^{-1} \frac{p_1 + p_2}{k}$$

$$\delta_1^{(2)} = \text{Sin}^{-1} \frac{p_1 + p_2}{k} \qquad\qquad \delta_2^{(2)} = \pi - \text{Sin}^{-1} p_2 + \text{Sin}^{-1} \frac{p_1 + p_2}{k}$$

$$\delta_1^{(3)} = \pi - \text{Sin}^{-1} \frac{p_1 + p_2}{k} \qquad\qquad \delta_2^{(3)} = \pi + \text{Sin}^{-1} p_2 - \text{Sin}^{-1} \frac{p_1 + p_2}{k}$$

$$\delta_1^{(4)} = \pi - \text{Sin}^{-1} \frac{p_1 + p_2}{k} \qquad\qquad \delta_2^{(4)} = - \text{Sin}^{-1} p_2 - \text{Sin}^{-1} \frac{p_1 + p_2}{k} \tag{8.4}$$

Figure 8.2 shows the location of these fixed points in the (δ_1, δ_2) plane for the example $p_1 = 0.1$ and $p_2 = 0.1$. Clearly fixed point 1 is close to the condition $\delta_1 = \delta_2 = 0$ with no power generated.

The stability of the fixed points is determined by linearizing the system (8.1) near a fixed point, so that small deviations from the fixed point are governed by the matrix

$$
\begin{bmatrix}
0 & 1 & 0 & 0 \\
-\dfrac{1}{m_1}\cos(\delta_1 - \delta_2) - \dfrac{k}{m_1}\cos\delta_1 & \dfrac{-d_1}{m_1} & \dfrac{1}{m_1}\cos(\delta_1 - \delta_2) & 0 \\
0 & 0 & 0 & 1 \\
\dfrac{1}{m_2}\cos(\delta_2 - \delta_1) & 0 & -\dfrac{1}{m_2}\cos(\delta_2 - \delta_1) & -\dfrac{d_2}{m_2}
\end{bmatrix}
\tag{8.5}
$$

Table 8.1 gives numerical values of the coordinates δ_1 and δ_2 of the four fixed points for the example $p_1 = 0.1$ and $p_2 = 0.1$ illustrated in Figure 8.2. Also given in Table 8.1 are the eigenvalues of the matrix (8.5) for each fixed point. The fixed point 1 has negative real part for all four eigenvalues and is therefore stable. The remaining three fixed points have at least one positive real eigenvalue and are therefore unstable.

The single positive eigenvalue of fixed points 2 and 3 indicates that near each of these points there is a three-dimensional subspace within which trajectories approach the fixed point, and a one-dimensional subspace in

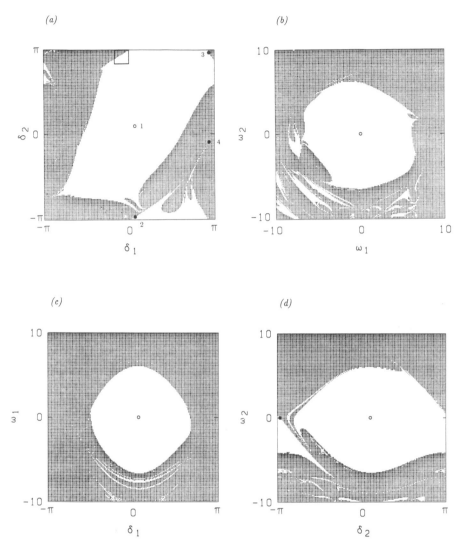

Figure 8.2. Partial basin portraits of coupled swing equations (8.1) with $p_1 = 0.1$ and $p_2 = 0.1$ in sections of the four-dimensional phase space taken at $\omega_1 = 0$, $\omega_2 = 0$, and δ_1 and/or δ_2 at the coordinates of fixed point 1; white regions are in the basin of attraction of the stable operating condition.

which trajectories diverge from the fixed point. Well-known theorems establish the fact that these subspaces can be extended to globally defined manifolds in phase space, a three-dimensional stable manifold, or *inset*, and a one-dimensional unstable manifold, or *outset*. The inset consists of trajectories that are asymptotic to the fixed point as $t \rightarrow \infty$. The inset can be located by starting from initial conditions near the fixed point in the three-dimen-

TABLE 8.1. Coordinates and Eigenvalues of Fixed Points for $p_1 = p_2 = 0.1$

	Fixed point 1	Fixed point 2	Fixed point 3	Fixed point 4
δ_1	0.20136	0.20136	2.94023	2.94023
δ_2	0.30153	−3.04040	3.04040	−0.30153
Eigenvalues	−0.0250 + j5.0930	−4.0427	−2.4827	−5.1181
	−0.0250 − j5.0930	−0.0250 + j2.4574	−0.0250 + j4.0176	−1.9638
	−0.0250 + j1.9385	−0.0250 − j2.4574	−0.0250 − j4.0176	1.9138
	−0.0250 − j1.9385	3.9927	2.4327	5.0681

sional subspace spanned by the three incoming eigenvalues and by integrating such initial conditions backwards in time. The outset consists of two trajectories asymptotic to the fixed point as $t \to -\infty$. That is, each of the fixed points 2 and 3 has an outset consisting of two branches. The local inset of each fixed point separates trajectories and might be part of the boundary of the basin of attraction of the stable operating condition.

In general the basin of attraction of the stable operating condition is expected to contain some singularity, either an unstable fixed point or an unstable periodic orbit. In the region of (p_1, p_2) control space satisfying inequalities (8.3), there are typically both stable and unstable periodic motions. For example, at $p_1 = 0.1$ and $p_2 = 0.1$, we found three stable periodic motions, corresponding to undesirable system operation (oscillation). It should be noted that a real generating system would not sustain these oscillations as our model does, because centrifugal forces would destroy the rotor.

By searching for fixed points in the surface of section $\delta_2 = 0$, we also found three unstable periodic motions. We have not yet tested whether the insets of these unstable periodic motions form part of any basin boundaries.

8.3. BASIN PORTRAITS

Figure 8.2 shows partial attractor-basin phase portraits of the coupled swing equations (8.1). In the upper left, Figure 8.2a is a section of the phase space in the (δ_1, δ_2) plane with $\omega_1 = \omega_2 = 0$. All initial conditions in the torus were integrated forward to determine whether the long-term behavior turned out to be the fixed point 1 or an oscillatory periodic attractor. An initial point leading to a periodic attractor is marked with a black dot, whereas an initial point in the basin of fixed point 1 is left unmarked. Another section of the same portrait in Figure 8.2b shows the (ω_1, ω_2) plane through the coordinates of fixed point 1, and the (δ_1, ω_1) and (δ_2, ω_2) sections in Figure 8.2c and 8.2d. We note that both fixed points 2 and 3 appear to lie in the basin boundary of the stable operating condition. We also checked that fixed point 4 lies in the basin boundary as well, although this is not so clear from Figure 8.2.

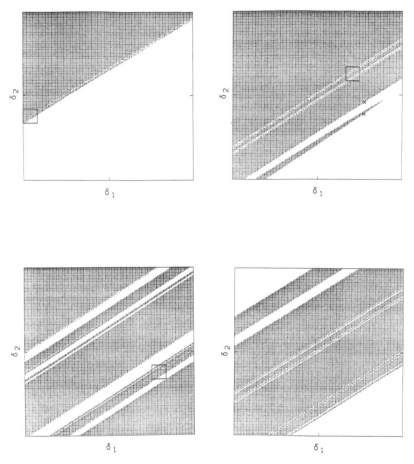

Figure 8.3. Magnification sequence of basin portrait section with $p_1 = 0.1$ and $p_2 = 0.1$, showing fractal structure.

The global structure of this basin boundary appears to be complicated. To check whether fractal structure exists, a zoom sequence of successively magnified basin portraits was obtained inside the small square in Figure 8.2a. The results are shown in Figure 8.3, which confirms that the structure of the basin boundary is roughly the product of a one-dimensional Cantor-like set with a smooth three-dimensional surface. The structure would be typical of a transverse homoclinic intersection of a three-dimensional inset. Note, however, that even though the insets of fixed points 2 and 3 are three-dimensional and they do intersect the two-dimensional outset of fixed point 4, they cannot have a transverse homoclinic intersection. It appears that unstable periodic motions generate tangles, and the insets of fixed points 2 and 3 become involved with those tangles, perhaps following a structurally unstable homoclinic connection of Shil'nikov type [18], or like that of the Lorenz system [19].

8.4. CONTROL SPACE PORTRAITS

Although the relationship between fixed points 2 and 3 and basin boundary of the stable operating condition is not simple, Figure 8.2 shows that the locations of these fixed points could be useful in devising approximate stability criteria. Indeed, such approximations could be more accurate than stability criteria based on energy-integral or Lyapunov function approaches. Thus it might be of interest to know whether fixed point 2, for example, lies on this basin boundary for typical values of p_1 and p_2.

This question was studied numerically by following the outset branches of fixed point 2 for a large number of (p_1, p_2) values. The result is illustrated in Figure 8.4. A point in control space is marked with a black dot if exactly one branch of the outset of fixed point 2 leads to the stable operating condition; a gray dot means both branches lead to the stable fixed point; no dot indicates that both branches lead away.

The control space is thus divided into three regions. It appears that the structure of these regions is complicated. In the small square near the center

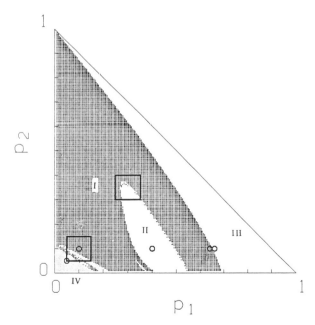

Figure 8.4. Control space portrait of position of fixed point 2 relative to the boundary of the basin of the stable operating condition, with black dots indicating regions where fixed point 2 is in the boundary.

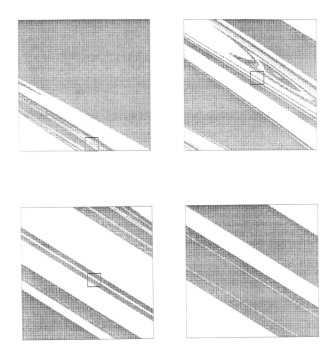

Figure 8.5. Magnification sequence of the control space portrait in the small box near the corner of Figure 8.4, showing fractal structure.

of the triangular region in Figure 8.4, a zoom sequence showed an infinitely layered but not fractal structure. In the other small square in Figure 8.4, the zoom sequence shown in Figure 8.5 reveals an apparently fractal structure. Like the structure of the basin boundary near fixed point 2, this control region boundary seems to be roughly the product of a Cantor-like set with a three-dimensional smooth manifold.

8.5. CONCLUSIONS

In the study of transient stability dynamical systems, the attractor-basin phase portrait provides the most direct information about system stability. For example, if the system is subjected to an impulsive disturbance that effectively displaces the phase space coordinates from a stable condition to some other point, the system will regain the stable operating condition if and only if the perturbed point is in the basin of attraction of the stable operating condition.

Thus a direct numerical attack on the question of stability requires a systematic trial of many initial conditions and parameter values. Although such studies may require substantial computer time, in principle they can be carried out with any existing numerical model of system dynamics.

In interpreting the results of systematic trials, geometric phase portraits are essential tools. The presence of fractal structure should be expected, and the theory of invariant manifolds helps to understand and explain such structure. This requires numerical methods for locating unstable fixed points and unstable periodic motions of system dynamics, such as straddle orbit algorithms [20] and push–pull algorithms [16].

In addition to attractor-basin phase portraits, it also may be useful to construct transient-time portraits as for example in Pezeshki and Dowell [21] and Thompson and Soliman [22]; these may in some cases reveal the influence of additional tangled structure in the interior of a basin.

ACKNOWLEDGMENT

The authors are grateful to the National Institute of Fusion Science at Nagoya for the use of the facilities of the Computer Center.

REFERENCES

1. T. Athay, R. Podmore, and S. Virmani, A practical method for the direct analysis of transient stability. *IEEE Trans. Power Apparatus and Systems* **PAS-98** 573–584 (1979).

2. N. Narasimhamurthi and M. T. Musavi, A generalized energy function for transient stability analysis of power systems. *IEEE Trans. Circuits and Systems* **CAS-31** 645 (1984).

3. A. N. Michel, A. A. Fouad, and V. Vittal, Power system transient stability using individual machine energy functions. *IEEE Trans. Circuits and Systems* **CAS-30** 266–276 (1983).

4. A. A. Fouad and S. E. Stanton, Transient stability of a multi-machine power system, Part I: Investigation of system trajectories. *IEEE Trans. Power Apparatus and Systems* **PAS-100** 3408–3416 (1981).

5. A. A. Fouad and S. E. Stanton, Transient stability of a multi-machine power system, Part II: Critical transient energy. *IEEE Trans. Power Apparatus and Systems* **PAS-100** 3417–3424 (1981).

6. H.-D. Chiang, F. Wu, and P. Varaiya, Foundations of direct methods for power system transient stability analysis. *IEEE Trans. Circuits and Systems* **CAS-34** 160–173 (1987).

7. V. Vittal, S. Rajagopal, M. A. El-Kady, and E. Vaahedi, Transient stability analysis of stressed power systems using the energy function method. *IEEE Trans. Power Systems*. **3** 239–244 (1988).

8. A. R. Bergen and D. J. Hill, A structure preserving model for power system stability analysis. *IEEE Trans. Power Apparatus and Systems* **PAS-100** 25–35 (1981).

9. M. A. Pai, *Energy Function Analysis for Power System Stability*. Kluwer Academic, Boston, 1989.

10. A. H. El-Abiad and K. Nagappan, Transient stability regions of multimachine power systems. *IEEE Trans. Power Apparatus and Systems* **PAS-85** 169–179 (1966).

11. P. D. Aylett, The energy-integral criterion of transient stability limits of power systems. *Proc. IEE Part C: Institution Monographs* **105** 527–536 (1958).

12. E. Carton and M. R-Pavella, Lyapunov methods applied to multimachine transient stability with variable inertia coefficients. *Proc. IEE* **118** 1601–1606 (1971).

13. J. M. T. Thompson, R. C. T. Rainey, and M. S. Soliman, Ship stability criteria based on chaotic transients from incursive fractals. *Philos. Trans. Roy. Soc. A* **332** 149–167 (1990).

14. S. W. McDonald, C. Grebogi, E. Ott, and J. A. Yorke, Fractal basin boundaries. *Physica D* **17** 125 (1985).

15. R. Thom, *Structural Stability and Morphogenesis*. Benjamin, Reading, MA, 1975.

16. R. H. Abraham, Dynasim: Exploratory research in bifurcations using interactive computer graphics. In *Bifurcation Theory and Applications in Scientific Disciplines*. New York Academy of Sciences, New York, 1979.

17. R. H. Abraham and C. D. Shaw, *Dynamics: The Geometry of Behavior, Parts I through IV*. Aerial Press, Santa Cruz, CA, 1982 (Part I)–1988 (Part IV).

18. J. Guckenheimer and P. Holmes, *Nonlinear Oscillations, Dynamical Systems, and Bifurcations of Vector Fields*. Springer-Verlag, New York, 1983.

19. J. M. T. Thompson and H. B. Stewart, *Nonlinear Dynamics and Chaos*. Wiley, Chichester, 1986.

20. P. M. Battelino, C. Grebogi, E. Ott, and J. A. Yorke, Multiple coexisting attractors, basin boundaries, and basic sets. *Physica D* **32** 296–305 (1988).

21. C. Pezeshki and E. H. Dowell, An examination of initial condition maps for the sinusoidally excited buckled beam modeled by the Duffing's equation. *J. Sound and Vibration* **117** 219–232 (1987).

22. J. M. T. Thompson and M. S. Soliman, Fractal control boundaries of driven oscillators and their relevance to safe engineering design. *Proc. Roy. Soc. London Ser. A* **428** 1–13 (1990).

CHAPTER 9

PROBABILISTIC ANALYSIS OF A CHAOTIC DYNAMICAL SYSTEM

SOLOMON C. S. YIM and HUAN LIN
Department of Civil Engineering
Oregon State University
Corvallis, OR 97331-2302

This study examines the chaotic behavior of a practical nonlinear dynamical system from the perspective of probability-based fatigue design. The stochastic characteristics of chaotic response of freestanding rigid objects subjected to horizontal harmonic base excitations are investigated. An approximate method based on the Melnikov function to predict analytically the existence of chaotic response is presented. Although the excitations to the rocking systems are simple and purely deterministic, some stochastic characteristics of the chaotic responses are detected using Poincaré maps and amplitude probability densities. It is found that although the chaotic time histories have a periodic time dependency (thus nonstationary), time series consisting of Poincaré points may be ergodic. These stochastic characteristics are useful in determining the fatigue life of nonlinear dynamical systems that operate frequently in chaotic states.

9.1. INTRODUCTION

The existence of chaotic response has been observed in many deterministic nonlinear physical and engineering systems [8, 6]. For some systems, the occurrence of chaotic responses is undesirable and should be avoided. However, for other systems, such as mixing of fluids and the control of heart rate variability, the occurrence of chaos may be desirable and should be promoted. In the design of such systems, the effects of long-duration steady-

Applied Chaos, Edited by Jong Hyun Kim and John Stringer.
ISBN 0-471-54453-1   © 1992 John Wiley & Sons, Inc.

state chaos on fatigue should be taken into account. Because many conventional fatigue design procedures are probability-based, it is important to determine whether chaotic responses possess stochastic properties commonly used in the design procedures. In particular, it is advantageous to examine the stochastic properties of chaotic responses, if they exist, in terms of probability density functions, even though chaotic responses are fully deterministic and repeatable. To demonstrate the stochastic characteristics of chaotic responses, a simple, freestanding rocking system is examined in this study.

The rocking behavior of rigid objects has been of interest to civil engineers for over a century. Many structures, from ancient historical minarets, monumental columns, and tombstones, to modern-day petroleum storage tanks, water towers, power transformers, nuclear reactors, and concrete radiation shields, can be considered to be freestanding objects. They may be subjected to support excitations due to earthquake ground motion and/or nearby machine vibrations. An understanding of the rocking behavior of these objects is of vital importance for safe operations, preservation of existence, and the design of new equipment.

Recent studies have demonstrated that the rocking response of rigid objects can be very sensitive to the system parameters and the excitation details. In a series of dynamic tests performed on the Berkeley shaking table, Aslam, Godden, and Scalise [1] showed that the rocking response was so sensitive that some of the experiments were deemed nonrepeatable. Using a single-degree-of-freedom (SDOF) model for the system, Yim, Chopra, and Penzien [13] examined the sensitivity of the rocking response by conducting a numerical study to identify, in a statistical sense, the parametric dependency of overturning stability on the size, slenderness ratio, coefficient of restitution, and ground motion intensity. The deterministic results confirmed that the rocking response can be sensitive to small changes in system parameters, and probabilistic trends can be established only with a large sample size.

Recognizing the insurmountable difficulties in the analysis of the complex behavior of the fully nonlinear rocking objects subjected to earthquake excitations, Spanos and Koh [7] and Tso and Wong [9] simplified the SDOF system by assuming the rigid objects to be slender and the base excitation to be harmonic, thus allowing linearization of the individual governing equations of motion and removing the randomness in the excitation. They were able to develop approximate analytical methods to predict the existence and stability of harmonic and subharmonic responses. Hogan [3] extended the stability analysis methods and developed a procedure to predict the existence of chaotic responses. He applied his prediction procedure to analyzing the experimental data of Wong and Tso [11] and was able to obtain a quantitative match. As a result of these studies, significant advances in the understanding of the rocking behavior were made.

Recently, Yang and Cheng [12] discovered new features of chaotic responses of nonlinear systems that fall in between the typical deterministic

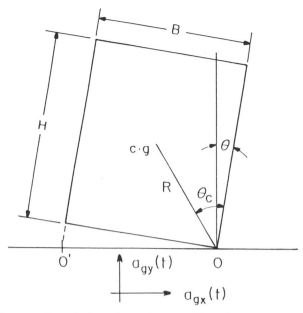

Figure 9.1. Freestanding rigid rocking object subjected to horizontal and vertical excitations.

and stochastic dynamics. They showed that, in a stochastic sense, the steady-state chaotic process is stationary and ergodic in terms of Poincaré time series. Kapitaniak [5] also developed a new indicator of chaotic response in a probabilistic sense. He pointed out that the chaotic response has a multimaxima curve in the amplitude probability density function.

This investigation focuses on identifying possible stochastic properties of chaotic responses of the simplified system. An approximate analytical method based on the Melnikov function to predict the existence of chaotic responses is first derived. The analysis techniques developed by Kapitaniak [5] and Yang and Cheng [12] are extended and employed to examine their stochastic properties.

9.2. SYSTEMS CONSIDERED

The freestanding slender object is modeled as a rectangular rigid body subjected to horizontal base motion excitation (Figure 9.1). Assuming that the coefficient of friction is sufficient that there will be no sliding between the object and the base, depending on the support accelerations, the object may move rigidly with the base or be set into rocking. If rocking occurs, it is assumed that the body will oscillate rigidly about the centers of rotation O and O'. The governing equation of motion for the rigid object with positive angular rotation about corner O is [13]

$$I_O\ddot{\theta} + MRa_{gx} + MgR(\theta_{cr} - \theta) = 0, \qquad \theta > 0 \qquad (9.1a)$$

where I_O is moment of inertia about O, M is mass, a_{gx} is the horizontal base acceleration, R is the distance from O to the center of mass, and $\theta_{cr} = \cot^{-1}(H/B)$ is the critical angle beyond which overturning will occur for the object under gravity alone (H and B are the height and width of the object). Similarly, the rocking about O' is governed by the equation

$$I_O\ddot{\theta} + MRa_{gx} - MgR(\theta_{cr} + \theta) = 0, \qquad \theta < 0 \qquad (9.1b)$$

Impact occurs when the angular rotation crosses zero approaching from either positive or negative, and the base surfaces recontact. Associated with the impact is a transition from rocking about one corner to rocking about the other and a finite amount of energy loss (or "damping") that can be accounted for by reducing the angular velocity of the object after impact. As in reference [13], it is assumed that the angular velocity before and after impact is related by a parameter e through the following equation:

$$\dot{\theta}(t^+) = e\dot{\theta}(t^-), \qquad 0 \le e \le 1 \qquad (9.2)$$

where e is the coefficient of restitution, t^+ is the time just after impact, and t^- is the time just before impact. In this study, the horizontal base excitation is assumed to be harmonic with constant amplitude and a single frequency,

$$a_{gx} = a_x \cos(\omega t + \phi) \qquad (9.3)$$

Note that there are two nonlinearities in the system associated with impact. The first results from the transition from one governing equation to the other as the center of rotation changes from one edge to the other. The second is due to the abrupt reduction (jump discontinuity) in the angular velocity caused by the impact (damping). In the limiting case, the damping nonlinearity can be removed if it is assumed that the material of the rigid object and the base support are stiff and that the rebound following impact is perfectly elastic so that there is no energy loss, that is, $e = 1$ (a conservative, or Hamiltonian, system).

9.3. METHOD OF ANALYSIS

Because the individual equations governing the rocking motion about each edge [equations (9.1a) and (9.1b)] are linear, exact analytical solutions for these equations exist and are used in conjunction with the nonlinear transition conditions at impact in determining the responses [7]. Let $\tau = \alpha t$ and $\Omega = \omega/\alpha$ [where $\alpha = (MgR/I_O)^{1/2}$] be the normalized time and frequency, respectively, the piecewise linear versions of (9.1a) and (9.1b) can be reduced to

$$\ddot{\Theta} - \Theta = \begin{cases} -A_x \cos(\Omega\tau + \phi) - 1, & \Theta > 0 \qquad (9.4a) \\ -A_x \cos(\Omega\tau + \phi) + 1, & \Theta < 0 \qquad (9.4b) \end{cases}$$

The solutions of (9.4a) and (9.4b) are

$$\Theta^+(\tau) = a^+ \sinh \tau + b^+ \cosh \tau + 1 + \beta \cos(\Omega\tau + \phi) \qquad (9.5a)$$

$$\Theta^-(\tau) = a^- \sinh \tau + b^- \cosh \tau - 1 + \beta \cos(\Omega\tau + \phi) \qquad (9.5b)$$

The superscript "+" indicates that the expression is valid for $\Theta > 0$ and the superscript "−" indicates that the expression is valid for $\Theta < 0$. The symbols a^+, b^+, a^-, and b^- denote constants of integration dependent on the initial conditions, and $\beta = A_x/(1 + \Omega^2)$. The response to harmonic excitations can be obtained by selecting the proper equation with the appropriate initial conditions and increasing the time variable successively. If at a particular time the magnitude of Θ exceeds 1.0 and continues to diverge, the rigid object will overturn. If Θ returns to zero, then the corresponding angular velocity can be determined and the solution given by the other equation can be applied. This procedure is then repeated until steady-state response is obtained.

9.4. CLASSIFICATION OF RESPONSES

When subjected to harmonic horizontal base excitation, depending on the amplitude and frequency of the excitation, the response of a slender object may either be *unbounded*, which leads to overturning, or it will eventually settle into a *bounded* motion. It has been shown that in addition to *periodic* response, there exist two additional types of bounded responses, namely, *quasiperiodic* and *chaotic* [4, 14]. Examples of these responses are shown in Figures 9.2, 9.3, and 9.4, respectively. In order to determine the regions of existence of chaotic responses, we derive analytical expressions of their boundaries.

9.5. CONDITIONS FOR EXISTENCE OF CHAOTIC RESPONSE

The existence conditions for chaotic response of the rocking system are derived following the methodology outlined by Melnikov [2], and the Melnikov integral function, which includes the effects of nonlinearity and excitations, is

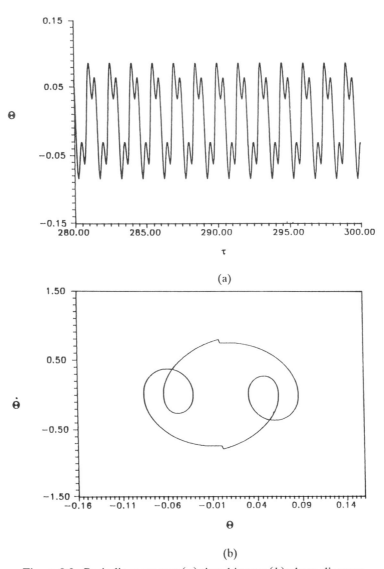

Figure 9.2. Periodic response: (a) time history; (b) phase diagram.

constructed. The integration is along a closed curve formed by heteroclinic orbits that are trajectories connecting two fixed points. The heteroclinic orbits are identified by examining the phase diagram of the associated (undamped, unforced) Hamiltonian system. Existence of zeros of the Melnikov function indicates the existence of chaotic response. The derivation of the existence conditions is shown as follows.

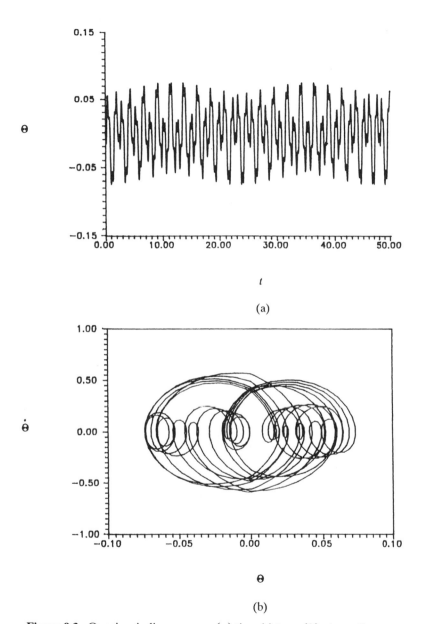

(a)

(b)

Figure 9.3. Quasiperiodic response: (a) time history; (b) phase diagram.

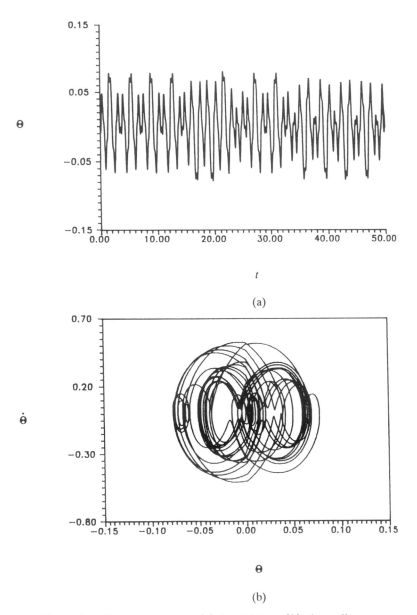

Figure 9.4. Chaotic response: (*a*) time history; (*b*) phase diagram.

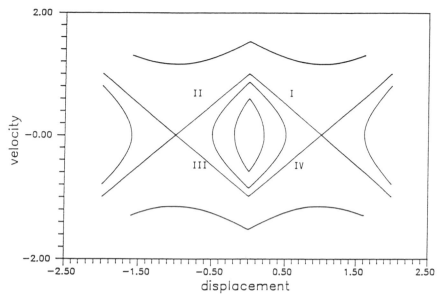

Figure 9.5. Heteroclinic orbit for rocking system.

First the equation of motion is rewritten in the following vector form:

$$\dot{x} = f(x) + \varepsilon g(x,t), \qquad x = (u \quad v)^* \tag{9.6}$$

where the asterisk indicates the transpose of vector (u, v). It is assumed that the unperturbed (i.e., undamped and unforced) system has a Hamiltonian H with $f_1 = \partial H/\partial u$ and $f_2 = \partial H/\partial v$, and the system satisfies the conditions [2] for the existence of two heteroclinic orbits connecting the two unstable hyperbolic singular points forming a closed curve. For the rocking system, it can be shown that these conditions are satisfied. The singular points for the rocking object are unstable equilibrium positions, $(1, 0)$ and $(-1, 0)$, where it stands on one edge (Figure 9.5). In the center of the closed curve is a stable singular point $(0, 0)$. Phase space trajectories near the stable singular point remain in its neighborhood, whereas trajectories near the hyperbolic singular points will diverge from them.

The external force and damping are then represented as perturbations to the Hamiltonian system. The damping effect in the rocking system, which causes sudden reduction in velocity by a factor e at displacement equal to 0, can be expressed analytically by a function $F(\theta, \dot{\theta})$. The integral of this function along the heteroclinic orbit is equal to the energy loss caused by impact in each oscillation. For the rocking system considered, the equation of motion for the general case with damping can be written as

$$I_O \ddot{\theta} \pm MgR(\theta_{\mathrm{cr}} \mp \theta) = -MRa_{gx}(t) + F(\theta, \dot{\theta}) \tag{9.7}$$

where $a_{gx} = \varepsilon g \cos \omega t$ is the perturbation caused by external force, and the equation with the upper signs represents equation of motion for $\theta > 0$ and the lower signs for $\theta < 0$. The perturbation due to damping is found to be

$$F(\theta, \dot{\theta}) = -\frac{1}{2} I_O \dot{\theta}^2 (1 - e^2) \left[\delta \left(\frac{\theta}{\theta_{cr}} \right) \right] \left(\frac{1}{\theta_{cr}} \right) \tag{9.8}$$

where $\delta(\theta/\theta_{cr})$ represents the Dirac delta function.

The equation of motion is then rewritten in the format used in (9.6) in the form of a first-order system as follows:

$$\dot{\theta} = v \tag{9.9a}$$

$$\dot{v} = \mp \frac{MgR}{I_O} (\theta_{cr} \mp \theta) - \frac{MgR}{I_O} \varepsilon \cos \omega t - \frac{1}{2} v^2 (1 - e^2) \delta \left(\frac{\theta}{\theta_{cr}} \right) \left(\frac{1}{\theta_{cr}} \right) \tag{9.9b}$$

The Hamiltonian function corresponding to the undamped, unforced system is then found by energy considerations:

$$H(\theta, v) = \frac{v^2}{2} - \frac{MgR}{2I_O} \theta^2 \pm \frac{MgR}{I_O} \theta \theta_{cr} \tag{9.10}$$

If the value of the Hamiltonian, which corresponds to the total (kinetic plus potential) energy of the system, exceeds a critical value ($MgR\theta_{cr}^2/2I_O$), the response will become unbounded, leading to overturning. Conversely, if the total energy is less than the critical value, the response will be bounded and stable.

Next, analytical expressions of the heteroclinic orbits are determined. By equating the Hamiltonian $H(\theta, v)$ and the critical value $MgR\theta_{cr}^2/2I_O$, the heteroclinic orbits can be expressed as functions of time. As mentioned previously, there are two heteroclinic orbits. Note that there is a discontinuity within each heteroclinic orbit and, therefore, four analytical expressions are required to describe the closed curve.

Section I (with $\theta > 0$ and $v > 0$) of the heteroclinic orbit can be expressed as

$$(\theta, v)^1 = (\theta_{cr} - e^{-p}, \alpha e^{-p}) \tag{9.11a}$$

where $p = \alpha t - \ln \theta_{cr}$. Similarly, section II (with $\theta < 0$ and $v > 0$) can be expressed as

$$(\theta, v)^2 = (e^p - \theta_{cr}, \alpha e^p) \tag{9.11b}$$

The expressions of the lower orbit (sections III and IV) can be obtained from (9.11a) and (9.11b) with opposite signs by symmetry. These expressions are then normalized by setting $\Theta = \theta/\theta_{cr}$, $\alpha t = \tau$, $V = v/\theta_{cr}$, and $\omega = \alpha\Omega$, and are rewritten as

$$(\Theta, V)^1 = (1 - e^{-\tau}, e^{-\tau}) \tag{9.12a}$$

$$(\Theta, V)^2 = (e^\tau - 1, e^\tau) \tag{9.12b}$$

$M^+(t_O)$ is used to denote the Melnikov's function for the upper orbit, and is defined by

$$M^+(t_O) = \int_{-\infty}^{+\infty} f(q_0(t)) \wedge g(q_0(t), t + t_O)\, dt$$

$$= \int_{-\infty}^{+\infty} (f_1 f_2) \wedge (g_1 g_2)^*\, dt$$

$$= \int_{-\infty}^{+\infty} (f_1 g_2 - f_2 g_1)\, dt \tag{9.13}$$

where the asterisk represents the transpose of a vector, and the f_is are the linearized Hamiltonians and the g_is are the perturbations. By symmetry, the expressions for $M^-(t_O)$ and $M^+(t_O)$ are identical. Therefore, the existence of solution to $M^+(t_O) = 0$ implies the existence of zero solution to $M^-(t_O)$. Hence, the condition for existence of chaotic response can be determined by examining just the upper curve. When no zeros exist, there is no chaotic response for this system.

The analytical methods for existence and stability of periodic response and the conditions for existence of chaotic response derived previously are applied to the undamped and damped systems in the following sections.

9.5.1. Undamped Systems

The Melnikov function for an undamped slender system can be obtained by letting $F(\theta, v) = 0$, thus (9.13) becomes

$$M^+(t_O) = \frac{-2\varepsilon\alpha^2 \cos \omega t_O}{\theta_{cr}(\alpha^2 + \omega^2)}$$

$$= -2A_x \frac{\cos \Omega \tau_O}{1 + \Omega^2} \tag{9.14}$$

where A_x and Ω are the normalized amplitude and frequency of excitation, respectively, with $A_x = \varepsilon/\theta_{cr}$ and $\Omega = \omega/\alpha$. The zero solutions, which yield the condition for the existence of chaotic response, are given by $\cos \Omega\tau_O = 0$. Because this condition can always be satisfied by $\tau_O = 2n\pi$ for all integer n, the condition for existence of chaotic response is satisfied by any arbitrary combination of system and excitation parameters.

9.5.2. Damped Systems

For damped systems, the Melnikov function (9.13) can be written as

$$
M^+(t_O) = \int_{-\infty}^{+\infty} v^{(2)}(t)\big(\varepsilon\alpha^2 \cos \omega(t + t_O) - F(\theta,\dot\theta)\big)\, dt
$$

$$
+ \int_{-\infty}^{+\infty} v^{(1)}(t)\big(\varepsilon\alpha^2 \cos \omega(t + t_O) - F(\theta,\dot\theta)\big)\, dt
$$

$$
= \frac{-2\varepsilon\alpha^2 \cos \omega t_O}{\alpha^2 + \omega^2} - \frac{\theta_{cr}(1 - e^2)}{2} \tag{9.15}
$$

$M^+(t_O)$ equal to zero is equivalent to

$$
\cos \Omega\tau_O = -\frac{(1 - e^2)(1 + \Omega^2)}{4A_x} \tag{9.16}
$$

where A_x and Ω are the normalized amplitude and frequency of excitation respectively with $A_x = \varepsilon/\theta_{cr}$ and $\Omega = \omega/\alpha$.

Equation (9.16) can always be satisfied by some τ_O if and only if the magnitude of the right-hand side is less than or equal to unity. Thus, the existence of chaotic response is given by the combination of system parameters e, Ω, and A_x that satisfies the following inequality:

$$
\frac{(1 - e^2)(1 + \Omega^2)}{4A_x} \le 1 \tag{9.17}
$$

It can be shown that (9.17) yields only the lower bound for existence of chaotic responses. In the given region, periodic and quasiperiodic responses coexist with chaotic responses [14].

9.6. STOCHASTIC CHARACTERISTICS

When a nonlinear dynamical system is in a chaotic state, precise prediction of the time history of the motion is impractical because small uncertainties in the initial conditions lead to diverging responses. However, probability density functions can provide a statistical measure of the chaotic dynamics. There is some mathematical and experimental evidence that such distributions do exist for chaotic responses [6]. Accordingly, an identification technique for chaotic response in a probabilistic setting, the amplitude probability density function, will be introduced.

Furthermore, in the probabilistic sense, under an almost unique set of initial conditions and a deterministic harmonic steady-state excitation, chaotic responses for certain deterministic structural systems are found to be stochastic [12]. Thus the chaotic process may serve well as a link between the traditional deterministic and stochastic processes. Two stochastic properties of chaotic responses—stationarity and ergodicity—also will be examined.

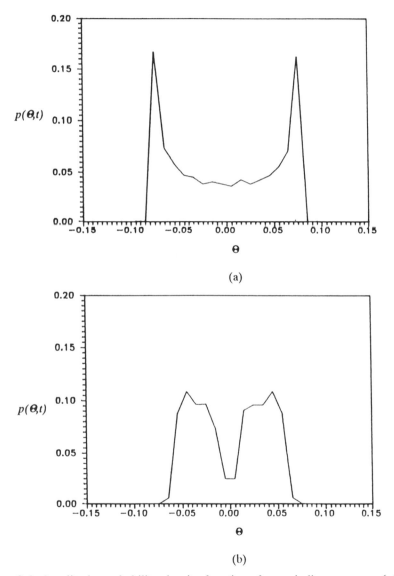

(a)

(b)

Figure 9.6. Amplitude probability density functions for periodic responses: (*a*) harmonic response, $A_x = 2.0$ and $T_x = 1.0$; (*b*) one-third subharmonic response, $A_x = 6.5$ and $T_x = 0.4$. ($H/B = 10$, $R = 290$, and $e = 0.925$.)

9.6.1. Amplitude Probability Density Function

The amplitude probability function of a response time history $\Theta(t)$ is defined as the normalized relative frequency of occurrence of the values between Θ and $\Theta + d\Theta$ (i.e., a normalized temporal average of Θ). This definition

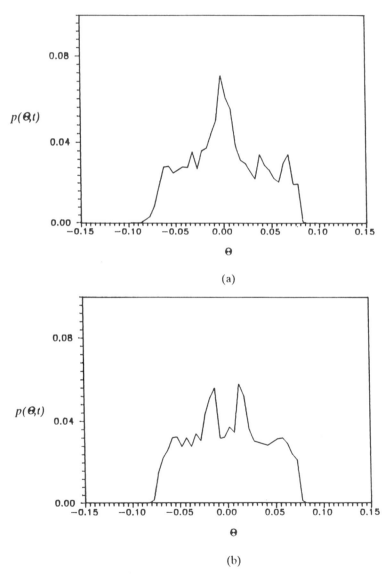

Figure 9.7. Amplitude probability density functions for nonperiodic responses: (*a*) chaotic response, $A_x = 3.0$ and $T_x = 0.4$; (*b*) quasiperiodic response, $A_x = 4.0$ and $T_x = 0.4$. ($H/B = 10$, $R = 290$, and $e = 1.0$.)

differs from the conventional definition of probability density function of $\Theta(t)$ for a given time t in stochastic analysis, which is a normalized ensemble average of the responses falling between Θ and $\Theta + d\Theta$ at time t.

The amplitude probability density functions of a harmonic response and a subharmonic response are shown in Figure 9.6a and 9.6b, respectively. They are characterized by symmetric maxima at the two extreme values. The maxima diffuse toward the center as the order of the subharmonic response increases.

However, for chaotic responses, the amplitude probability density function is characterized by a multimaxima curve [5], as shown in Figure 9.7a. The multimaxima nature of the density function is due to the fact that in the oscillatory waveform of chaotic responses there exist values of the amplitude that are more probable than neighboring values. Although a multimaxima probability density curve is useful in identifying possible chaotic responses, it is not a sufficient indicator. In particular, it cannot differentiate between quasiperiodic and chaotic responses. As shown in Figure 9.7b, quasiperiodic response can also have a multimaxima curve in the amplitude probability density function.

9.6.2. Stochastic Properties of Chaotic Response

Stochastic properties for some chaotic systems are known to exist based on "ensemble experiment" [12]. To identify possible stochastic properties of the rocking systems, time histories and Poincaré maps of their chaotic responses are examined. The focus will be on two fundamental properties, stationarity and ergodicity.

Prior to examining the stationarity and ergodicity properties of chaotic responses, it is essential to determine the required sample length of a chaotic time history to ensure stable probability properties. For this we examine the amplitude probability function of a typical chaotic response. To avoid the effects of transience, the response corresponding to the first 3000 cycles of excitation period was discarded. Different segments of the same time history were then sampled. Two representative amplitude probability density functions, each segment equal to 1000 times the excitation period, are shown in Figure 9.8a and 9.8b. As indicated, the amplitude probability density functions are practically identical, indicating that the segment length is sufficient for convergence of the amplitude probability density function.

A *chaotic process* is defined as an ensemble of chaotic responses that have identical system parameters, but with infinitesimally small perturbations in the initial conditions about a known chaotic attractor, and each sample response is itself chaotic. Numerically, ensembles of responses are generated by computing the time histories of responses of the system with initial conditions corresponding to the grid points of an infinitesimal neighborhood containing the known chaotic attractors. The reference time t is chosen far enough from the initial time that the correlations between each pair of

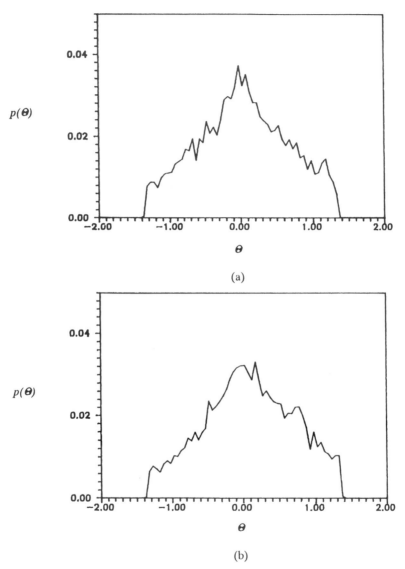

Figure 9.8. Convergence of amplitude probability density functions of chaotic process: (*a*) density function of first 1000-excitation-cycle segment of a chaotic time history; (*b*) density function of second 1000-excitation-cycle segment of the same time history. ($A_x = 4.6$, $T_x = 2.327$, $H/B = 100$, $R = 290$, and $e = 0.5$.)

responses have vanished. In this study, time histories and Poincaré points are generated by computing approximately 1000 time series with initial conditions corresponding to the grid points of a small square. The sizes of the squares are carefully chosen to ensure the samples are representative of the chaotic processes.

Time History. A stochastic process $\Theta(t)$ is stationary if its statistical properties are invariant to a shift of the time reference, that is, the process $\Theta(t)$ and $\Theta(t + \phi)$ have identical distributions for any value of ϕ. To test stationarity

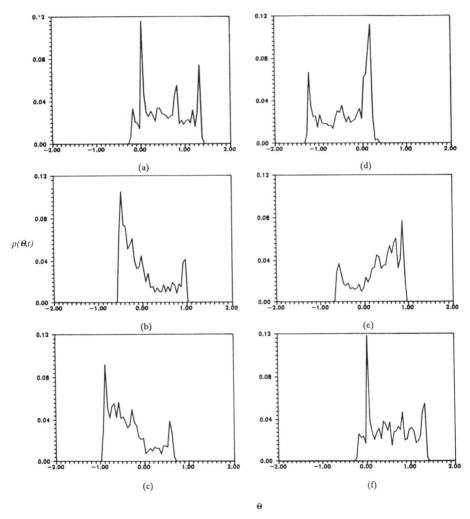

Figure 9.9. Probability density functions with sampling shifts (ϕ): (a) $\phi = 0.0 \times T_x$; (b) $\phi = 0.2 \times T_x$; (c) $\phi = 0.4 \times T_x$; (d) $\phi = 0.6 \times T_x$; (e) $\phi = 0.8 \times T_x$; (f) $\phi = 1.0 \times T_x$. $(A_x = 4.6, T_x = 2.327, H/B = 100, R = 290,$ and $e = 0.5.)$

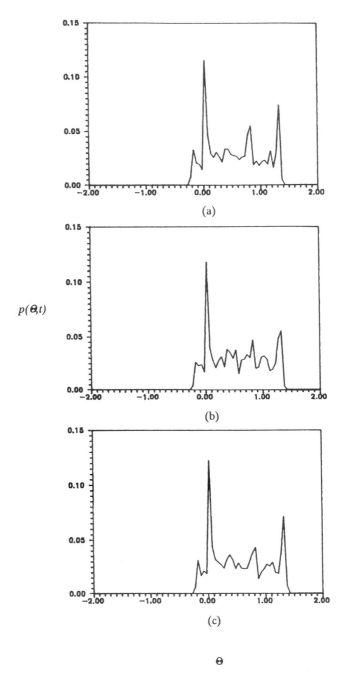

Figure 9.10. Stationarity and ergodicity of Poincaré time series: (a) probability density function at $t = 100T$; (b) probability density function at $t = 150T$; (c) amplitude probability density function of a typical time series. ($A_x = 4.6$, $T_x = 2.327$, $H/B = 100$, $R = 290$, and $e = 0.5$.)

of chaotic responses, the probability density functions of a (5000-sample) chaotic process corresponding to six values of time shifts (ϕ = 0.0, 0.2, 0.4, 0.6, 0.8, and 1.0 times the excitation period) are presented in Figure 9.9. The time t is chosen as the end of the 1000th cycle of the excitation. Note that the probability density function is dependent on the phase shift; thus chaotic time histories are nonstationary. However, when the phase shift is an integer multiple of the excitation period, the probability density functions are nearly identical (Figures 9.9a and 9.9f). In addition, the probability density function appears to be symmetric about the half-cycle time point, that is, the function at 0.4 and 0.6 (Figure 9.9c and 9.9d) and the function at 0.2 and 0.8 (Figure 9.9b and 9.9e) are practically equal. Thus the time dependence appears to be periodic with period equal to the excitation period, and there may be an invariant structure when periodic dependency is removed.

Poincaré Maps. The Poincaré points generated from the 1000-sample time histories may themselves be considered as time series with a sampling rate equal to the excitation period. The probability density functions at two representative times t_1 and t_2 of the Poincaré time series of a chaotic process are shown in Figure 9.10a and 9.10b. It is observed that the two functions are practically identical, thus indicating that the Poincaré processes of chaotic responses are stationary. This property is anticipated because the trajectory of chaotic response settles into a strange attractor; hence, after transient response has died out, the shape of the Poincaré map will not change with time.

Yang and Cheng [12] observed an interesting behavior of the Poincaré points of chaotic responses. In their study of nonlinear structures with hysteresis and degradation, they found that the asymptotic distribution of the Poincaré points originating from the infinitesimal square appear to be identical to that of the Poincaré map of a single time history. In particular, for small multiples of the excitation period T, the Poincaré points stay close together. As time increases, the Poincaré points keep stretching and folding (Smale horseshoe effect [8]). Eventually, the Poincaré points settle down into the same attractor as the one created by a single time history. This phenomenon is demonstrated in Figure 9.11 for the chaotic rocking responses. Numerical results indicate that the convergent rate of ergodicity depends on the chosen small square of initial conditions and the system sensitivity. If the small square of initial conditions falls in the sensitive region of the chaotic state, the strange attractor can be achieved by fewer loading cycles. Furthermore, the more sensitive the response is, the fewer loading cycles are needed for the ensemble Poincaré points to reach the distribution of the strange attractor. Thus the rate of convergence of the Poincaré process to the chaotic attractor may also be used as a measure of the sensitivity of the system response.

Poincaré processes of the rocking systems are further examined for ergodicity in this study by examining the probability density functions. A stochastic

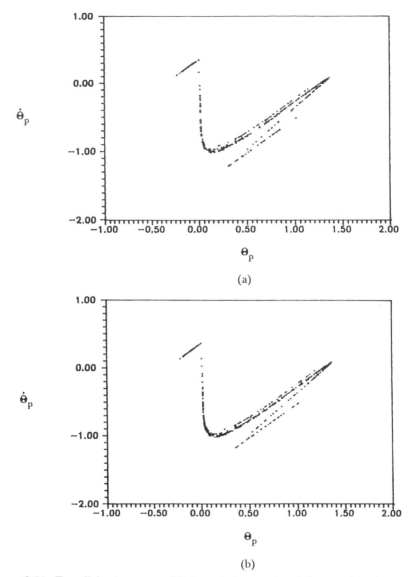

Figure 9.11. Ergodicity in terms of Poincaré time series: (a) ensemble time series; (b) one time history. ($A_x = 4.6$, $T_x = 2.327$, $H/B = 100$, $R = 290$, and $e = 0.5$.)

process $\Theta(t)$ is ergodic if its ensemble average equals appropriate time average, that is, with probability 1, the statistics of $\Theta(t)$ can be determined from a single sample $\Theta(t, \zeta)$. To examine the ergodic property of the Poincaré process, the amplitude probability density function constructed from a single time series (Figure 9.10c) is compared to those obtained from the 1000 sample Poincaré process discussed (Figure 9.10a and 9.10b). Observe that the two types of distributions are practically identical, which indicates that the Poincaré process may be ergodic. The properties of the correlation function of the Poincaré process will be examined in detail in a future paper to confirm ergodicity.

9.7. CONCLUSIONS

This investigation deals with the identification of stochastic properties of chaotic response of a dynamical system describing the rocking response of rigid objects to horizontal harmonic base excitation. The existence periodic, chaotic, and quasiperiodic responses are demonstrated. An analytical technique based on the Melnikov function is derived for the prediction of existence of chaotic response. The stochastic dynamics of the rocking system are examined by merging and extending the numerical analysis techniques developed by Kapitaniak [5] and Yang and Cheng [12]. It is found that in a stochastic sense, probability density functions are time dependent; hence, a process formed by time histories of chaotic responses is not stationary. However, a process consisting of Poincaré points of chaotic responses as time histories may be ergodic. These stochastic properties are useful for fatigue design of nonlinear dynamical systems with frequent chaotic motions.

Finally, the excitations considered in this study are purely deterministic. For future research, it would be interesting to apply the stochastic techniques to examine the behavior of nonlinear dynamical systems subjected to excitations with significant stochastic components.

ACKNOWLEDGMENT

The authors gratefully acknowledge the financial support for this research by the Office of Naval Research Young-Investigator Award under Grant No. N00014-88-K-0729.

NOTATION

a_{gx}, a_{gy}	horizontal and vertical ground accelerations
a_x, a_y	amplitude of horizontal and vertical excitations
a_s	coefficient of periodic steady-state response
A_x	normalized amplitude of horizontal excitation, $A_x = a_x/(g\theta_{cr})$
A_y	normalized amplitude of horizontal excitation, $A_y = a_y/g$
b_s	coefficient of periodic steady-state response
d_c	fractal dimension
e	coefficient of restitution
g	acceleration of gravity

r_a	ratio of vertical versus horizontal excitation amplitude, $r_a = a_y/a_x$
B, H	width and height of block
I_O	mass moment of inertia
M	mass of object
$p(\Theta, t)$	amplitude probability density function
R	radius of rotation
T_x	period of horizontal excitation
θ	rotation of angle of rocking block
θ_b	bracing angle
θ_{cr}	critical angle
$\dot{\theta}$	angular velocity
Θ	normalized angle, $\Theta = \theta/\theta_{cr}$
$\dot{\Theta}$	normalized angular velocity, $\dot{\Theta} = \dot{\theta}/\theta_{cr}$
α	$(MgR/I_O)^{1/2}$
β	$A_x/(1 + \Omega^2)$
τ	initial time, $\tau = \alpha t$
τ_i	the instant of occurrence of impact
Ω	normalized frequency of horizontal excitation
λ	Lyapunov exponent
ω_x	frequency of horizontal excitation
ϕ	sampling shift
ϕ_s	shift between excitation and steady-state response

REFERENCES

1. M. Aslam, W. G. Godden, and D. T. Scalise, Earthquake rocking response of rigid bodies. *J. Structural Eng. Div. ASCE* **106** (ST2) 377–392 (1980).

2. J. Guckenheimer and P. Holmes, *Nonlinear Oscillations, Dynamical Systems, and Bifurcations of Vector Fields.* Springer-Verlag, New York, 1983.

3. S. J. Hogan, On the dynamics of rigid-block motion under harmonic forcing. *Proc. Roy. Soc. London A* **425** 441–476 (1989).

4. S. J. Hogan, The many steady state responses of a rigid block under harmonic forcing. *Earthquake Eng. and Structural Dynam.* **19** 1057–1071 (1990).

5. T. Kapitaniak, Quantifying chaos with amplitude probability density function. *J. Sound and Vibrations* **114** 588–592 (1987).

6. F. C. Moon, *Chaotic Vibrations.* Wiley, New York, 1987.

7. P. D. Spanos and A. S. Koh, Rocking of rigid blocks due to harmonic shaking. *J. Eng. Mech. Div. ASCE* **110** 1627–1642 (1984).

8. J. M. T. Thompson and H. B. Stewart, *Nonlinear Dynamics and Chaos.* Wiley, New York, 1986.

9. W. K. Tso and C. M. Wong, Steady state rocking response of rigid blocks, Part I: Analysis. *Earthquake Eng. and Structural Dynam.* **18**(106) 89–106 (1989).

10. A. Wolf, J. B. Swift, H. L. Swinney, and J. A. Vastano, Determining Lyapunov exponent from a time series. *Physica D* **16** 285–317 (1985).

11. C. M. Wong and W. K. Tso, Steady state rocking response of rigid blocks, Part II: Experiment. *Earthquake Eng. and Structural Dynam.* **18**(106) 107–120 (1989).

12. C. Y. Yang, A. H-D. Cheng, and R. V. Roy, Chaotic and stochastic dynamics for a nonlinear structural system with hysteresis and degradation. *Int. J. Probabilistic Eng. Mech.* **6**(3/4) 193–203 (1991).

13. C. S. Yim, A. K. Chopra, and J. Penzien, Rocking response of rigid blocks to earthquakes. *Earthquake Eng. and Structural Dynam.* **8**(6) 565–587 (1980).

14. S. C. S. Yim and H. Lin, Chaotic behavior and stability of free-standing offshore equipment. *Ocean Eng.* **18**(3) 225–250 (1991).

PART II

APPLICATIONS IN PHYSICAL SCIENCES

CHAPTER 10

CHAOTIC BEHAVIOR OF COUPLED DIODES*

HILDA A. CERDEIRA[†]
International Centre for Theoretical Physics
34100 Trieste
Italy

A. A. COLAVITA[‡] and T. P. EGGARTER[§]
Microprocessor Laboratory
International Centre for Theoretical Physics
34100 Trieste
Italy

We study the behavior of a reverse-biased *pin* diode structure with both junctions in avalanche mode. The density of the oscillating charge clouds presents a very nonlinear behavior. Here we study the stability of the solutions that present bifurcations, chaos, and coexistence of attractors of different periods similar to those found in coupled logistic maps.

Coupled maps have been the subject of intensive study in connection with many physical, chemical, and biological systems, such as the following: ac-driven superconducting quantum interference devices (SQUIDs); convection in conducting fluids; instabilities in solids; lasers; different vibrational degrees of freedom in polyatomic systems coupled to each other and also separately coupled to strong electromagnetic fields; neurodynamics, biologi-

*This work is supported in part by the project ALUD of the Instituto Nazionale di Fisica Nucleare (INFN).
[†]Also, Instituto de Física, Universidade Estadual de Campinas, 13081 Campinas, S. P., Brazil, and Associate Member, INFN.
[‡]Also, Facultad de Ciencias, Universidad Nacional de San Luis, San Luis, Argentina, and Associate Member, INFN.
[§]Associate Member of the Instituto Nazionale di Fisica Nucleare.

Applied Chaos, Edited by Jong Hyun Kim and John Stringer.
ISBN 0-471-54453-1 © 1992 John Wiley & Sons, Inc.

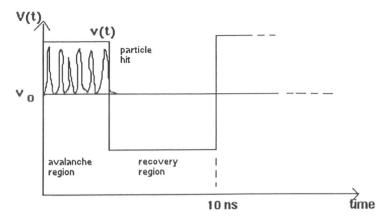

Figure 10.1. Reverse bias voltage as a function of time. $V(t)$ is in synchronism with the mechanism that produces the change in the density of carriers.

cal systems in general; and so on (see, e.g., reference [1]; [2–10]). All of these examples have the characteristic feature of being coupled oscillatory systems. It is then natural to represent them by coupled logistic maps, which have been shown to have a very rich structure in parameter space, although still belonging to the same universality class of period-doubling bifurcations [11]. Even globally coupled maps, which are important in studying neural networks, have only been investigated using logistic maps [12]. In this chapter we address a problem of coupled maps of a *different nature*, presenting many of the features of logistic maps although not following the same route to chaos.

We consider a device consisting of a *pin* diode structure [13]—two oppositely doped silicon regions separated by a layer of intrinsic, or undoped, silicon. When the *pin* diode is reverse-biased the two regions of fixed charges are separated by the relatively long intrinsic Si region, which acts as an insulator. The structure supports a high reverse voltage and greatly reduces the capacitance of the diode. The *pin* diode under consideration is kept at low temperature, with a reverse bias voltage $V_0 + V(t)$, where $V(t)$ is in synchronism with the mechanism that produces the change in the density of carriers (see Figure 10.1). The peaks of $V(t)$ make the field at both junctions slightly above the critical value for an avalanche process to develop. Therefore a slight change in the density of carriers will trigger an avalanche. During the lower half-cycle of $V(t)$ the field is subcritical, stopping the avalanche and sweeping away the electron and hole charge clouds. Such a device produces a current with an oscillating component in the following way: A carrier crossing the left junction produces two charge clouds, one of which moves to the right (see Figure 10.2). As it reaches the second junction it again produces two charge clouds, one of them moving toward the left, and so on. This produces an oscillation in the current; its frequency depends on

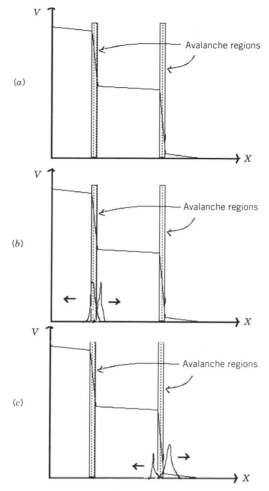

Figure 10.2. (*a*) Voltage profile across the device; (*b*) cloud of charges leaving the first junction; (*c*) cloud of charges leaving the second junction.

the distance between the junctions and should be much higher than the frequency of the bias voltage.

To solve the problem completely is outside the scope of this chapter and shall be published elsewhere. Here we consider a simplified one-dimensional structure, with two slabs (representing the junctions) with an applied electric field strong enough to produce an avalanche, separated by a region where the field is weak enough that particles just drift with a constant velocity (the intrinsic region). Before addressing the problem of the two slabs, we consider that of one slab at $-L \leq x \leq L$, under the action of an electric field, with the following simplifying assumptions: The electric field is not modified by

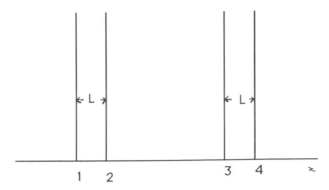

Figure 10.3. Sketch of the simplified model for the *pin* structure.

the pairs that are generated during the avalanche; electrons and holes have equal mobilities ($\mu_+ = \mu_- = 1$) as well as pair-generation constants; and we neglect the diffusion current compared to the drift current, which is equivalent to letting the temperature go to zero [14]. With these assumptions the equations for the density of positive and negative carriers, are

$$\frac{\partial \rho_+}{\partial t} = -\frac{\partial \rho_+}{\partial x} + \alpha(\rho_+ + \rho_-)$$

$$\frac{\partial \rho_-}{\partial t} = \frac{\partial \rho_-}{\partial x} + \alpha(\rho_+ + \rho_-) \tag{10.1}$$

where the last term represents pair generation. Without the first assumption, equations (10.1) together with Poisson's equation should be solved self-consistently. After Laplace-transforming, equations (10.1) can be easily solved for $\rho_+(x, s)$, with the appropriate boundary conditions; in this case we use $\rho_+(-L, t) = \delta(t)$, $\rho_+(L, 0) = 0$, and $\rho_-(-L, 0) = 0$. Assuming that this result is known, we go on to solve the model of the device. Consider the system shown in Figure 10.3. Our aim is to find the density of charges at the boundaries of the slabs. To do that let us define g_s to be the output coming out on the same side of an incoming $\delta(t)$ pulse, for an isolated avalanche region. Let g_0 be the output at the opposite side. Therefore, g_s refers to particles of the opposite sign as the incoming pulse; g_0 refers to those of the same sign. Using the simplifying notation $\rho_+^{(2)} = \rho_+(2, s)$, and so on, we have

$$\rho_+(2, s) = g_0(\alpha)\rho_+(1, s) + \rho_-(2, s)g_s(\alpha) \tag{10.2}$$

Because $\rho_+(1, s) = 1$,

$$\rho_+(2, s) = g_0(\alpha) + g_s(\alpha)e^{-s\tau}\rho_-(3, s) \tag{10.3}$$

and

$$\rho_-(3, s) = \rho_+(2, s) g_s(\alpha) e^{-s\tau} \tag{10.4}$$

Notice that the value of $\rho_-^{(3)}$ depends on the previous value of $\rho_+^{(2)}$. It can be shown by substitution of the results of a single slab and extensive numerical calculation, that these densities have oscillatory behavior of the type needed for detection. The functions g as function of the electric field are highly nonlinear [14], which makes it interesting to study the stability of the solutions.

Because the behavior of the density of carriers at a certain slab end depends linearly on the density of carriers of opposite sign at the slab end where it bounced before, we may represent these successive densities as the coupled maps:

$$x_{n+1} = cy_n \exp\left\{-\frac{\alpha}{|V - x_n|}\right\}$$

$$y_{n+1} = cx_{n+1} \exp\left\{-\frac{\alpha}{|V - y_n|}\right\} \tag{10.5}$$

with the correspondence

$$x_{n+1} \leftrightarrow \rho_-^{(3)}$$

$$y_{n+1} \leftrightarrow \rho_+^{(2)} \tag{10.6}$$

where V is the potential applied across the junction, α and c are parameters associated with the characteristics of the sample whose particular dependence is of no interest at this moment [14], and we have made a shift of variables to simplify the maps. But the important physical feature is that the exponential behavior on the potential is typical of the density of a cloud of charges leaving a junction where a voltage is applied. Notice that with that definition of x_n and y_n, n represents successive "bounces" of clouds of charges at each junction.

The instabilities of this map should not be interpreted as the avalanche process being a chaotic one, but that the device, in order to work as such, should be stable under changes in controllable variables—in this case the external potential, which is designed to be controllable to prevent the avalanche process from destroying the device. However, it is within the range of so-called stable values of the potential where we want to study the instabilities of this system. To analyse the behavior of coupled maps we shall study the Lyapunov exponent, which is a measure of how rapidly two nearby orbits in an attracting region converge or diverge [15]. Define the set of coupled maps as

$$\mathbf{x}_{n+1} = \mathbf{F}(\mathbf{x}_n), \qquad n = 1, 2, \ldots \tag{10.7}$$

the Lyapunov exponent is given by

$$\lambda = \lim_{n \to \infty} \left\{ \frac{\ln\left[\|\mathbf{D}^{(n)}(\mathbf{x}_0)\|\right]}{n} \right\}, \tag{10.8}$$

where $\|\mathbf{D}^{(n)}\|$ is the norm of the derivative matrix. The exponent allows us to recognize periodic, quasiperiodic, and chaotic motion; thus, being negative, zero, or positive indicates the type of orbit under description. The derivative matrix of order n is given by

$$\mathbf{D}^{(n)}(\mathbf{x}_0) = \mathbf{D}(\mathbf{x}_{n-1}) \cdots \mathbf{D}(\mathbf{x}_1)\mathbf{D}(\mathbf{x}_0) \tag{10.9}$$

where $\mathbf{D}(x_n)$ represents the derivative matrix for the nth iteration. In our case,

$$\mathbf{D}(x_{n+1}, y_{n+1})$$

$$= \begin{pmatrix} \alpha x_{n+1} \dfrac{\text{sgn}(V - x_n)}{|V - x_n|^2} & c \exp\left(-\dfrac{\alpha}{|V - x_n|}\right) \\ \alpha y_{n+1} \dfrac{\text{sgn}(V - y_n)}{|V - y_n|^2} & c^2 \exp\left\{-\left(\dfrac{\alpha}{|V - x_n|} + \dfrac{\alpha}{|V - y_n|}\right)\right\} + \alpha y_{n+1} \dfrac{\text{sgn}(V - y_n)}{|V - y_n|} \end{pmatrix}$$

$$\tag{10.10}$$

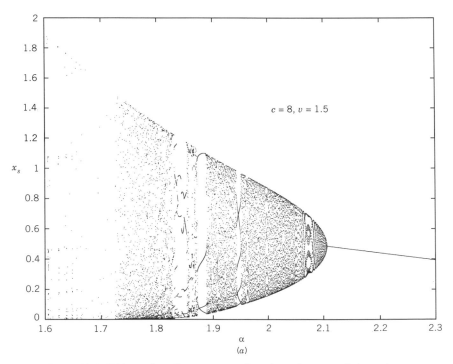

Figure 10.4. Bifurcation diagrams as a function of α, c, and the voltage.

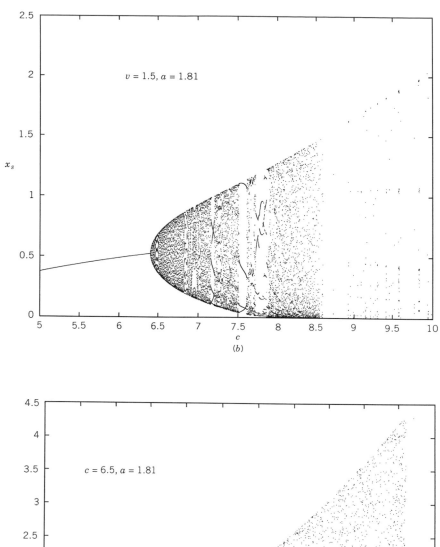

Figure 10.4. (*Continued*)

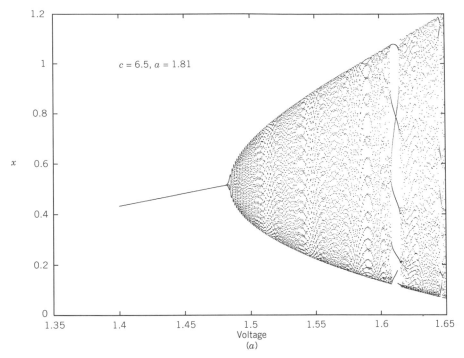

Figure 10.5. (*a*) Detailed patterns on the region of periodic motion.

We now study the behavior of the system as a function of the different parameters. The bifurcation diagrams have been represented only for the variable *x*, as a function of α, *c*, and *V*, and they can be seen in Figures 10.4*a*, 10.4*b* and 10.4*c*, respectively. We notice that these diagrams are quite complex and that they all show the well-known structure belonging to chaotic systems. A closer look at the first point where the number of periods change (shown in Figure 10.5*b* and 10.5*c* only as a function of c) shows that this is a continuous bifurcation and that the fixed points oscillate as function of *c*, thus giving rise to complicated, although periodic, patterns (see Figure 10.5*a*).

Because it is clear that the parameter we should be able to control is the external potential, we have done a more exhaustive study of these diagrams as a function of *V*, and we have kept the values of α and *c* within the values showing chaotic behavior. From Figure 10.6*a*, we have selected two regions to study in detail—one is the empty zone around *V* = 1.7 and the other some closely packed points that can be seen around *V* = 2. Figure 10.6*b–e* shows successive enlargments of these apparently well behaved regions—around *V* = 1.7—which show a phenomenon previously observed in coupled logistic maps—the coexistence of attractors of different periods. This effect is not

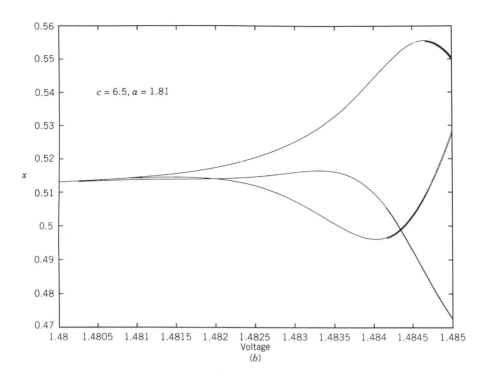

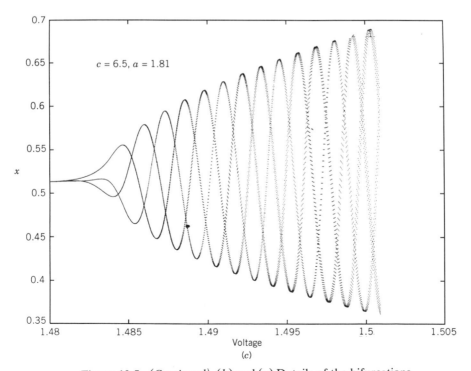

Figure 10.5. *(Continued).* (*b*) and (*c*) Details of the bifurcations.

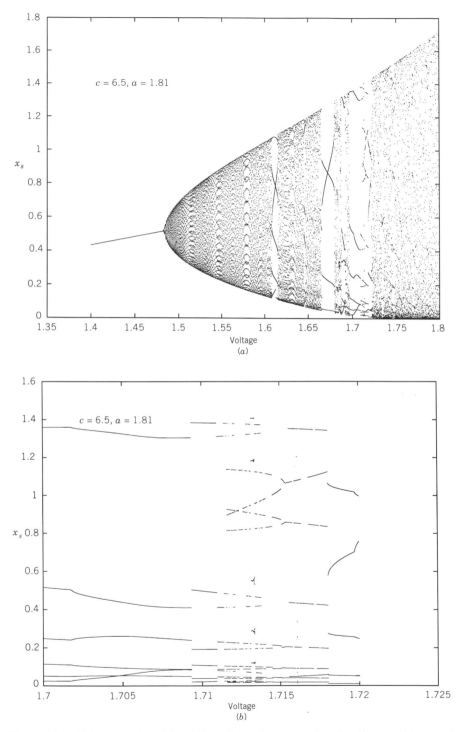

Figure 10.6. Enlargements of the bifurcations diagrams, showing the coexistance of attractors of different periods.

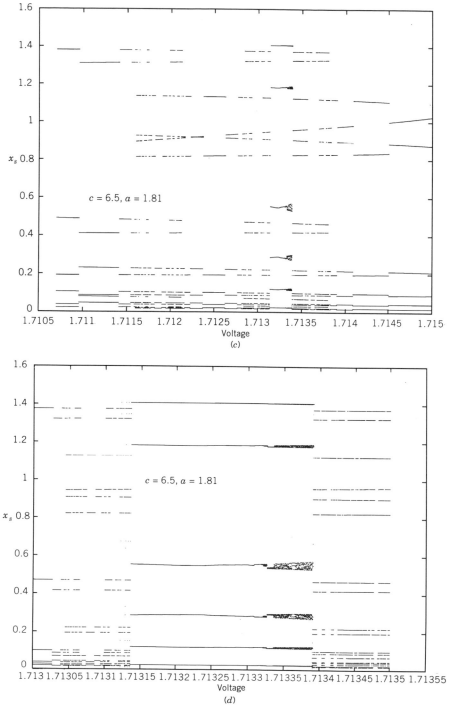

Figure 10.6. (*Continued*)

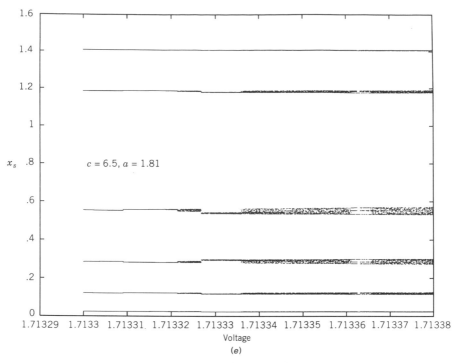

Figure 10.6. *(Continued)*

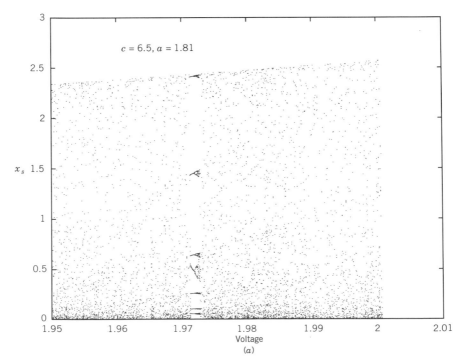

Figure 10.7. Windows of periodicity.

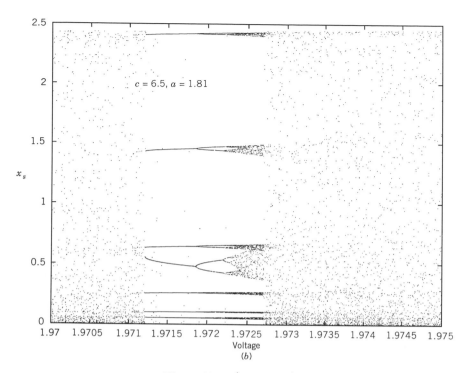

$c = 6.5, a = 1.81$

Figure 10.7. (Continued)

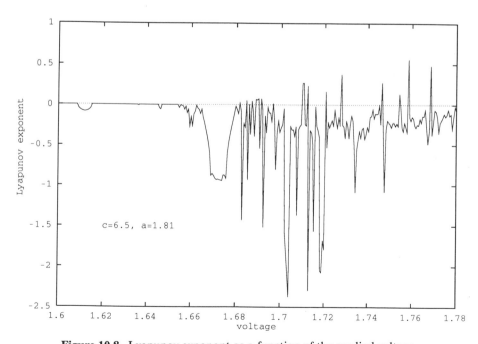

c=6.5, a=1.81

Figure 10.8. Lyapunov exponent as a function of the applied voltage.

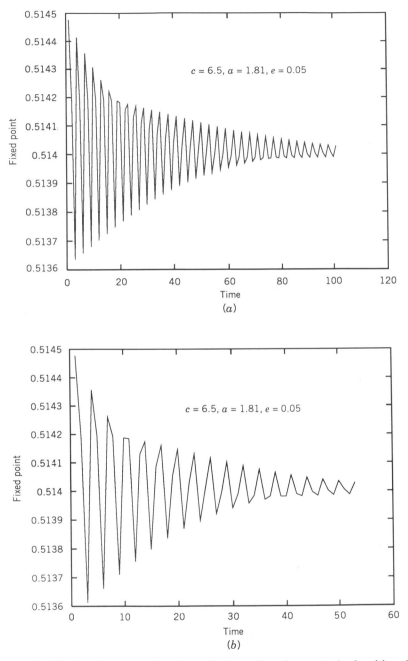

Figure 10.9. Fixed point as a function of time after the control algorithm $V = \varepsilon(x - x_s)$ has been applied.

superimposed on the other stable solutions—they simply disappear to give way to these new bifurcation sequences, which evolve by period doubling all the way to chaos. At a certain point they again stop and the system "remembers" the stable solutions. Whether this is a universal feature of coupled maps and, in that case, what are the complications due to the multiplicity of parameters need an exhaustive study, which we are carrying out separately. On the other hand, Figure 10.7a and 10.7b show that the region around $V = 2$ has the well-known windows of periodicity. We have also plotted (Figure 10.8) the largest Lyapunov exponent as a function of the parameter V.

It is then clear that, if we want to use a system of coupled diodes as a detector, we must obtain from it a univocal response. Hence, the system must be controlled. The ideas put forward by Huberman and Lumer [16] and later elaborated by Sinha, Ramaswamy, and Subba Rao [17] of a simple adaptive control algorithm utilizing an error signal proportional to the difference between the actual output of the signal and the desired one can be applied here, where the controllable parameter is indeed something that can be done electronically with a feedback process. This result of a simple adaptive control of this kind can be seen in Figure 10.9a and 10.9b.

Summarizing, we have studied a simplified model of a device that consists of a system of coupled diodes and can be used as a detector of charged particles when near avalanche breakdown. This model shows chaotic behavior, which can be controlled if necessary by a simple feedback process. A study of a more-realistic model of this device is currently under way.

REFERENCES

1. A. C. Rose-Innes and E. H. Roderick (eds.), *Introduction to Superconductivity.* Pergamon, Oxford, 1978.

2. A Libchaber and S. Fauve, Dissipative dynamical systems and Rayleigh–Bénard experiments. In *Melting, Localization and Chaos* (R. Kalia and P. Vashista, Eds.), p. 195. North Holland, New York, 1982.

3. C. D. Jeffries, *Phys. Scripta* **T9** 11 (1985).

4. F. T. Arecchi, *Phys. Scripta* **T9** 85 (1985).

5. A. Kittel, W. Clauss, M. Knoop, U. Ran, J. Parisi, J. Peinke, and R. P. Huebner, Possible experimental verification of self organized critical behavior in extrinsic germanium. *20th International Conference on the Physics of Semiconductors* (E. M. Anastassakis and J. D. Joannopulos, eds.), p. 2213. World Scientific, Singapore, 1990.

6. A. Ferretti and N. K. Rahman, *Chem. Phys. Lett.* **140** 71 (1987).

7. A. V. Holden (ed.), *Chaos.* Manchester University Press, 1986.

8. A. T. Winfree, *The Geometry of Biological Time.* Springer, New York, 1980.

9. L. Glass and M. C. Mackey, *From Clock to Chaos.* Princeton University Press, 1988.

10. W. Freeman, *Brain Res. Rev.* **11** 259 (1986).

11. T. Hogg and B. A. Huberman, *Phys. Rev. A* **29** 275 (1984).

12. K. Kaneko, *Phys. Rev. Lett.* **65** 1391 (1990).

13. S. M. Sze, *Physics of Semiconductor Devices*. Wiley-Interscience, New York, 1981.

14. M. Schur, *Physics of Semiconductor Devices*. Prentice-Hall, Englewood Cliffs, NJ, 1990.

15. A. J. Lichtenberg and M. A. Lieberman, *Regular and Stochastic Motion*. Springer-Verlag, New York, 1983.

16. B. A. Huberman and E. Lumer, *IEEE Trans. Circuits Systems.* **37** 547 (1990).

17. S. Sinha, R. Ramaswamy, and J. Subba Rao, *Physica D* **43** 118 (1990).

CHAPTER 11

REAL-TIME IDENTIFICATION OF FLAME DYNAMICS

MICHAEL GORMAN
Department of Physics
University of Houston
Houston, TX 77204-5506

KAY A. ROBBINS
Division of Mathematics, Computer Science, and Statistics
The University of Texas at San Antonio
San Antonio, TX 78249

Our experiments on premixed flames in a number of experimental settings have revealed an extensive variety of dynamical behavior. We have found that high frequency fall-off of power spectra can be an effective, real-time tool for classifying and organizing dynamical behavior when a vast parameter space must be surveyed experimentally. We use the example of a Bunsen flame to illustrate the important issues involved in implementing this technique and show the correspondence between power spectra and a variety of dynamical modes found in flat, premixed flames.

11.1. INTRODUCTION

Our experiments on premixed flames on circular burners have resulted in the discovery of many new modes of propagation with an astonishing variety of dynamical behaviors. The complexity of these modes and the vastness of the parameter space has forced us to develop analysis tools that can be used in real-time. A central issue in these investigations is the question "How does

Applied Chaos, Edited by Jong Hyun Kim and John Stringer.
ISBN 0-471-54453-1 © 1992 John Wiley & Sons, Inc.

one determine that the time dependence of an experimental system is described by low-dimensional chaotic dynamics?" This question is important because there is an arsenal of techniques that have been developed over the last 15 years to analyze and quantify chaotic behavior.

Standard approaches that have been used to address this question include observations of time series, reconstruction of attractors from time series, dimension calculations, and Liapunov exponent calculations. Mere observation of time series or visual observations of experiments may not even be able to distinguish periodic behavior from nonperiodic behavior, and attractor reconstruction can produce yarn balls, even for experimental systems of low to moderate dimension. Unfortunately, both Liapunov exponent and dimension calculations require large, stable data sets and significant computation times. Osborne [1] has pointed out the difficulties in using dimension calculations to identify deterministic chaos in experimental systems and has shown that these calculations can lead to erroneous conclusions identifying stochastic systems as deterministic.

It is our central thesis that exponential and power-law fall-offs in the power spectrum distinguish between low-dimensional deterministic chaos, on one hand, and high-dimensional deterministic chaos and stochastic dynamics on the other hand. We describe an explicit prescription for the implementation of the power spectra technique and apply the technique to the Bunsen flame, a system that has both exponential and power-law fall-offs due to the presence of both diffusion and premixed flames. Finally we show observations of spectra for a variety of premixed flame modes for the case of a flat flame on a circular burner at low pressure (which eliminates the diffusion flame).

11.2. BACKGROUND

It is instructive to review the history of the relationship between power spectra and chaotic dynamics. It is well known and can be easily verified that the power spectrum of a chaotic time series generated by any of the "standard" sets of ordinary differential equations (Lorenz, Rossler, Silnikov, etc.) has an exponential fall-off at high frequencies. Frisch and Morf [2] first argued that the power spectrum of a deterministic system should fall off faster than any power law.

Later Greenside, Ahlers, Hohenberg, and Walden [3] were puzzled by their experiments on the dynamics of convection when they found that their time series of a supposedly deterministic system (low Rayleigh number convection in a small container) had a power-law fall-off in the power spectrum. They concluded that their supposedly deterministic system was best described by a stochastic differential equation.

Sigeti and Horsthemke [4] argued that for dynamical systems that are C^{∞}, the power spectrum must fall off faster than any power of the frequency.

They also showed that some stationary stochastic processes have power spectra that fall off asymptotically like an inverse power of the frequency. They further argued that these differences can be used to distinguish between deterministic and stochastic processes—the first time this claim had been made even though it was implicit in the work of Greenside et al.

Sigeti and Horsthemke showed power spectra from two experimental systems, a log-log plot from Rayleigh–Benard data showing power-law fall-off and a semilog plot from circular Couette flow showing exponential fall-off. They did not show log-log and semilog plots for *the same* data set, they did not quantitatively compare the fits for the two kinds of plots, and they did not indicate how to choose the frequency range of the high frequency fall-off.

Our work is a synthesis of the results of Sigeti and Horsthemke with those of Greenside et al. However, we think that it is implicit in this latter work that deterministic systems that are high dimensional have power-law fall-offs. Such a conclusion is supported by the work of Keeler and Farmer [5] on the dynamics of coupled-lattice maps. The power spectra of these systems had a power-law behavior at high frequencies. The distinguishing characteristic of power-law fall-offs seems to be the cascade of information from short time scales to long time scales. This cascade, which is present in both stochastic processes and in processes such as turbulent diffusion, makes it impossible to model the system by a low-dimensional set of modes. An exponential fall-off, on the other hand, is indicative of low-dimensional behavior.

11.3. DYNAMICS OF A BUNSEN FLAME

The Bunsen flame, because of the coexistence of the chaotic premixed flame and the more complicated diffusion flame, represents an extremely stringent test of this technique. We have chosen it as a pedagogical example because it exhibits both exponential and power-law fall-offs. A schematic diagram of a Bunsen burner is shown in Figure 11.1. In typical operation there are two flames, an inner conical premixed flame and an outer diffusion flame. The gas enters the Bunsen tube and it entrains air through holes in the bottom. A premixed flame front is an equivelocity selecting front that has a velocity that is equal in magnitude but opposite in direction to the flow velocity of the premixed gas. The inner premixed flame is conical because the velocity profile is approximately parabolic. If there is excess fuel (if the mixture is rich), some of the fuel that travels through the flame front without combining mixes with the surrounding air and ignites, forming the outer diffusion flame.

To determine the dynamics of these flames we measure the light intensity emitted at a point. This intensity is, to a very good approximation, proportional to the local temperature, which is a dynamical variable that is calculated in theoretical studies of the dynamics of flames. The motion of the

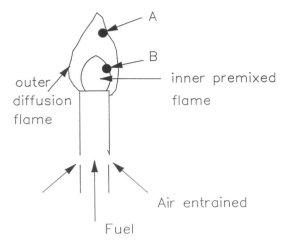

Figure 11.1. A schematic diagram of a Bunsen flame.

inner cone becomes unsteady because of the thermodiffusive instability. This instability, which results from a competition between mass diffusivity and thermal diffusivity, has been the subject of a large number of theoretical studies by Sivashinsky and Michelson [6], [7], and Margolis and Matkowsky [8], who have described the dynamic modes of propagation that replace the steady flame front. A principal part of our experimental effort [9, 10] has been directed toward studying the dynamics of flat circular flames on porous plug burners. Some of these results are discussed in Section 11.4.

In Figure 11.2 we show power spectra computed from time series taken at points A and B in Figure 11.1. The top two plots are log-log and semilog plots of the power spectrum at point A in the diffusion flame; the bottom two plots are the log-log and semilog plots of the power spectrum at point B in the premixed flame. The straight lines are least squares fits to the data. A visual inspection of these plots agrees with a quantitative analysis, which shows that the best fit is the log-log plot for the diffusion flame and the semilog plot for the premixed flame. We interpret these results to indicate that the dynamics of the inner cone of a Bunsen flame exhibits deterministic chaos. We cannot distinguish between high-dimensional deterministic chaos and stochasticity in the case of the diffusion flame.

Our measurement of intensity of the premixed flame also contains contributions from the outer diffusion flame. Because the intensity of the diffusion flame is less than that of the premixed flame, the power spectrum still decays exponentially. We suspect that the log-log plot of the power spectrum of the premixed flame would show a greater deviation if the contributions from the diffusion flame were absent.

In fitting the power spectrum, a high and low frequency cutoff must be chosen. We have chosen the high frequency cutoff to be the point where the

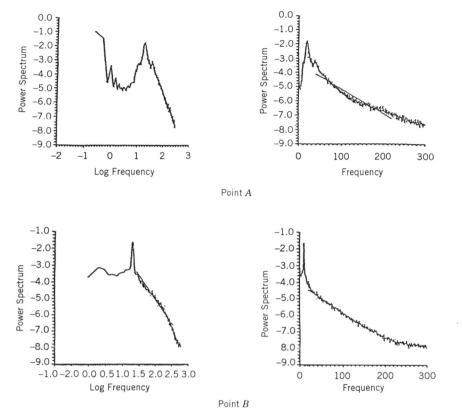

Figure 11.2. Log-log and semilog plots of power spectra computed from time series taken at points A and B in Figure 11.1.

spectrum meets the digitizing noise due to the analog-to-digital (AD) converter. We have chosen the low frequency cutoff to be just higher than any characteristic frequency of the system. In the Bunsen flame this frequency is about 40 Hz because there is an 11 Hz oscillation (and its harmonic) that is caused by the periodic shedding of vortices.

Real-time spectrum analyzers with log-log and semilog displays are commercially available so that a qualitative visual determination of the high frequency behavior can be made in the laboratory as the external parameters are being changed. In the operation of the experiment it is relatively easy to distinguish (by eye) between a power law and an exponential decay. Such analyzers automatically compute the digitizing noise, allowing one to easily determine the high frequency cutoff. Time series records of arbitrary length are stored in a computer for later quantitative analysis. This analysis includes a comparison of residuals for a least squares fit on both the log-log and the semilog scales and the application of sliding window fits to detect concavity trends. A more thorough explanation of the details of the power spectra technique is presented in reference [11].

11.4. CHAOTIC DYNAMICS OF FLAT FLAMES

Although a Bunsen burner is a relatively simple laboratory system, the conical shape of its premixed flame makes it relatively complicated to study theoretically. Most theoretical studies concentrate on the dynamics of freely propagating flat flames in circular and square geometries. Our experimental studies have been conducted using bronze and stainless steel porous plug burners in a circular geometry.

The velocity profile of the premixed gas is made uniform by forcing the flow through the small holes that are randomly arrayed in the porous plug. A schematic diagram of our porous plug burner (McKenna Products, Pittsburg, CA) is shown in Figure 11.3. A cooling coil (not shown) is wound in a spiral

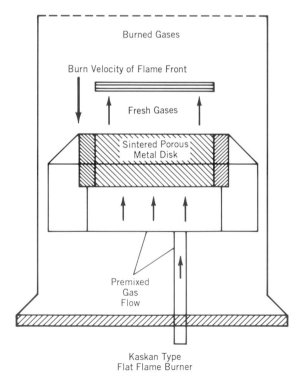

Figure 11.3. A schematic diagram of the porous plug burner used in our experiments on circular flat flames.

shape and imbedded in the porous plug in order to produce a temperature gradient from the center of the plug to its edge, which counteracts the gradient caused by the back conduction of heat from the flame through the flowing premixed gas. A coflow of inert gas is introduced around the outer edge of the burner to suppress any instabilities occurring at the edge of the flame front.

Our experiments are conducted at $\frac{1}{3}$ atm in a combustion chamber made from process glass pipe. We use a silicon intensified target camera to record the dynamics on videotape. The frames of videotape are then digitized, stored in a microcomputer, contrast-enhanced, and printed on a laser printer. The results that we present here concentrate on periodic modes in the pulsating regime of methane–air flames.

A steady circular flame appears as a luminous disk, $2\frac{3}{4}$ in. in diameter. It sits about 5 mm from the surface of the burner. A flame front is stable when the burning velocity of the flame front is equal in magnitude and opposite in

Figure 11.4. Computer enhanced prints of frames of videotape depicting the axial mode.

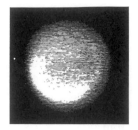

Figure 11.5. Computer enhanced prints of frames of video-tape depicting the radial mode.

direction to the flow of the premixed gases. In all of the prints we will show in this paper the view is from the top, the burner is below the page, the premixed gases are flowing out of the page, and the flame front is burning into the page.

In the axial mode the entire flame front pulsates along its axis. Figure 11.4 shows a top view of the flame at three axial positions. In the top frame the flame front is closest to the burner where the flame is dim because of its significant heat loss due to its proximity to the burner. In the middle frame the flame front is brightest at the center of its oscillation, and in the top frame the flame front is again dim, but now because of a decreased burning velocity.

In the radial mode the flame front periodically changes its radial extent. The three frames in Figure 11.5 show half a cycle of oscillation in which the

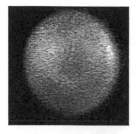

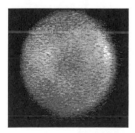

Figure 11.6. Computer enhanced prints of frames of video-tape depicting the drumhead mode.

flame front decreases from near its maximum to its minimum size. In the drumhead mode only a central region of the flame front moves perpendicular to the burner surface. In Figure 11.6 three frames show the flame front intensity at three different positions.

The spiral mode is a two frequency mode in which the flame front moves as a point in a complicated spiral-like path. The six successive frames of videotape shown in Figure 11.7 are arranged in columns. This sequence may give the reader a sense of the pointlike motion behavior of the flame front. Our best guess from direct visual observation and careful examination of videotape is that the front motion is similar to the motion of a toy train that moves between upper and lower circular tracks.

Figure 11.8 is an experimentally determined schematic diagram, showing *only* the periodic modes as a function of flow rate and equivalence ratio. A lean (rich) flame has a mixture with stochiometry less (greater) than 1. At the

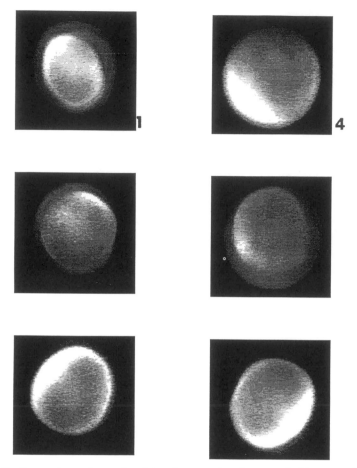

Figure 11.7. Computer enhanced prints of successive frames of videotape depicting the spiral mode. The sequence is arranged in columns of 3.

boundaries between the modes very complicated dynamics can occur. Figure 11.9a–c shows power spectra taken at three points (labeled 1, 2, and 3 on the diagram). The axial mode is labeled 1 and the radial mode is labeled 3. At a point intermediate between the two regimes both modes, their combinations, and harmonics are present. Figure 11.10a–c shows power spectra at three points (4, 5, and 6) between the axial and spiral modes. The power spectrum of the axial mode taken at point 4 is shown in Figure 11.10a. The power spectrum of the spiral mode taken at point 6 contains two frequencies labeled A and B. At a point intermediate between the two modes the power spectrum has broad components indicative of deterministic chaos. In order to demonstrate that this mode exhibits deterministic chaos, we show in Figure 11.11, a semilog plot of a power spectrum that was computed from a time

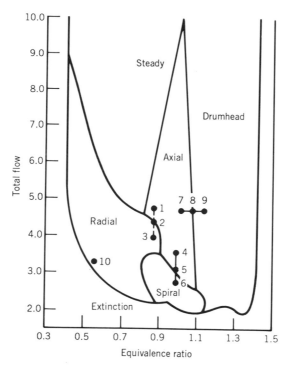

Figure 11.8. A schematic diagram of the occurrence of certain periodic modes of methane–air pulsating flames at $\frac{1}{3}$ atm.

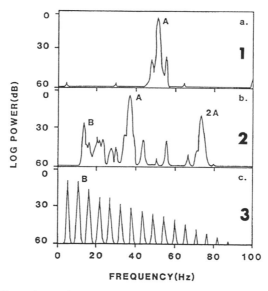

Figure 11.9. Semilog plots of power spectra obtained from a real-time spectrum analyzer taken at points 1, 2, and 3 in the stability diagram corresponding to the transition from the axial mode to the radial mode.

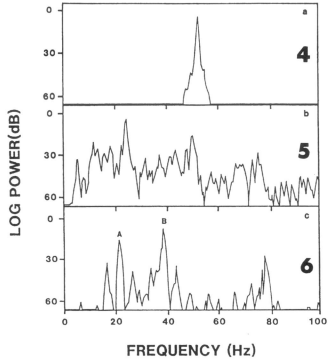

Figure 11.10. Semilog plots of power spectra obtained from a real-time spectrum analyzer taken at points 4, 5, and 6 in the stability diagram corresponding to the transition from the axial mode to the spiral mode.

series. The solid line is a guide for the eye. We interpret this result to indicate that the axial-spiral mode exhibits low-dimensional deterministic chaos.

Prints of frames of videotape of the dynamics of the flame taken at points 2 and 5 would not reveal any easily identifiable features such as the modes shown in Figures 11.4–11.7. Furthermore, it would not be obvious which mode corresponded to a chaotic state and which to a (multiply) periodic state. In order to give the reader a feel for the chaotic motion of pulsating flames, we have included a sequence of six successive frames in Figure 11.12 that correspond to the radial mode near the extinction limit, at parameter values at location 10 on the stability boundary diagram. Figure 11.13 shows a semilog plot of the power spectrum of this mode, and we interpret this exponential fall-off to be indicative of low-dimensional deterministic chaos.

These modes represent only a small fraction of the dynamics that we have discovered. If we use other fuels and oxidizers, we obtain rotating spots and

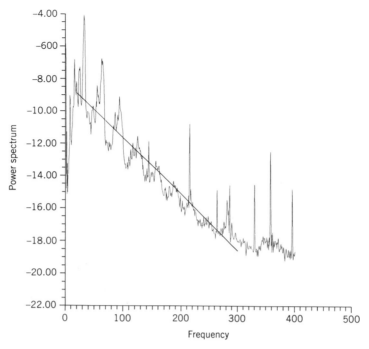

Figure 11.11. A semilog plot of a power spectrum, computed from a time series stored in a microcomputer, taken at point 5 on the stability boundary diagram corresponding to the mixed axial-spiral mode.

spirals. But most of the parameter regime that corresponds to pulsating flames is filled with very complicated dynamics in which modes are not readily identified. In a separate region of parameter space cellular flames form and we have discovered a number of periodic and chaotic modes of propagation.

In a given experiment, such as the one described in the preceding text, we typically vary the flow rate and the equivalence ratio. There are a number of parameters we have *not* yet varied that we expect to affect the dynamical modes we have observed. We have not systematically investigated another flame geometry (possibilities include square, annular, linear, triangular, spherical); we have not changed the size of the circular porous plug; we have not extensively used an oxidizer other than air; we have not used exotic neutrals, such as helium or argon; we have not used a nonhydrocarbon fuel; we have not inverted the burner, reversing the direction of the buoyant force; and we have not systematically studied variations of the pressure. There is an enormous parameter space to be investigated, and it would be surprising if there were not significant dynamics yet to be discovered.

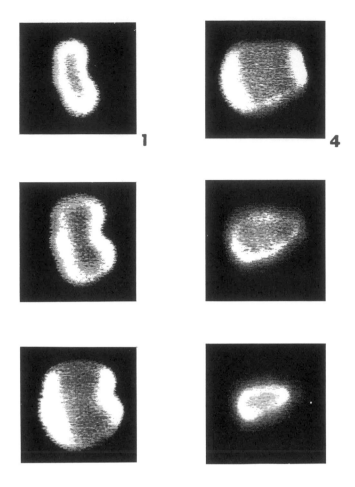

Figure 11.12. Computer enhanced prints of successive frames of videotape depicting the chaotic mode near the boundary between the radial mode and the extinction limit. This mode was observed at parameter values labeled 10 on the stability boundary diagram.

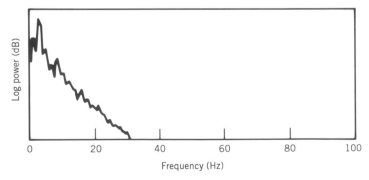

Figure 11.13. Power spectrum obtained from a real-time spectrum analyzer taken at point 10 in the stability boundary diagram corresponding to the mode depicted in Figure 11.12.

Such periodic and chaotic modes of propagation are not confined to premixed flames; they are present in any kind of propagating front. There are indications from the literature that periodic and chaotic modes of propagation also occur in solid state combustion (the fuel is a solid that vaporizes; this process is the basis of most propellants) and in detonations.

11.5. SUMMARY

The purpose of our research is to show that many of the nonperiodic modes of propagation in premixed flames are chaotic and can, therefore, be analyzed by the techniques of nonlinear dynamics. These nonperiodic modes had not been systematically investigated by the combustion community, and the probable reason is that there was no framework in which to "understand" such modes. It was assumed that the nonperiodicities were due to stochastic processes.

In the past 10 years the papers of Sivashinsky, Margolis, and Matkowsky have shown that nonperiodic modes can occur in the solutions of equations that model flame front propagation. It was generally thought that such solutions corresponded to parameter regimes that could not be easily produced in the laboratory. Our work has shown that these modes can easily be produced, that they are ubiquitous, and that power spectra can be used to distinguish among the modes and to identify chaotic behavior.

We should point out the current limitations that govern the use of power spectra. If the fall-off of the power spectrum is *not* exponential, that does *not* mean that the system is *not* deterministic. For instance, we believe that the diffusion flame is deterministic even though its power spectrum falls off as a power law. Rather, a power law indicates a level of cascade of information from short time scales to longer time scales, thereby preventing low-dimensional behavior. Currently we have no technique that can distinguish between stochastic processes and high-dimensional chaos.

We do not yet understand how or why the fall-off of the power spectrum changes from exponential to power law, and it is not clear how to apply power spectra in the case of an open flow in which the dynamical system occurs in a moving frame of reference.

In spite of these limitations, we have found power spectra to be essential to the real-time identification of deterministic chaos of the dynamics of flames. But power spectra must be used in conjunction with other techniques in order to provide as clear and complete a picture of the dynamics as possible. Most of the techniques of nonlinear dynamics are difficult to implement on real experimental data; the real importance of power spectra is that it may help us to identify the best candidates for testing and applying these techniques.

The research community in chaotic dynamics is still developing tools to understand the spatial as well as the temporal characteristics of chaotic systems. We hope that our investigations of the chaotic dynamics of flames will be a contribution to that effort.

ACKNOWLEDGMENTS

Mohamed el-Hamdi conducted these experiments and discovered much of what we have presented. This research has benefited greatly from the support of the Office of Naval Research, the Energy Laboratory at the University of Houston, and the Thermosciences Section of the Engineering Directorate of the National Science Foundation. Julio Reyes helped to prepare this manuscript. This research was initiated by James Blackshear and Jerry Mapp. We have benefited greatly from conversations with Martin Golubitsky, John Guckenheimer, Gregory Sivashinsky, and especially, Stephen Margolis and Bernard Matkowsky.

REFERENCES

1. A. R. Osborne and A. Provenzale, Finite correlation dimensions for stochastic systems with power-law spectra. *Physica D* **35** 357 (1989).
2. U. Frisch and R. Morf, Intermittency in nonlinear dynamics and singularities in complex times. *Phys. Rev. A* **23** 2673 (1981).
3. H. H. Greenside, G. Ahlers, P. C. Hohenberg, and R. W. Walden, A simple stochastic model for the onset of turbulence in Rayleigh–Benard convection. *Physica D* **5** 322 (1982).
4. D. Sigeti and W. Horsthemke, High frequency power spectra for systems subject to noise. *Phys. Rev. A* **35** 2276 (1987).
5. J. Keeler and J. D. Farmer, Robust space-time intermittency in $1/f$ noise. *Physica D* **23** 413 (1986).
6. G. I. Sivashinsky and D. M. Michelson, Nonlinear analysis of hydrodynamic instabilities in laminar flames. *Acta Astronautica* **4** 1207 (1977).
7. G. I. Sivashinsky, Instabilities, pattern formation, and turbulence in flames. *Ann. Rev. Fluid Mech.* **15** 179 (1983).
8. S. B. Margolis and B. J. Matkowsky, Flame propagation in channels. *Combustion Science & Technology* **34** 45 (1983).
9. J. W. Mapp, J. I. Blackshear, and M. Gorman, Nonsteady, nonplanar modes of propagation in premixed burner-stabilized flames. *Combustion, Science & Technology* **43** 217 (1985).
10. M. el-Hamdi, M. Gorman, J. W. Mapp, and J. I. Blackshear. Stability boundaries of periodic modes of propagation in burner-stabilized methane–air flames. *Combustion, Science & Technology* **55** 33 (1987).
11. M. el-Hamdi, M. Gorman, and K. A. Robbins, Power spectra as a guide for identification of deterministic chaos. To be submitted to *Physica D* (1992).

CHAPTER 12

A QUANTITATIVE ASSESSMENT OF THREE METAL-PASSIVATION MODELS BASED ON LINEAR STABILITY THEORY AND BIFURCATION ANALYSIS

ALAN J. MARKWORTH and J. KEVIN McCOY
Metals and Ceramics Department
Battelle Memorial Institute
Columbus, OH 43201-2693

ROGER W. ROLLINS and PUNIT PARMANANDA
Department of Physics and Astronomy
Condensed Matter and Surface Science Program
Ohio University
Athens, OH 45701-2979

Three models for the kinetics of metal passivation, one by Sato and two by Talbot and Oriani, are assessed by using the formalism of linear stability theory and bifurcation analysis, and the models are evaluated to determine their suitability as bases for future models that predict chaotic behavior. The Sato model is described in terms of its stable fixed points, and the results are shown to yield a physically realistic description of anodic current density as a function of electrode potential. A model by Talbot and Oriani is shown to exhibit no fixed points, and its rate equations are integrated in closed form for a limiting case. A global description of another Talbot–Oriani model is obtained by numerical integration of its rate equations. For all three models, it is found that the application of linear stability theory and bifurcation analysis results in major contributions to the understanding of system behavior. The second Talbot–Oriani model shows the greatest promise as a basis for future chaotic models.

Applied Chaos, Edited by Jong Hyun Kim and John Stringer.
ISBN 0-471-54453-1 © 1992 John Wiley & Sons, Inc.

12.1. INTRODUCTION

Metallic passivation is a subject that has been investigated for over a century. Basically, it consists of the loss of chemical reactivity of a metal surface, exposed to a gaseous or aqueous environment, within a range of values for the surface potential. The phenomenon is of considerable technological importance, particularly as it applies to the prevention of metallic corrosion.

Passivation kinetics is quite complex and must generally be described in terms of a system of coupled, nonlinear, chemical rate equations. Time-dependent solutions of such equations can usually be obtained in closed form only for highly simplified passivation models, although solutions are not required if one wishes only to understand the nature of the fixed points (i.e., steady states) of the system. However, the nonlinearities inherent in the rate equations often give rise to very interesting and important behavior. This includes multiple fixed points, continuous or discontinuous bifurcations, and periodic or even chaotic oscillations. Behavior such as this has long been known to exist; Centnerszwer [1] observed temporal oscillations associated with the oxidation of phosphorus in 1898. However, detailed theoretical understanding, based on concepts of stability theory and bifurcation analysis, has only recently begun to emerge, as in the work of Talbot, Oriani, and DiCarlo [2], where these approaches were applied to several passivation models.

The work presented here consists of a quantitative assessment, based on stability theory and bifurcation analysis, of three passivation models, one developed by Sato [3] and two by Talbot and Oriani [4–6]. For the first model [3], we emphasize the nature of the stable fixed points and their dependence upon the magnitudes of the various rate constants. These results are then related to observed passivation behavior, that is, the anodic current density as a function of electrode potential. For the second [4], we demonstrate the absence of any fixed points and integrate the coupled rate equations in closed form for a limiting case. For the third [5, 6], we develop a global understanding of its dynamic behavior, including the application of stability analysis as well as numerical integration of the rate equations. In each case, we obtain considerable information relative to the overall dynamics and/or steady-state behavior of the system using straightforward analytical procedures.

The models described here apply to general corrosion of a homogeneous material. More complex models are necessary to describe the corrosion of an inhomogeneous material or corrosion geometries in which mass transport significantly affects the corrosion rate.

12.2. THE SATO MODEL

The first model that we consider was developed by Sato [3]. It is an analog of a theoretical treatment of Gelain, Cassuto, and LeGoff [7] for the passivation of silicon in a low-pressure, high-temperature oxygen atmosphere. We restrict our treatment mainly to certain aspects of its fixed points. The model is summarized by the following set of reactions, which describe the passivation of a metal M exposed to an aqueous environment:

$$M + H_2O \xrightarrow{k_1} MOH_{ad} + H^+ + e^- \tag{12.1}$$

$$MOH_{ad} \xrightarrow{k_2} MOH_{aq}^+ + e^- \tag{12.2}$$

$$MOH_{ad} \xrightarrow{k_3} MO_{ad} + H^+ + e^- \tag{12.3}$$

$$M + MO_{ad} + H_2O \xrightarrow{k_4} 2MOH_{ad} \tag{12.4}$$

$$MO_{ad} + H^+ \xrightarrow{k_5} MOH_{aq}^+ \tag{12.5}$$

These reactions can be represented, in turn, by the following set of rate equations:

$$\dot{\theta}_M = -k_1\theta_M - k_4\theta_M\theta_O + k_2\theta_{OH} + k_5\theta_O \tag{12.6}$$

$$\dot{\theta}_{OH} = k_1\theta_M - (k_2 + k_3)\theta_{OH} + 2k_4\theta_M\theta_O \tag{12.7}$$

$$\dot{\theta}_O = k_3\theta_{OH} - k_4\theta_M\theta_O - k_5\theta_O \tag{12.8}$$

with

$$\theta_O + \theta_{OH} + \theta_M = 1 \tag{12.9}$$

where θ_M, θ_{OH}, and θ_O are the respective fractional surface coverages of bare metal and the two adsorbed species MOH_{ad} and MO_{ad}, the k_i are the rate constants for the various reactions, and the dot represents differentiation with respect to time. In the analysis that follows, we treat all the k_i values as being independent of θ_O or θ_{OH}.

The fixed points are those sets of values for θ_M, θ_{OH}, and θ_O for which the rates of change with respect to time (i.e., the left-hand sides of equations 12.6–12.8) are zero. One can use equations 12.6–12.9 to show that the fixed points satisfy the relations

$$\theta_M = \frac{(k_2 + k_3)(1 - \theta_O)}{k_1 + k_2 + k_3 + 2k_4\theta_O} \tag{12.10}$$

$$\theta_M = \frac{k_3 - (k_3 + k_5)\theta_O}{k_3 + k_4\theta_O} \tag{12.11}$$

which are the same as those given by Sato.

We concentrate further attention upon one specific case, $k_5 = 0$, for which the adsorbed oxide species, once it is formed, remains stable. This case, which is of particular interest because there may be *two* fixed points, was also considered by Sato (he actually discussed $k_3 \gg k_5$). It is clear that one of the fixed points for this special case is

$$\theta_O = 1 \tag{12.12}$$

$$\theta_{OH} = \theta_M = 0 \tag{12.13}$$

and one can show that the other is

$$\theta_O = \frac{k_1 k_3}{k_4(k_2 - k_3)} \tag{12.14}$$

$$\theta_M = \frac{k_4(k_2 - k_3) - k_1 k_3}{k_4(k_2 - k_3) + k_1 k_4} \tag{12.15}$$

$$\theta_{OH} = \frac{k_1}{k_2 - k_3} \theta_M \tag{12.16}$$

Of course, the latter fixed point is physically attainable only if none of the θ values is negative. (We consider only the physically attainable solutions here.) The former fixed point clearly represents a completely passive state, whereas the latter represents one for which at least a fraction of the surface remains active. As we now demonstrate, only one of these fixed points is a stable steady state for a given set of values for the rate constants.

Conditions for fixed-point stability for sets of autonomous equations are well known. Basically, a fixed point represents a stable steady state if the real parts of the eigenvalues of the Jacobian matrix J, evaluated at the fixed point, are negative. A sufficient condition for local stability of a fixed point for a two-dimensional autonomous system is tr $J < 0$ and det $J > 0$.

In general, an element J_{ij} of the Jacobian matrix is given by

$$J_{ij} = \partial \dot{v}_i / \partial v_j \tag{12.17}$$

where v_i is one of the variables in question. Thus, for the Sato model, we may choose $v_1 = \theta_M$ and $v_2 = \theta_O$ and obtain

$$J = \begin{bmatrix} -(k_1 + k_2) - k_4 \theta_O & k_5 - k_2 - k_4 \theta_M \\ -k_3 - k_4 \theta_O & -(k_3 + k_5) - k_4 \theta_M \end{bmatrix} \tag{12.18}$$

It is clear, from equation 12.18, that tr $J < 0$ under any set of conditions. However, the behavior of det J is more complex. Again considering the case for which $k_5 = 0$, we find that, for the fixed point corresponding to full passivation,

$$\det J > 0 \quad \text{for } k_1 k_3 > k_4(k_2 - k_3) \tag{12.19}$$

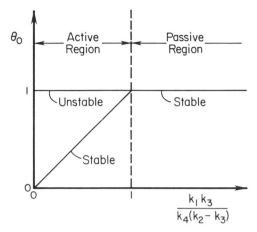

Figure 12.1. Illustration of fixed-point behavior for Sato model.

Conversely, for the fixed point corresponding to the active state,

$$\det J > 0 \quad \text{for } k_1 k_3 < k_4(k_2 - k_3) \tag{12.20}$$

The boundary between these two cases thus occurs at

$$\det J = 0 \quad \text{for } k_1 k_3 = k_4(k_2 - k_3) \tag{12.21}$$

As discussed by Talbot, Oriani, and DiCarlo [2], a fixed point for which $\det J < 0$ has eigenvalues of J that are real and opposite in sign, specifying a saddle or unstable steady state.

Consequently, only one of the two fixed points represents a stable steady state for a given set of rate constants. Thus, for example, by varying the potential of the metal surface one would vary the magnitude of those rate constants that are potential dependent. This, in turn, could induce a transition from one stable steady state to the other (that is, from the active to the passive state, or vice versa), the point of transition being given by equation 12.21.

Figure 12.1 provides a convenient summary of the model. Here, the fixed-point values of θ_O (actually, only the physically attainable values) are plotted as a function of the quantity $k_1 k_3 / [k_4(k_2 - k_3)]$. As can be seen when the abscissa has values greater than unity, a single fixed point exists: $\theta_O = 1$. This fixed point is stable and corresponds to the passive state. Then, when the abscissa is reduced to a value of unity, a bifurcation is exhibited; below this value there are two fixed points, one stable and one unstable. The stable fixed point for this latter regime corresponds to the active state, and the stable value for θ_O linearly approaches zero as the value of the abscissa approaches zero. The bifurcation is a transcritical bifurcation [8, 9], or, in the terminology of Lyberatos, Kuszta, and Bailey [10], a D_1 bifurcation.

To illustrate some additional characteristics of the Sato model, we consider the anodic current density j, which can be expressed as

$$j = j_0(k_1\theta_M + k_2\theta_{OH} + k_3\theta_{OH}) \tag{12.22}$$

including each of the reactions equations 12.1–12.3, that involves electron transfer away from the surface into the metal. The quantity j_0 in equation 12.22 is a constant of proportionality. For the passive-state fixed point, we obtain

$$j = 0 \tag{12.23}$$

whereas for the active-state fixed point

$$j = j_0\left(\frac{2k_1k_2}{k_2 - k_3}\right)\frac{k_4(k_2 - k_3) - k_1k_3}{k_4(k_2 - k_3) + k_1k_4} \tag{12.24}$$

To illustrate the dependence of j upon potential, we consider a specific numerical example: Using dimensionless units and assuming that the rate constants k_1, k_2, and k_3 vary exponentially with the potential of the metal surface, we take

$$j_0 = 1, \quad k_1 = k_2/2 = k_3 = e^\eta, \quad k_4 = 1 \tag{12.25}$$

where η represents a dimensionless surface potential and where all pertinent units are consistent but arbitrary. For this case, equation 12.24 becomes

$$j = 2e^\eta(1 - e^\eta) \tag{12.26}$$

In addition, the surface potential, η_t, corresponding to the point of transfer between active and passive states is seen, using equation 12.21, to be given by

$$\eta_t = 0 \tag{12.27}$$

Equations 12.23 and 12.26 are plotted in Figure 12.2, each over its appropriate range of potential. The interesting feature of this plot is that the general manner in which current density varies with potential is indeed similar to that

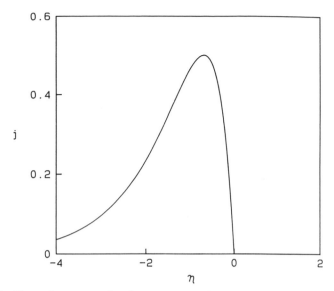

Figure 12.2. Plot of current density versus surface potential, both expressed in dimensionless units for Sato model [3], after equations 12.23 and 12.26.

observed experimentally for real systems, that is:

1. At lower values of the applied potential, the current density increases sharply (approximately exponentially) with increasing potential, this being the active state.
2. Within a rather narrow range of potential, the current density reaches a maximum value and then drops abruptly (to zero for this model) as the potential is increased further, indicating a transition to the passive state.
3. The passive state obtains for all higher values of potential.

In summary, the Sato model can exhibit interesting and physically realistic behavior, which can be readily understood through the use of elementary stability theory and bifurcation analysis. However, the dynamics of this model are so simple that it is not promising as a basis for a higher-dimensional chaotic model.

12.3. EARLY TALBOT – ORIANI MODEL

In one of their earlier studies, Talbot and Oriani [4] considered a simple three-step model for coupled dissolution and passivation of a metal surface exposed to an aqueous environment:

$$M \xrightarrow{k_1} M^+ + e^-$$ (12.28)

$$M^+ + H_2O \underset{k_{-2}}{\overset{k_2}{\rightleftharpoons}} MOH_{aq} + H^+$$ (12.29)

$$MOH_{aq} \underset{k_{-3}}{\overset{k_3}{\rightleftharpoons}} MOH_{ad}$$ (12.30)

Assuming that the concentration of H^+ is unaffected by the second reaction, the kinetic equations for this model can be expressed (in somewhat different notation from Talbot and Oriani's) as

$$\dot{C}_1 = k_1(1 - \theta) - k_2C_1 + k_{-2}C_2$$ (12.31)

$$\dot{C}_2 = k_2C_1 - k_{-2}C_2 - k_3C_2(1 - \theta) + k_{-3}\theta$$ (12.32)

$$\dot{\theta} = k_3C_2(1 - \theta) - k_{-3}\theta$$ (12.33)

where C_1 and C_2 are the respective concentrations of M^+ and MOH_{aq}, and θ is the fractional surface coverage of MOH_{ad}. Effects of H^+ activity are assumed to be included in the pertinent rate constants k_2 and k_{-2}.

It is convenient, in equations 12.31–12.33, to regard C_1 and C_2 as dimensionless quantities, equal to the total number of molecules of each respective species in solution divided by the total number of adsorption sites on the metal surface exposed to the solution. In this manner, the concentrations are analogous in meaning to θ, except that they are not restricted to values less than or equal to unity.

The authors incorrectly concluded that, because the sum of the right-hand sides of equations 12.31–12.33 is equal to $k_1(1 - \theta)$, the steady-state condition is given by $\theta = 1$. In fact, inspection of this set of equations leads us to the conclusion that there is *no* set of conditions for which a fixed point can exist, that is, for which the right-hand side of each equation is zero.

A particularly interesting limiting case for this model is that for which the hydrolysis reaction, equation 12.29, is always in virtual equilibrium, so C_1 is proportional to C_2. We continue to assume that the concentration of H^+ is unaffected by the reaction. For this case, equations 12.31–12.33 reduce to

$$\dot{C}_1 = k_1(1 - \theta)$$ (12.34)

$$\dot{\theta} = k'_3C_1(1 - \theta) - k_{-3}\theta$$ (12.35)

where k'_3 is a new rate constant. Equations 12.34–12.35 form a set of two autonomous nonlinear equations. This set is interesting because a solution can be obtained in closed form, as has been demonstrated by Glasser [11].

The procedure is straightforward but lengthy, and is presented in the Appendix.

Rather than consider the general solution for C_1 (equation A.8 of the Appendix) in detail, we shall focus upon its asymptotic behavior. Toward this end, we note that the Airy functions and their derivatives, in terms of which this solution is expressed, approach the following forms [12] for large values of the argument, which in our case is real:

$$\text{Ai}(x) \sim \tfrac{1}{2}\pi^{-1/2}x^{-1/4}e^{-z} \tag{12.36}$$

$$\text{Ai}'(x) \sim -\tfrac{1}{2}\pi^{-1/2}x^{1/4}e^{-z} \tag{12.37}$$

$$\text{Bi}(x) \sim \pi^{-1/2}x^{-1/4}e^{z} \tag{12.38}$$

$$\text{Bi}'(x) \sim \pi^{-1/2}x^{1/4}e^{z} \tag{12.39}$$

where

$$z = \tfrac{2}{3}x^{3/2} \tag{12.40}$$

from which it follows, by using equation A.8, that, for long times,

$$C_1 \sim \sqrt{2\alpha\gamma t/\beta} \tag{12.41}$$

For equation 12.41 to be consistent with equation A.1, we must have

$$\theta \sim 1 \tag{12.42}$$

It is interesting to observe that equation 12.41 can be obtained in a very simple manner, merely from inspection of equation A.7.

In summary, the Talbot–Oriani model [4] does *not* possess a fixed point. For the limiting case considered here, the total amount of metal in solution diverges at asymptotic times, and the coverage of the surface by the passivating species approaches unity. Interestingly, the system thus actually does asymptotically approach a condition for which both $\dot{C}_1$ and $\dot{\theta}$ are zero. This model predicts unbounded trajectories, and that defect must be remedied before chaos can be produced. Talbot and Oriani have themselves provided the remedy, as discussed in the next section.

12.4. RECENT TALBOT–ORIANI MODEL

Recently, Talbot and Oriani [6] generalized their earlier model to include a homogeneous reaction in solution:

$$n\text{M}^+ + \text{A}^{n-} \xrightarrow{k_r} \text{M}_n\text{A} \tag{12.43}$$

where the anion A^{n-} is not OH^-. The reactions equations 12.28, 12.29, 12.30, and 12.43 thus form the basis for the model.

It is assumed that the concentrations of H_2O, H^+, and A^{n-} are in excess and that the hydrolysis reaction, equation 12.29, is always in equilibrium so that the concentration of MOH_{aq} is proportional to the concentration of M^+ in solution. These assumptions, together with the further limitation to a univalent anion A^-, lead to the rate equations

$$\dot{C} = k_1(1 - \theta) - k'_r C \tag{12.44}$$

$$\dot{\theta} = k'_3 C(1 - \theta) - k'_{-3} e^{-\beta\theta}\theta \tag{12.45}$$

where C is the concentration of metal ions in solution, θ is the fraction of the surface covered by metal hydroxide, k'_r is proportional to k_r of equation 12.43, k'_3 is proportional to k_3, and $k'_{-3} e^{-\beta\theta} = k_{-3}$ of equation 12.30.

Following Talbot and Oriani [6] we introduce the reduced concentration $Y \equiv (k'_3/k'_{-3})C \equiv KC$ and reduced time $\tau \equiv k'_{-3}t$ so that equations 12.44 and 12.45 become

$$\dot{Y} = p(1 - \theta) - qY \tag{12.46}$$

$$\dot{\theta} = Y(1 - \theta) - \theta e^{-\beta\theta} \tag{12.47}$$

where

$$\dot{Y} \equiv dY/d\tau, \qquad \dot{\theta} \equiv d\theta/d\tau, \qquad p \equiv Kk_1/k'_{-3}, \quad \text{and} \quad q \equiv k'_r/k'_{-3} \tag{12.48}$$

Ultimately, we are interested in investigating the possibility of deterministic chaos in the dynamics of the passivation process, and chaos is not possible in a two-dimensional system. Thus, we must eventually add a third dimension, by adding a second adsorbed species on the surface for example, but in order to obtain a clear understanding of the much simpler two-dimensional system we carefully extended the work of Talbot and Oriani.

We obtain a global understanding of the dynamical behavior of the two-dimensional system in two steps. We first apply stability analysis to the fixed points determined by $dY/d\tau = 0$ and $d\theta/d\tau = 0$. Then, we numerically integrate the system for many initial conditions at points of interest in parameter space (p, q, β). Most of the numerical integration was performed using the highly interactive environment provided by a program by one of the authors [13], modified to solve our set of equations.

The set of solution trajectories forms a state space portrait revealing the global behavior for a particular set of parameters. Stable limit cycles and fixed points are found by integrating forward in time, whereas unstable limit cycles and unstable fixed points are located by integrating backward in time. The location and stability of limit cycles is important in understanding the

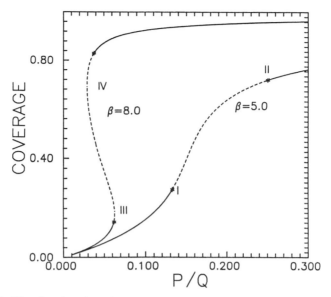

Figure 12.3. The fractional surface coverage θ for all fixed points as a function of p/q, $q = 0.001$, at two values of β. The unstable or saddle fixed points are indicated by the dashed curves. The regions at I, II, III, and IV are displayed at a larger scale in Figures 12.4–12.7.

global behavior of the system as a function of the parameters p, q, and β. The behavior of the limit cycles can lead to bifurcations in the state space portrait of solution trajectories that are nonlocal in origin and hence cannot be predicted by an analysis of the stability of the fixed points. The existence of stable limit cycles is of interest for finding chaos in a modified system. We might expect to find chaotic behavior in the neighborhood (in parameter space) of these limit cycles when we modify the model to add the necessary third dimension to the state space.

Building on the work of Talbot and Oriani, we chose to study carefully the global behavior of the system as a function of the ratio p/q for $q = 0.001$ and for $\beta = 5.0$ and $\beta = 8.0$. Figure 12.3 traces the behavior of all the fixed points as a function of p/q for these two values of β. The dashed portions of these curves represent unstable fixed points. The stability of the fixed points is determined by the sign of the real part of the eigenvalues of the Jacobian matrix, evaluated at the corresponding fixed points. The regions marked I and II for $\beta = 5.0$ and regions marked III and IV for $\beta = 8.0$ are expanded in Figures 12.4, 12.5, 12.6, and 12.7, respectively, showing transitions from one type of fixed point to another.

Numerical integration of the equations, both forward and backward in time, was used to obtain the global behavior, including global bifurcations, described below. The cases $\beta = 5.0$ and $\beta = 8.0$ are discussed separately,

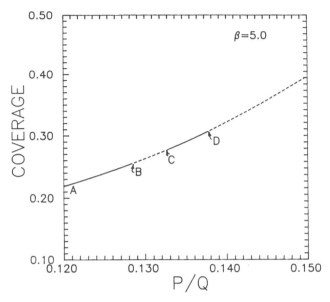

Figure 12.4. Region I of Figure 12.3 shown at an expanded scale. The points marked B–D are local bifurcation points (see text).

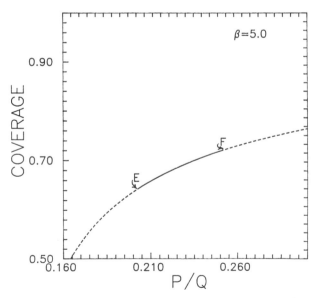

Figure 12.5. Region II of Figure 12.3 shown at an expanded scale. The points marked E and F are local bifurcation points (see text).

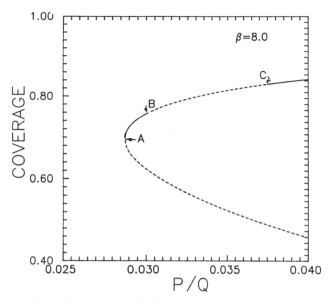

Figure 12.6. Region III of Figure 12.3 shown at an expanded scale. The points marked A–C are local bifurcation points (see text).

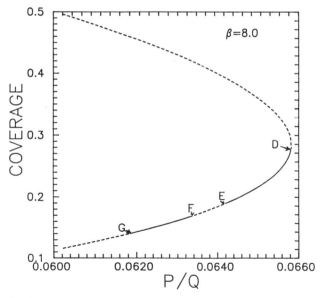

Figure 12.7. Region IV of Figure 12.3 shown at an expanded scale. The points marked G–D are local bifurcation points (see text).

TABLE 12.1. Summary of Results for $\beta = 5.0$, $q = 0.001$

Region in Figs. 12.4 and 12.5	p/q	Attractors or Bifurcations
A–B	0.0–0.128	SN
B–C	0.128–0.132776	SF
C	0.132776	Supercritical Hopf
C–	0.132776–0.1328313	UF, SLC
	0.1328313	Global jump in SLC size
–D	0.1328313–0.13730	UF, SLC
D–E	0.13730–0.20390	UN, SLC
E–F	0.20390–0.250035	UF, SLC
F	0.250035	Subcritical Hopf
F–	0.250035–0.2509286	SF, SLC, ULC
	0.2509286	Global LC annihilation
	0.2509286–0.3000	SF

and each fixed point is characterized as a stable node (SN), stable focus (SF), unstable node (UN), unstable focus (UF), or a saddle point (SP) by evaluating the eigenvalues of the Jacobian matrix. The periodic attractors are either stable limit cycles (SLC) or unstable limit cycles (ULC).

Table 12.1 summarizes the results for $\beta = 5.0$. For $p/q < 0.132776$ we find only a single stable fixed point. At $p/q = 0.132776$ a supercritical Hopf bifurcation takes place where the stable focus changes to an unstable focus and there emerges a stable limit cycle surrounding it. The existence of the stable limit cycle is consistent with the Poincaré–Bendixson theorem, which requires the existence of at least one stable limit cycle surrounding an unstable node or an unstable focus when no other fixed points exist in a region of state space where all trajectories are bounded. This unstable focus persists (region $C–D$) as we increase the ratio p/q, changing to an unstable node at $p/q = 0.13730$ (point D). Meanwhile, the stable limit cycle (SLC), surrounding the unstable focus, undergoes a global bifurcation at $p/q = 0.1328313$ where it suddenly increases in size. The precise nature of this reversible global bifurcation is as yet undetermined. The (now large) limit cycle increases in size very slowly as we increase p/q further. The existence of this important large limit cycle is not mentioned by Talbot and Oriani [6]. Figure 12.8 shows a state space portrait including this large stable limit cycle. As can be seen from the squarish shape of the limit cycle, the system is undergoing relaxation oscillations.

A global bifurcation occurs at $p/q = 0.2509286$, where this large stable limit cycle suddenly disappears upon annihilation with an unstable limit cycle. The unstable limit cycle (ULC) is born (as p/q increases) at a subcritical Hopf bifurcation that occurs at $p/q = 0.250035$. For a short range of p/q, the ULC coexists in the lower right corner of the large SLC as shown in Figure 12.9. In this range, $0.2500 < p/q < 0.2509$, the ULC is the bound-

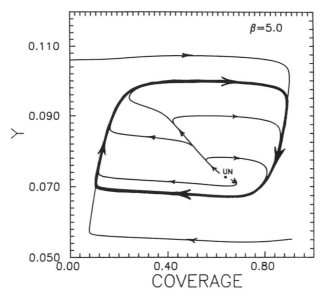

Figure 12.8. A state space portrait showing the global behavior of the system at $\beta = 5.0$ and $p/q = 0.2$, $q = 0.001$. There is an unstable node (UN) inside the large stable limit cycle indicated by the heavy curve. Several trajectories starting at different initial conditions and converging on the stable limit cycle are shown.

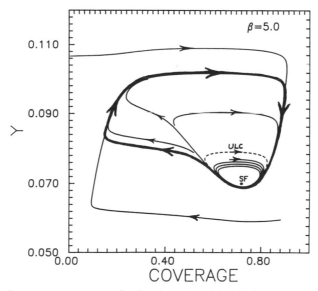

Figure 12.9. A state space portrait showing the global behavior of the system at $\beta = 5.0$ and $p/q = 0.25036$, $q = 0.001$. A stable focus (SF) is shown surrounded by an unstable limit cycle (ULC) all inside the large stable limit cycle (SLC) indicated by the heavy curve. Several trajectories starting at different initial conditions are shown. At a slightly higher value of p/q the SLC and ULC suddenly annihilate, leaving only the stable focus.

TABLE 12.2. Summary of Results for $\beta = 8.0$, $q = 0.001$

Region in Figs. 12.6 and 12.7	p/q	Attractors or Bifurcations
−A	0.0–0.02890	SN
A	0.02890	Saddle-node bifurcation
A–B	0.02890–0.0302	SN, UN, SP
B–C	0.0302–0.03750	SN, UF, SP
C	0.03750	Subcritical Hopf
C−	0.03750–0.03980	SN, SF, SP, ULC
	0.03980	Saddle-loop bifurcation
−G	0.03980–0.06170	SN, SF, SP
G−	0.06170–0.062313	SF, SF, SP
	0.062314	Saddle-loop bifurcation
−F	0.062314–0.062325	SF, SF, SP, ULC
F	0.062326	Subcritical Hopf
F–E	0.062326–0.06280	UF, SF, SP
E–D	0.06280–0.0633	UN, SF, SP
D	0.0634	Saddle-node bifurcation
D−	0.0634–2.0	SF

ary between two regions of state space. The region inside the ULC is the basin of attraction for the stable focus; the region outside the ULC is the basin of attraction for the SLC.

The sudden change in the global behavior of the system in this range of p/q shows hysteresis. This would be observed physically as the sudden collapse of a large amplitude oscillatory response to a fixed point steady-state behavior at $p/q = 0.2509$ as p/q increases and the sudden transition from fixed point steady-state behavior to a large amplitude oscillatory behavior at $p/q = 0.2500$ as p/q decreases. Our numerical results do not show evidence of a slowing down of the oscillation (i.e., no lengthening of the period) as the upper bifurcation point is approached from below. Such slowing down is present in other types of global bifurcations involving limit cycles such as the saddle-loop bifurcation [14].

Now we turn briefly to the case $\beta = 8.0$. The S shaped curve shown in Figure 12.3 traces the behavior of all fixed points for $\beta = 8.0$, $q = 0.001$, and $0.0 < p/q < 0.3$. The S shape leads to a range of p/q where multiple fixed points coexist. Multiple fixed points occur because there are vertical tangents to the S shaped curves. Saddle-node bifurcations occur at these two tangent points. We completed a numerical study of the global behavior of the system, and Table 12.2 summarizes the results.

As p/q is increased, only a stable node exists until $p/q = 0.02890$. At $p/q = 0.02890$ a saddle-node bifurcation takes place (marked IV in Figure 12.3), giving birth to a saddle point and an unstable node. The saddle-node bifurcation point is marked as point A in Figure 12.6. The lower branch of

Figure 12.6 is the saddle point and the upper branch is an unstable node from A to B. The stable node, which is not shown in Figure 12.6, persists. At $p/q = 0.03020$, marked by point B in Figure 12.6, this unstable node changes to an unstable focus, which persists until $p/q = 0.037500$ (C in Figure 12.5), where a subcritical Hopf bifurcation takes place, giving rise to an unstable limit cycle surrounding a stable focus.

Talbot and Oriani [6] also analyzed the state space for $\beta = 8.0$ and showed the existence of a limit cycle for $p/q = 0.037500$ (see Figure 7 of reference [12]), but they do not specify whether the limit cycle is stable or unstable. We find that this limit cycle is unstable and originates from the subcritical Hopf bifurcation at C. This unstable limit cycle, which is the boundary of the basin of attraction for the stable focus inside, increases in size rapidly as we increase p/q. At $p/q = 0.03980$ this unstable limit cycle hits the saddle point and disappears (saddle-loop bifurcation), leaving behind the stable focus and saddle point.

As can be seen from Table 12.2, several interesting bifurcations occur, but there are no stable limit cycles for the system at all for $\beta = 8.0$. Thus, there is less of a possibility for chaos here by modifying the model to add a third dimension to the phase space as mentioned before. We should point out that the S shaped curve shown in Figure 12.3 does provide a possibility for chaos if the added reaction could couple to the parameters p and q in such a way as to drag the system back and forth between the vertical tangent points on the S curve.

12.5. CONCLUSIONS

Methods of linear stability theory and bifurcation analysis have been applied to three models for aqueous corrosion of metals. Each of these models has a two-dimensional state space, so none of them can give rise to chaos. Although the models describe closely related mechanisms for corrosion, they have strikingly different dynamic behaviors. The Sato model exhibits stable and unstable fixed points and a transcritical bifurcation. The early Talbot–Oriani model produces trajectories that diverge at long times. The recent Talbot–Oriani model, with suitable choices of kinetic parameters, exhibits saddle points, stable and unstable nodes, foci, and limit cycles, and several types of bifurcations.

The remarkable variety of dynamics for corrosion models with two-dimensional state spaces strongly suggests that models with three-dimensional state spaces will have even more complex dynamics, including chaotic dynamics. Compared with the other two models examined here, the recent Talbot–Oriani model has dynamics that are markedly richer, and so it is undoubtedly the most promising of the three as the basis for a future model with a three-dimensional state space. The complexity of the dynamics for the current models also clearly indicates that methods of nonlinear dynamics are essential for analyzing even simple models of corrosion.

ACKNOWLEDGMENTS

Financial support of this research by the Electric Power Research Institute (EPRI), Palo Alto, California, under contract RP 2426-25, is gratefully acknowledged. Particular thanks go to Dr. John Stringer of EPRI for his support and encouragement.

APPENDIX. SOLUTION OF EQUATIONS 12.34 AND 12.35

In this Appendix, we outline the procedure, originally developed by Glasser [11], for obtaining a closed-form solution of the coupled rate equations given by equations 12.34 and 12.35. We begin by rewriting these equations as

$$\dot{C}_1 = \alpha(1 - \theta) \tag{A.1}$$

$$\dot{\theta} = \beta C_1(1 - \theta) - \gamma\theta \tag{A.2}$$

where $\alpha = k_1$, $\beta = k'_3$, and $\gamma = k_{-3}$. Substituting equation A.1 into A.2, we obtain

$$\dot{\theta} = \frac{\beta}{\alpha}C_1\dot{C}_1 + \frac{\gamma}{\alpha}\dot{C}_1 - \gamma \tag{A.3}$$

Equation A.3 can be integrated, subject to initial conditions, which we take to be

$$C_1(t = 0) = 0 \tag{A.4}$$

$$\theta(t = 0) = 0 \tag{A.5}$$

where t represents time. We thus find that

$$\theta = \frac{\beta}{2\alpha}C_1^2 + \frac{\gamma}{\alpha}C_1 - \gamma t \tag{A.6}$$

Next, equation A.6 can be substituted into equation A.1 to yield

$$\dot{C}_1 + \gamma C_1 + \tfrac{1}{2}\beta C_1^2 = \alpha\gamma t + \alpha \tag{A.7}$$

Equation A.7 is a special form of the Riccati equation. Its solution is straightforward [11], but the procedure is rather lengthy. Briefly, the approach is to make successive transformations of variables that linearize equation A.7 and ultimately yield the Airy equation. We omit the details and simply state the solution, which is

$$C_1 = -\frac{\gamma}{\beta} + \left(\frac{4\alpha\gamma}{\beta^2}\right)^{1/3}\left[\frac{Ai'(\xi) + \rho\,Bi'(\xi)}{Ai(\xi) + \rho\,Bi(\xi)}\right] \tag{A.8}$$

where Ai and Bi are Airy functions [12] and

$$\xi \equiv \left(\frac{4\alpha\gamma}{\beta^2}\right)^{1/3}\left(\frac{\beta t}{2} + \frac{\beta}{2\gamma} + \frac{\gamma}{4\alpha}\right) \tag{A.9}$$

$$\rho \equiv -\frac{Ai(\xi_0) - \left(4\alpha\beta/\gamma^2\right)^{1/3}Ai'(\xi_0)}{Bi(\xi_0) - \left(4\alpha\beta/\gamma^2\right)^{1/3}Bi'(\xi_0)} \tag{A.10}$$

with

$$\xi_0 = \xi(t = 0) \tag{A.11}$$

Given this solution for C_1, the explicit dependence of θ upon time can be obtained from equation A.6.

REFERENCES

1. M. Centnerszwer, *Z. Phys. Chem.* **26** 1 (1898).
2. J. B. Talbot, R. A. Oriani, and M. J. DiCarlo, *J. Electrochem. Soc.* **132** 1545 (1985).
3. N. Sato, In *Passivity of Metals*, (R. P. Frankenthal and J. Kruger, eds.). p. 29. The Electrochemical Society, Princeton, NJ, 1978.
4. J. B. Talbot and R. A. Oriani, In *Proceedings of the International Congress on Metallic Corrosion*, vol. 3, p. 440. National Research Council of Canada, Ottawa, Ontario, 1984.
5. J. B. Talbot and R. A. Oriani, In *Electrochemical Engineering Applications* (R. E. White and R. F. Savinell, eds.). AIChE Symposium Series No. 254, vol. 83, p. 64. American Institute of Chemical Engineers, New York, 1987.
6. J. B. Talbot and R. A. Oriani, *Electrochim. Acta* **30** 1277 (1985).
7. C. Gelain, A. Cassuto, and P. LeGoff, *Oxid. Metals* **3** 139 (1971).
8. J. Dorning, In *Heat Transfer–Philadelphia 1989* (S. B. Yilmaz, ed.). AIChE Symposium Series No. 269, vol. 85, p. 13. American Institute of Chemical Engineers, New York, 1989.
9. J. M. T. Thompson and H. B. Stewart, *Nonlinear Dynamics and Chaos*. Wiley, Chichester, 1986.
10. G. Lyberatos, B. Kuszta, and J. E. Bailey, *Chem. Eng. Sci.* **40** 1177 (1985).

11. M. L. Glasser, personal communication. Department of Mathematics and Computer Science and Department of Physics, Clarkson University, Potsdam, NY.

12. H. A. Antosiewicz, In *Handbook of Mathematical Functions With Formulas, Graphs, and Mathematical Tables* (M. Abramowitz and I. A. Stegun, eds.). Applied Mathematics Series **55**, p. 435. National Bureau of Standards, Washington, DC, 1968.

13. A modified version of *Chaotic Dynamics Workbench*, R. W. Rollins, Physics Academic Software. American Institute of Physics, NY, 1990.

14. V. Gáspár and K. Showalter, *J. Chem. Phys.* **88** 778 (1988).

PART III

APPLICATIONS IN PHYSIOLOGICAL SCIENCES

CHAPTER 13

DYNAMICAL SIGNATURES IN ELECTROCARDIOGRAPHIC DATA

ROBERT DE PAOLA and WILLIAM I. NORWOOD
Division of Cardiothoracic Surgery
Children's Hospital of Philadelphia
Philadelphia, PA 19104

LEON GLASS
Department of Physiology
McGill University
Montreal, Canada

13.1. INTRODUCTION

Unexpected premature death resulting from cardiac disease is one of the leading health problems confronting developed countries. Although a large number of different factors have been associated with a higher incidence of sudden cardiac death [1], none can be used to predict with any degree of certainty who will suffer a fatal arrhythmia or when this will occur.

The underlying hypothesis of this research is that hidden in the electrocardiogram (ECG), which is a measure of the electrical activity of the heart, is information reflecting details of the dynamics of the heart that can complement techniques conventionally used in the diagnosis of cardiac rhythm disturbances. In this preliminary report we investigate transformations of electrocardiographic data that reveal dynamical features underlying complex cardiac rhythms. We propose that these features, "dynamical signatures," are related to physiological mechanism and can thus be useful in obtaining a more precise clinical characterization of cardiac rhythms.

Applied Chaos, Edited by Jong Hyun Kim and John Stringer.
ISBN 0-471-54453-1 © 1992 John Wiley & Sons, Inc.

Two cardiac rhythms of different physiological origin are treated:

1. An atrial rhythm that results from variations in the levels of neurological tone.
2. A ventricular rhythm, which may result from the competition between two cardiac oscillators.

Large scale analysis of clinical data reveals subtle dynamical features not evident using traditional analytical methods.

13.2. THE BASICS OF CARDIAC PHYSIOLOGY AND THE CONNECTION WITH NONLINEAR DYNAMICS

In the normal heart the rhythm is governed by a small group of cells in the right atrium called the sinoatrial or sinus node. Electrical excitation spreading from the sinoatrial node passes through the atria, then through the atrioventricular (AV) node separating the atria and ventricles, then through specialized conducting tissue called the Purkinje fibers, and finally into the ventricles. The contraction of the ventricles following excitation leads to the profusion of blood to the body. The firing rate of the sinoatrial node is controlled by the interplay of the two antagonistic branches of the autonomic nervous system. Activity of the sympathetic nerves increases the heart rate and activity of the parasympathetic (vagus) nerves decreases the heart rate. The activity of these nerves is controlled by a variety of feedback loops that monitor cardiac function. Thus, viewed from the perspective of a physicist, the heart is a complex nonlinear system involving oscillators and wave propagation in an excitable medium with complex feedback mechanisms regulating its frequency.

In the past decade there has been increasing interest in the theoretical analysis of complex electrical and mechanical events in the heart [2, 3]. To simplify matters to the extreme, two complementary approaches are taken. In one, an attempt is made to develop realistic models of the heart using stiff ordinary differential and partial differential equations for the various electrical currents involved in cardiac excitation. This offers insights into the various ionic currents that help shape the action potential and propagation properties of cardiac excitation. A second approach approximates the complex electrophysiological and anatomical structure of the heart using simplified mathematical models, such as low-dimensional finite difference equations. This work has demonstrated that some rhythms that appear to be haphazard or difficult to interpret are at least consistent with low-dimensional models.

In searching for dynamical signatures, we take advantage of the perspective introduced from the study of nonlinear dynamics and develop new ways to display data based on low-dimensional finite difference equations. In particular, because many problems involving cardiac function involve interactions of two or more oscillators, we focus on methods that have been employed to study the dynamics of interacting nonlinear oscillations [4]. Detailed information about the interactions between the rhythms can sometimes be derived by examining the timing of one oscillation relative to the second oscillation. Let one rhythm be used to set the scale of time and measure the phase of the nth activation, denoted ϕ_n, of the second rhythm relative to a fixed event in the first oscillation. Then in some circumstances we have

$$\phi_n = \phi_{n-1} + \Omega/T_o + f(\phi_{n-1}), \quad \text{mod } 1 \tag{13.1}$$

where T_o is the period of the first oscillation, Ω is the period of the second oscillation, and f reflects the interaction (coupling) between the two rhythms. In equation (13.1), ϕ_n is normalized to lie between 0 and 1. It is sometimes convenient to consider the actual times of the nth activation t_n, where $\phi_n = t_n/T_o \pmod 1$. From equation (13.1) we obtain

$$t_n = t_{n-1} + \Omega + T_o f(\phi_{n-1}). \tag{13.2}$$

The duration of the nth interval, denoted h_n, where $h_n = t_n - t_{n-1}$, corresponds to the time interval between successive activations of the second oscillator. If $f = 0$, there is no interaction between the oscillations. It is also possible that f depends on the values of several preceding terms.

In cardiac physiology there are many circumstances in which there is coupling between oscillations. For example, the respiratory rhythm affects the timing of cardiac impulses so that in normal circumstances the heart rate speeds up during inspiration. This effect, called respiratory sinus arrhythmia, is mediated principally by parasympathetic nerves innervating the heart. Another arrhythmia that can sometimes be generated by the interaction of coupled oscillators is frequent premature ventricular contractions (PVCs). In this arrhythmia, the ventricle contracts prematurely. Following each PVC, the next expected normal (sinus) beat is usually blocked, leading to a longer than normal interval, called a compensatory pause, before the next activation of the ventricles. In some situations, frequent PVCs may be generated by an autonomous pacemaker located in the ventricles (an ectopic pacemaker) competing for control of the heart with the sinus pacemaker. This is called parasystole. However, there are other mechanisms that can generate frequent PVCs, and it is frequently difficult to determine the underlying mechanism. Several theoretical studies of the dynamics of parasystole have been undertaken [5–8]. Because of the readily identifiable mathematical structure of the hypothesized mechanisms of respiratory sinus arrhythmia and parasystole, analysis of data that may display these rhythms is fertile ground for the application of techniques of nonlinear dynamics to cardiology.

13.3. METHODS

Data were collected from patients in the cardiac intensive care unit of the Children's Hospital of Philadelphia and from the cardiac step-down clinic of the Hospital of the University of Pennsylvania. All data were collected using a hardware interface composed of a Mac II (68020 CPU operating at 16 kHz) and a National Instruments NB-MIO-16 interface board. The ECG signals were preamplified and digitized using the NB-MIO-16 board at rates determined by software drivers. High frequency filtering and R wave selection (the R wave is the peak of the electrical activity associated with ventricular contraction) were performed in software that was written in Think C. The raw data set employed is the sequence of time intervals between successive R-wave detections, that is, the heartbeat intervals (h_n). The 80-kHz clock used in the computer interface produces a theoretical time resolution of 25 μs. This data set is downloaded to a Sun workstation for further analysis. Characteristic artifacts resulting from patient movement represent less than 0.2% of all collected data. In some patients, ventilatory data were obtained in order to measure accurate ventilatory rates. The data consist of a time series of chest expansion measurements digitized sampled at 4 Hz with an 8 bit analog to digital converter (ADC). Ventilatory rate was determined via autocorrelation of a 1000 point segment.

This data acquisition methodology is designed to store data from many subjects over long times in a readily analyzable format within the constraints of modern data storage technology. Although short samples of digitized ECG are also recorded for each patient, due to storage limitations, no beat to beat concordance between measured heartbeat intervals and ECG traces is maintained. Further, no image-processing of the ECG morphology is attempted.

13.4. RESULTS

13.4.1. Respiratory Sinus Arrhythmia

In normal, spontaneously breathing individuals, the fluctuation of the period of respiration complicates the analysis of respiratory sinus arrhythmia. In order to eliminate this source of variability, we consider data from a mechanically ventilated sleeping infant in the pediatric intensive care unit of Children's Hospital of Philadelphia. In Figure 13-1, we plot the successive heartbeat intervals for approximately 18 min of recording during ventilation at about 12 breaths/min. During this record the mean heartbeat interval increases from about 336 ms to 346 ms. Superimposed is a fluctuation with the periodicity of the ventilator, approximately 5 s, at an amplitude above baseline of about 3 ms (see inset). There is also a much slower, irregular oscillation with a period of several hundred heartbeats. Because there is a drift in the heart rate over the time of this recording, a plot of h_n as a

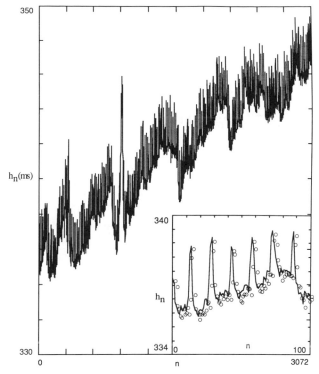

Figure 13.1. A time ordered sequence of 3072 consecutive heartbeat intervals obtained from a 1-month-old infant during repose following cardiac surgery. The duration of each interval is plotted on the ordinate against the position within the sequence plotted on the abscissa. The ordinate is scaled from 330 to 350 ms. The inset below to the right is a closeup of the first 100 intervals plotted in the main body. The ordinate of the inset is scaled from 334 to 340 ms.

function of ϕ_{n-1} would not give a clear picture of the respiratory modulation evident in the inset of Figure 13-1. In order to compensate for this drift, we plot the successive difference of the heartbeat intervals $\Delta h_n = h_n - h_{n-1}$ as a function of the ventilator phase of the R-wave of the nth heartbeat interval ϕ_n.

The plot of Δh_n versus ϕ_n is presented in Figure 13.2a. The central region of the figure during which the heartbeat interval is increasing corresponds to the expiratory portion of the breathing phase.* There appear to be

---

*Unfortunately, no beat to beat concordance of ventilatory and RR-interval measurements is currently available, although the two data sets were sampled within minutes of each other. The drift in ventilator frequency (Model 900C, Siemens-Elema Ventilator Systems, 2360 N. Palmer Dr., Schaumberg, IL) was estimated to be less than 1 part in 100,000 over a period of several hours.

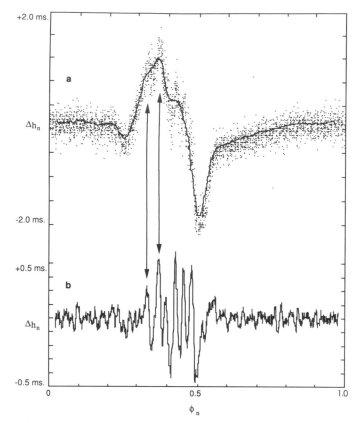

Figure 13.2. (a) The successive differences of the data of Figure 13.1 (Δh_r) mapped onto the phase of ventilation ϕ_r. The horizontal phase axis ranges from phase 0 to phase 1 (mod 1) and the vertical axis ranges from -2 to $+2$ ms. Expiration occurs near phase 0.5. The data of (a) have been subjected to digital bandpass filtering to illustrate the repetitive "fine structure" features at the center that occur during the initial phases of expiration. The ordinate ranges from -0.5 to $+0.5$ ms. Simultaneous chest expansion measures were used to determine the fundamental frequency of ventilation to be $\nu = 12.1356$ breaths/min. Using ν, the phase of each R wave was obtained using an arbitrary choice of phase origin (i.e., the first heartbeat was assumed to be initiated at phase 0.0) via the expression $\phi_n = \nu \langle \sum_{m=n}^{m=n} h_n \rangle_{\text{mod}(1/\nu)}$, so that ϕ_n ranges between 0 and 1.

two components of this function:

1. A slowly varying envelope.
2. A superimposed fine structure.

The superimposed solid line estimates the slowly varying component. It was obtained by sorting the transformed data by ventilator phase and performing

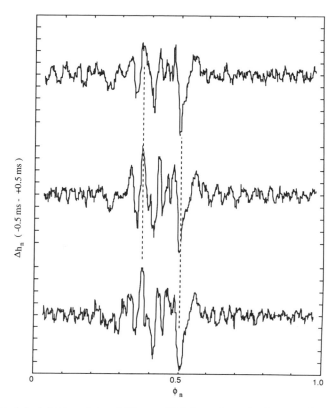

Figure 13.3. (*a*)–(*c*) The data of Figure 13.1 have been broken up into three 1024 point segments, each of which has been subjected to the same bandpass filtering. The "fine structure" appears to be stable (dotted lines) over time.

a 150 point running average. This smoothing digitally filters out components of frequency less than 1/20 of a ventilator cycle. To view the fine structure (Figure 13-2*b*), the residue of the original data and the smoothed data are treated with a 30 point running average. This procedure results in a bandpass filter between frequencies corresponding to 20 and 100 times that of the ventilator frequency. The arrows in Figure 13.2 highlight the close correspondence between fine structure features visible directly in Figure 13.2*a* and extracted in the smoothed data in Figure 13.2*b*. Although it may appear that less pronounced fine structure is present in the rest of the smooth plot as well, it should be kept in mind that smoothed *random* data will also show structure. One can discriminate random features from systematic fluctuations because the random fluctuations will not be stable in time, whereas systematic fluctuations will persist across time in successive plots. In Figure 13.3 we plot "fine structure" maps of three successive 1024 point data streams that comprise the single map presented in Figure 13.2. This figure demonstrates the stability over time of the fine structure. In each trace, a 30 point running average has been subtracted from the original remapped data and a 10 point running average has been performed on the residues. The positions of the dominant fine structure remain stable throughout this time, even though the heart rate varies from 178 to 185 bpm. Figure 13.1*b* demonstrates that the fine structure is confined to ventilator phases corresponding to 0.3–0.5, representing about 20% of the ventilator cycle. Based on the distance between the five most prominent sequential features, we estimate an average spacing of 0.017 phase units. At the measured ventilation rate, this spacing corresponds to fluctuations on the order of 85 ms.

13.4.2. Frequent PVCs

Premature ventricular contractions are a common clinical finding. Although in many instances, this arrhythmia is associated with a benign prognosis, in some cases, such as post myocardial infarction [9], there is elevated risk of sudden death associated with frequent PVCs. In this section we plot clinical data from two patients, using techniques suggested by the hypothesis that the frequent PVCs are due to a parasystolic mechanism, and can therefore be well described by the theoretical formulation in equations 13.1 and 13.2. We consider two patients with frequent PVCs. Patient A is a 57-year-old male who had suffered an acute myocardial infarction two years prior to data collection. Patient B is a 9-year-old female who was recovering from cardiac surgery in the pediatric intensive care unit.

A representative sample of the heartbeat interval data is shown in Figure 13.4. For each patient, the durations of 250 consecutive heartbeat intervals (ordinate) are plotted against their ordinal numbers (abscissa). Circles are drawn indicating the duration of each heartbeat interval and for clarity, lines have been drawn between adjacent circles. The center horizontal rows of circles indicate normal heartbeat intervals, which result from two consecutive discharges of the sinus node. The normal heartbeat interval of patient A is gradually increasing from 860 to 900 ms. That of patient B is fluctuating about 550 ms. The circles lying below indicate intervals that have been shortened due to premature ectopic discharges. These shortened heart beat intervals are clinically referred to as *coupling intervals*. The long intervals (termed *compensatory pauses*) occur because the sinus discharge immediately following the expressed ectopic discharge are concealed due to refractory effects. Although a clinical analysis of the ECG morphology suggests premature ventricular contractions in both patients, the patterns of occurrence of the ectopy and evolution of the coupling interval durations are different. In patient A, the coupling interval durations appear to vary continuously throughout the range 500–800 ms; those of patient B are fixed at very near 375 ms. A close examination of the figure indicates that the coupling intervals experienced by patient A are separated from the previous compensatory

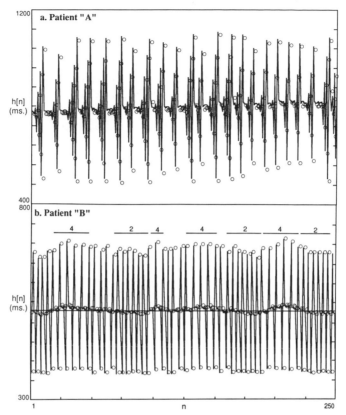

Figure 13.4. The durations of 250 consecutive heartbeat intervals (ordinate) are plotted against their ordinal numbers (abscissa) for two patients. These short data segments are excerpted from a 5000 point collection sequence obtained from an adult patient measured in the step-down clinic at the hospital of the University of Pennsylvania (patient A) and a pediatric patient measured in the intensive care unit at Children's Hospital of Philadelphia (patient B). Circles are drawn indicating the duration of each heartbeat interval and, for clarity, lines have been drawn between adjacent circles. The horizontal scalings are (*a*) 400 to 1200 ms and (*b*) 300 to 800 ms. In panel (*b*) the number of intervening sinus beats between ectopic beats are indicated for different regions of the sequence. Because the sinus beat immediately following the ectopic beat is always blocked due to refractory effects, the number of intervening sinus *R waves* is always one less than the numbers indicated.

pauses by either 0,* 2, 3, or 4 sinus beats. In the case of patient B separations of only 2 or 4 sinus intervals are observed: The longer separations apparently

---

*That is, only a single sinus beat intervenes between ectopic beats and thus no normal sinus intervals intervene.

occur when the sinus intervals are longer. Regions of 2 and 4 intervening sinus beats are indicated on Figure 13.4b.

We will explore the origins of the apparently different behaviors of the cardiac rhythms of these two patients through treatment of the 5000 point data sets from which the time sequences of each plate of Figure 13.4 were drawn. In the four data treatments that follow, we examine:

1. The relationship between the time interval between successive ectopic discharges (the interectopic time) and the phase within the sinus cycle of each ectopic discharge.
2. The distribution of phases of the ectopic discharges within the sinus cycle.
3. The distribution of interectopic times.
4. The distribution of the ratios interectopic times to the sinus interval.

To simplify the discussion, we introduce the following notation in which all times are measured with respect to an arbitrary initial time at which the data set (h_m) begins.

T_n^e: the time of the peak of the nth ectopic R wave within the data set.*
T_n^s: the time of the peak of the sinus R wave that immediately precedes the time of the nth ectopic R wave within the data set.
$I_n^e = T_{n+1}^e - T_n^e$: the nth interectopic time.
$C_n = T_n^e - T_n^s$: the coupling interval (i.e., the shortened heartbeat interval) preceding I_n^e.
H_n: the sinus interval that most nearly precedes T_n^e (in the case that T_n^s terminates a sinus interval, H_n is the interval preceding C_n):

$$\phi_n(\text{mod } 1) = C_n/H_n$$

In Figure 13.5, we plot the I_n^e (interectopic times) on the ordinate versus ϕ_n on the abscissa. Because no ectopic beat was observed to fall at sinus phases less than 0.5 in the data of these two patients, the abscissas are scaled from 0.5 to 1.0. Both ordinates are scaled from 0 to 8000 ms. The data of patients A and B contained 900 and 1163 premature beats, respectively. In both patients, interectopic times are grouped into horizontal strata. In the data obtained from patient A (Figure 13.5a), the lowest-lying locus of points, corresponding to the shortest duration interectopic times of approximately

---

*Premature beats, and thus the time of the ectopic R waves, are readily detected in the h_n sequence due to the fact that a premature beat typically conceals the following sinus beat. In the case of pure *parasystole* this leads to the relationship $h_{n+1} + h_{n+2} = 2h$, where h_n is the interval of normal duration prior to the interval shortened due to the premature beat, h_{n+1} is the premature beat, and h_{n+2} is the beat following the premature beat, which has been lengthened due to refractory effects.

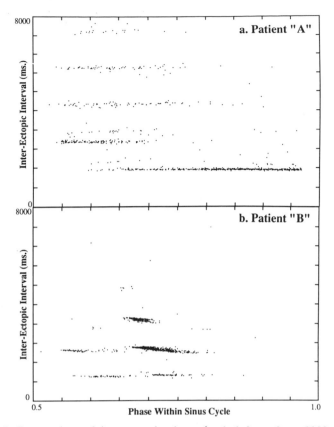

Figure 13.5. Scatterplots of interectopic times (scaled from 0 to 8000 ms on the ordinate) against phase within the sinus cycle of the initial ectopic discharge. The phase axes span 0.5 to 1.0 in phase units modulus 1. In both cases, the original data consisted of 5000 heartbeat intervals. The data of patient A contained 900 premature ventricular contractions; patient B's data contained 1163.

1600 ms, spans the phase region 0.65 to 1.0.* Another horizontal locus of points lies at approximately 3000 ms and spans the phase region 0.55 to 0.7. Nearly equally spaced horizontal lines appear above that span the early phase region of Figure 13.5*a*. The plot of patient B is distinct in that:

1. The points are more localized along the phase axis.
2. At least in the upper two clusters of points, there is a distinct negative slope that indicates that as the phase of the sinus interval sampled by

---

*To discrimate shortened ectopic intervals from the normal variability from, for example, respiratory effects, a shortening of any interval by greater than 50 ms with respect to its predecessor was used to trigger the ectopic interval detection algorithm.

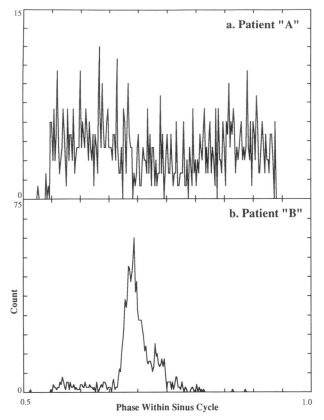

Figure 13.6. A histogram of the phases of the sinus cycle at which ectopic R waves were detected, obtained from the same data shown in Figure 13.5. The frequency axes are scaled from 0 to 15 (panel *a*, patient A) and from 0 to 75 (panel *b*, patient B). The phase axes are scaled from 0.5 to 1.0 in phase units modulus 1. The ectopic intervals of patient B fall at nearly fixed phase (near 0.7) but over a wide phase region for patient A.

the initial ectopic R wave *increases*, the duration of the following interectopic interval *decreases*.

To clarify the analysis of Figure 13.5, the next two figures plot distributions of the points of Figure 13.5 when projected onto the horizontal (Figure 13.6) and vertical (Figure 13.7). Figure 13.6 is the histogram of all ϕ_n scaled, as in Figure 13.5 from 0.5 to 1.0. Although the sinus phases sampled by ectopic R waves are widely dispersed in the phase region 0.55 to 0.95 in the case of patient A, in patient B's data they are localized and centered at phase

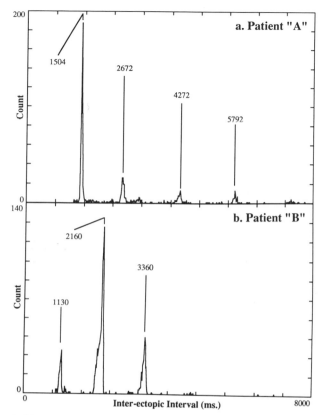

Figure 13.7. A histogram of the time intervals between successive ectopic R waves. The frequency axes on the ordinates are scaled from 0 to 200 (panel *a*, patient A) and from 0 to 140 (panel *b*, patient B). In both cases, the interectopic interval is scaled feom 0 to 8000 ms on the abscissa. Ectopic beats resulting from a pure parasystolic mechanism would produce lines at multiples of the ectopic period. The marked deviations from equal spacing in both plates argue that this simple model cannot describe the observed dynamics.

0.7. The histograms of interectopic times show several unequally spaced peaks in patient A. The peaks in the histogram of patient B are more nearly equally spaced with separations of approximately 1100 ms.

Finally, in Figure 13.8, we present the histogram of R_n, the ratio of the interectopic time (I_n^e) to the sinus period (H_n). In both Figure 13.8*a* and *b*, this ratio is scaled from 0 to 10. In addition to prominent lines centered at 1.7, 2.8, and 4.5, the data of patient A contains a considerable background contribution throughout the ratios 2.0 to 10. The histogram of patient B is essentially free of background. It contains three dominant lines, which to one decimal point precision lie at 2.0 4.0, and 6.0.

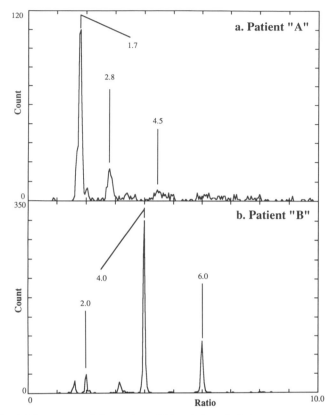

Figure 13.8. A histogram of the ratios of time intervals between successive ectopic R waves to the period of the preceding sinus interval. The frequency axes on the ordinates are scaled from 0 to 120 (panel *a*, patient A) and from 0 to 350 (panel *b*, patient B), respectively. In both histograms, the ratio is scaled from 0 to 10 on the abscissa.

13.5. DISCUSSION

The appropriate display of clinical electrocardiographic data cannot only yield new insight into the dynamics underlying the cardiac rhythm, but can reveal subtle features that are not observable using conventional analyses. We will first discuss the implications of this approach to the study of the respiratory sinus arrhythmia and frequent ectopy before addressing its broader implications in the conclusion.

13.5.1. Respiratory Sinus Arrhythmia

Heart rate variability may provide important information concerning autonomic activity in the body, and several previous studies have dealt with the quantitative analysis of heart rate variation in normal and pathological circumstances [10, 11]. The usual way to quantitate heart rate variability is to compute the power spectrum of the heart rate as interpolated from the sequence of successive heartbeat intervals. Such plots usually show several peaks, with one of the peaks at approximately the respiratory frequency. This peak is therefore taken as a measure of the respiratory sinus arrhythmia. This peak is markedly reduced (or absent) in patients who have had heart transplants [12], indicating the importance of neural activity in the generation of the intrinsic fluctuations.

The current study attempts a finer dissection of respiratory sinus arrhythmia by determining the heartbeat interval relative to the phase of the respiratory rhythm. The periodic sharp increase and slower decay of the heartbeat interval shown in the inset in Figure 13.1 is associated with a definite phase of the respiratory cycle. The magnitude of this effect is comparatively small (a change of about 3 ms in each respiratory cycle), and it is difficult to exclude the possibility that artifact may be responsible for some or all of the variability observed. In particular, the small changes in the shape of the QRS complex associated with lung inflation may change timing of the R wave detector, even in the presence of an absolutely regular rhythm. Similarly, the various mechanical and electrical events associated with the ventilator may also affect the algorithms for R wave detection in subtle ways.

Bearing these caveats in mind, the plots in Figures 13.1–13.3 provide important methods for quantifying respiratory sinus arrhythmia. In the absence of any drift or any long period oscillations of the heart rate, the plot of h_n versus ϕ_{n-1} would show fluctuations reflecting the modulation of the sinus rate by the respiratory cycle, that is, f in equation 13.1. In the current (and typical) case, a plot of h_n versus ϕ_{n-1} would not exhibit significant structure because the magnitude of the drift in heart rate is greater than the magnitude of f. One way to minimize the effects of drift is to plot Δh_n versus ϕ_{n-1} as in Figures 13.2 and 13.3. Assuming that equation 13.2 holds and that T_o is constant, we obtain the theoretical expression

$$\Delta h_n = T_o[f(\phi_n) - f(\phi_{n-1})] \qquad (13.3)$$

Equation 13.3 shows why the upper trace in Figure 13.2 appears as a "crude" first derivative of the inset in Figure 13.1, because the right-hand side contains the difference between $f(\phi_n)$ and $f(\phi_{n-1})$. What is totally unexpected, and is not presently understood, is the appearance of the high frequency (about 12 Hz), low amplitude (about 1 ms) oscillations in Figures 13.2 and 13.3. This frequency is different from high frequency oscillations that have been reported in diaphragmatic electromyographic recordings [13]. Similar oscillations can sometimes be found in experimental studies of mechanically ventilated piglets [14] and it is tempting to speculate that these high frequency oscillations may reflect delicate details of the interactions between the autonomic system and the heart.

Several clinicians have emphasized the importance of autonomic activity as a significant factor in the onset of cardiac arrest. For example, there are reports showing that increased heart rate, indicating increased sympathetic activity, or reduced heart rate variability may sometimes precede cardiac arrest [15]. However, systematic prospective clinical studies probing these questions have not yet been carried out. The plots in Figures 13.2 and 13.3 represent a novel way to characterize heart rate variability, and further studies are needed to assess their significance.

13.5.2. Frequent PVCs

It is often difficult to determine the mechanism of frequent PVCs. Standard methods of analysis give some idea of the number of PVCs during the observation period, but little in the way of detailed analysis of the dynamics. If the PVCs were generated by a ventricular pacemaker competing with the sinus rhythm (ventricular parasystole), and if the rates of both pacemakers were fixed, aspects of the dynamics could be well described by circle maps [5–8].

The simplest model of parasystole results from the presence of an ectopic pacemaker that is discharging independently of the sinus pacemaker. In this approximation (termed *pure parasystole*) the timing of successive discharges (t_n) of either pacemaker can thus be described using

$$t_{n+1} = t_n + \Omega, \tag{13.4}$$

where Ω is the period of that pacemaker. However, refractory effects complicate the *observed* behavior because discharges are expressed only when the cardiac tissue is nonrefractory. This model of parasystole is understood in considerable generality [16, 8]. Ideal pure parasystolic rhythms are characterized by:

1. Varying coupling intervals.
2. Interectopic intervals that are multiples of a common divisor: the ectopic cycle time.

Neither of the two patient data sets studied is consistent with both of these results. Although the data of patient A, as observed in Figures 13.4*a* and 13.6*a*, are consistent with result 1, the unequal spacings of lines in Figure 13.7*a* reveals significant inconsistencies with result 2. Although the lines in the interectopic interval histogram of patient B (Figure 13.7*b*) are more closely spaced, the observation of a preferred phase at which ectopic discharges sample the sinus cycle (Figure 13.6*b*) is inconsistent with result 1.

However, is it possible that the data sets obtained from these two patients can be understood if one considers the possibility that discharges emanating from the sinus focus can reset the phase of the ectopic rhythm? This dynamics is called *modulated* parasystole [5]. By converting to a phase coordinate, the influence of the phase response curve f of an ectopic focus can be incorporated into Equation 13.4 to obtain the circle map of Equation 13.1, which describes the phase within the ectopic cycle of successive sinus discharges (ϕ_n). In modulated parasystole, one would generally find some variability of coupling intervals, but the interectopic times are no longer multiples of a common divisor.

Unfortunately, obtaining the form of $f(\phi)$ is greatly complicated by refractory effects. However, some qualitative aspects of the form of the ectopic pacemaker phase response can be drawn from Figure 13.5. Assuming that all the parameters describing the parasystolic rhythm were to remain fixed, the most basic prediction of the parasystole mechanism is that the plots in Figure 13.5 should be single-valued functions with discontinuities resulting from refractory effects. This means that for any given phase within the sinus cycle there should be a single value for the interectopic time. In the ideal case of pure parasystole, the graph of the interectopic times as a function of phase in the sinus cycle would be composed, in general, of discontinuous branches, where the spacings between the branches are multiples of a common divisor [7]. Due to the absence of resetting, each of these branches must be of slope 0. In the cases of modulated parasystole one would also expect single-valued functions in the plots in Figure 13.5, but due to phase dependent resetting effects, the slopes of the branches will not, in general, be zero. The observation that neither of the plots of Figure 13.5 is a single-valued function indicates that either additional theoretical mechanisms must be considered to account for the data or that changing values of parameters, such as refractory period or ectopic rate, are confounding this analysis. We reserve for a future report a more complete time-resolved analysis of these effects.

Despite difficulties in interpreting Figure 13.5 based solely on equation 13.1, one feature distinguishes the two records. Although the slope of all branches of Figure 13.5a have a slope near 0, the upper two branches of Figure 13.5b have a noticeable slope suggestive of appreciable phase resetting. Thus the cumulative effects of the sinus discharges lying between the expressed ectopic discharges, which gives rise in the middle row of points in Figure 13.5b to a change of $300/2160 \simeq 15\%$ in the observed interectopic time when the phase relationships of the sinus and ectopic depolarizations, are shifted only 10% with respect to the sinus phase.

Studies of circle maps with sinusoidal coupling functions have demonstrated that one manifestation of such phase resetting is the entrainment of one oscillator by another [4, 17, 18]. It is possible that we observe such an entrainment in Figure 13.8b. This histogram of the ratio of interectopic period to sinus period contains three narrow features centered at 2.0, 4.0, and 6.0, which correspond to successive expressed ectopic discharges that are separated by 0, 1, and 2 concealed ectopic discharges and thus to 1, 2, and 3 ectopic cycle lengths. This apparent 1:2 entrainment of the periods of the sinus and ectopic pacemaker, despite variations in sinus period on the order of 15%, is consistent with the observation in Figure 13.5*b* that there exists a coupling between the sinus and ectopic pacemakers. In contrast, the more weakly coupled pacemakers of patient A (as inferred from Figure 13.5*a*) exhibit no obvious low integer ratios of sinus to interectopic interval (see Figure 13.8).

13.6. CONCLUSIONS

The topic of this conference is chaos, and until three words ago we have studiously avoided this term. The reason for this is that we believe that it is most important to try to identify the mechanisms underlying the fluctuations present in both normal and abnormal conditions, rather than to determine whether the observed dynamics are chaotic. Because definitions for chaos in biological data have many ambiguities, different individuals have reached differing conclusions concerning the presence or absence of chaos during normal and abnormal cardiac rhythms [19]. Although we do not attempt a complete summary of the debate, we comment briefly on several key questions.

13.6.1. Variability during Normal Rhythm

It is well known that the normal heart rhythm shows significant fluctuation [20]. This variability reflects the response of the individual's cardiorespiratory system to the changing environment via the intrinsic feedback mechanisms regulating heart rate. Although there have been claims that these fluctuations represent deterministic chaos [19], the analysis supporting these conclusions is based on application of subtle measures, such as dimension, Lyapunov number, and power spectra, that are difficult to interpret in the absence of a detailed understanding of the underlying physiological mechanism or of the inherent biological and measurement noise contributions. Therefore, we believe that the hypothesis that the intrinsic heart rate variability reflects deterministic chaos must be viewed as interesting conjecture rather than established fact. Normal neural control systems are nonlinear systems with time delays due to sensory input, neural information processing, and motor output. Although it is known that chaotic dynamics can be found in such systems, [21, 22], theoretical models for heart rate control that display deterministic chaos have not yet been proposed.

13.6.2. Lack of Variability as an Indication of Disease

A number of serious medical problems have been associated with a loss of sinus rate variability [23]. For example, there can be loss of sinus rate variability in high risk fetuses, during heart failure, and in diabetic neuropathy. There are also reports that low sinus rate variability is associated with higher risk of sudden cardiac death in post myocardial infarct patients. Certainly the heart rate variability is a window on the sympathetic and *parasympathetic* neurological systems and can be an important measure of well-being. Yet the significance of this variability or absence of variability in sudden cardiac death needs further analysis. Coumel and colleagues report sudden acceleration of heart rate just prior to cardiac arrest [15], and Goldberger and colleagues described loss of variability immediately preceding cardiac arrest [24, 23]. Systematic prospective studies are needed to assess the sinus rate variability and to correlate this with pathology.

13.6.3. Chaos and Cardiac Arrhythmia

Many individuals with serious heart disorders display one or more cardiac arrhythmias reflecting abnormalities in cardiac conduction and rhythm generation. Theoretical mechanisms that can give rise to deterministic chaos during periods of cardiac arrhythmia have been described [25] and there is a considerable body of research demonstrating deterministic chaos in model experimental systems. Because these mechanisms involve periodic forcing of oscillatory [26] or excitable [27] systems, individuals can have a low sinus rate variability and a ventricular response reflecting deterministic chaos at the same time. But laymen should be aware that there can be fantastic variability of the ventricular rate in diseased hearts, reflecting one or combinations of many different classes of cardiac arrhythmias including atrial fibrillation, multifocal atrial tachycardia, frequent ectopy originating from foci in the atria, ventricular, or AV node, blocked conduction at the AV node, and repetitive paroxysmal tachycardia. It is a medical problem to assess the risk of these arrhythmias, and a scientific problem to understand the source and nature of the variability.

In summary, there exists a wide spectrum of dynamics in the cardiac rhythm, ranging from clocklike regularity to extreme fluctuation and complexity. It is inaccurate to give the impression that there is now compelling evidence establishing deterministic chaos in any of these data, or that simple measures of health can be readily extracted from the regularity of the heart rate.

We believe that it is essential at this stage to develop novel methods of data analysis that may help to elucidate the underlying mechanism of cardiac variability. In pursuing this goal, we have developed data transformation and display techniques that reveal structure in the cardiac rhythm that can be used to develop dynamical classifications that may be used to further refine traditional diagnostic techniques. The analyses in this paper are feasible on microcomputers provided with the appropriate data interface software. Thus, they are not beyond the means of any well-equipped cardiology clinic. Although the analysis has shown certain striking *dynamical signatures* present in the timing of the sinus rhythm and ectopic beats, we are still not able to correlate these with underlying physiological mechanisms and the clinical state. Indeed the inverse cardiac arrhythmia problem—of posing a theoretical model that agrees with observed dynamics of complex arrhythmias—is still before us.

ACKNOWLEDGMENTS

We wish to thank Kevin Elliott and the nursing staff of the pediatric intensive care unit, Childrens Hospital of Philadelphia for their assistance in the collection of cardiac data and Jeffrey Breen for help with the interface software. The research has been partially supported by the Quebec Heart and Stroke Foundation.

REFERENCES

1. W. D. Kannel and A. Schatzkin, Sudden death: lessons from subsets in population studies. *J. Am. Coll. Cardiol.* **5** 141B–149B (1985).

2. L. Glass, P. Hunter, and A. McCulloch (eds.), *Theory of Heart*. Springer-Verlag, Berlin, 1991.

3. J. Jalife (ed.), *Mathematical Approaches to Cardiac Arrhythmias*, vol. 591, Annals of the New York Academy of Sciences, New York, 1990.

4. Guckenheiner and P. Holmes, *Nonlinear Oscillations, Dynamical Systems, and Bifurcations of Vector Fields*. Springer-Verlag, New York, 1986.

5. G. Moe, J. Jalife, W. Mueller, and B. Moe, A mathematical model of *parasystole* and its application to clinical arrhythmias. *Circulation* **56** 968–979 (1977).

6. N. Ikeda, S. Yoshizawa, and T. Sato, Difference equation model of ventricular *parasystole* as an interaction of pacemakers based on the phase response curve. *J. Theor. Biol.* **103** 439–465 (1983).

7. M. Courtemanche, L. Glass, J. Belair, D. Scagliotti, and D. Gordon, A circle map in a human heart. *Physica D* **40** 299–310 (1989).

8. R. dePaola, The dynamics of competing cardiac pacemakers, Ph.D. dissertation. University of Pennsylvania, 1989.

9. S. Goldstein, J. R. Landis, and R. Leighton, Predictive survival models for resuscitated victims of out-of-hospital cardiac arrest with coronary heart disease. *Circulation* **71** 873–880 (1985).

10. R. I. Kitney and O. Rompelman (eds.), *The Study of Heart Rate Variability*. Clarendon Press, Oxford, 1980.

11. S. Akselrod, D. Gordon, F. A. Ubel, D. C. Shannon, and R. J. Cohen, Power spectral analysis of heart rate fluctuation: A quantitative probe of beat-to-beat cardiovascular control. *Science* **213** 220–223 (1981).

12. L. Bernardi, F. Keller, M. Sanders, P. Reddy, B. Griffith, F. Meno, and M. Pinsky, Respiratory sinus arrhythmia in the denervated human heart. *J. Appl. Physiol.* **67** 1447–1455 (1989).

13. E. Bruce and L. Ackerson, High-frequency oscillations in human electromyograms during voluntary contractions. *J. Neurophysiol.* **56** 542–553 (1986).

14. R. de Paola, unpublished results.

15. J. F. Leclerq, P. Maisonblanche, B. Cauchemez, and P. Coumel, Respective role of sympathetic tone and of cardiac pauses in the genesis of 62 cases of ventricular fibrillation. *Eur. Heart J.* **9** 1276–1283.

16. L. Glass, A. L. Goldberger, and J. Belairk, Dynamics of pure *parasystole*. *Am. J. Physiol.* **251** 4841–4847 (1986).

17. M. Guevara and L. Glass, Phase locking, period-doubling bifurcations and chaos in a mathematical model of a periodically driven oscillator: A theory for the entrainment of biological oscillators and the generation of cardiac dysrythmias. *J. Math. Biol.* **14** 1–23 (1983).

18. S. Ostlund, J. Sethnia, and E. Siggia, *Physica D* **8** 303 (1983).

19. L. Glass, Is cardiac chaos normal or abnormal? *J. Cardiovasc. Electrophysiol.* **1**: 481–482 (1990).

20. J. Skinner, A. Goldberger, G. Mayer-Kress, and R. Ideker, Chaos in the heart: Implications for clinical cardiology. *Biotechnology* **8** 1018–1024 (1990).

21. M. Guevera, L. Glass, M. Mackey, and A. Shrier, Chaos in neurobiology. *IEEE Trans. Syst. Man. Cyb.* **SMC-13** 790–798 (1983).

22. L. Glass, A. Beuter, and D. Larocque, Complex oscillationsw in a human motor control system. *J. Motor Behavior* **21** 277–289 (1988).

23. D. Kaplan, M. Furman, S. Pincus, S. Ryan, L. Lipsitz, and A. Goldberger, Aging and the complexity of cardiovascular dynamics. *Biophys. J.* **59** 945–949 (1991).

24. A. Goldberger, D. Rigney, J. Mietus, E. Antman, and S. Greenwald, Nonlinear dynamics in sudden cardiac death syndrome: Heart rate oscillations and bifurcations. *Experientia* **44** 983–987 (1988).

25. M. Courtemanche, L. Glass, M. Rosargarten, and A. Goldberger, Beyond pure *parasystole*: Promises and problems in modeling complex arrhythmias. *Am. J. Physiol.* **257** H693–H706 (1989).

26. M. Guevara, L. Glass, and A. Shrier, Phase locking, period doubling bifurcations and irregular dynamics in periodically stimulated cardiac cells. *Science* **214** 1350–1353 (1981).

27. D. Chialvo, R. Gilmour, and J. Jalife, Low dimensional chaos in cardiac tissue. *Nature* **343** 653–657 (1990).

CHAPTER 14

APPLICATIONS OF CHAOS TO PHYSIOLOGY AND MEDICINE

ARY L. GOLDBERGER, M.D.
Associate Professor of Medicine
Harvard Medical School
Director of Electrocardiography
Beth Israel Hospital
Boston, MA 02215

The complex type of variability generated by healthy systems may be due in part to deterministic chaos. A reduction in this apparent chaos characterizes certain diseases, as well as drug toxicity and aging. This phenomenon represents an example of the principle of the *decomplexification of disease*. Quantitative nonlinear analysis of physiologic data sets based on concepts from chaos theory promises to provide new diagnostic and prognostic indices in a variety of different clinical conditions, including the prediction of sudden cardiac death.

14.1. INTRODUCTION

Over the past decade my colleagues and I have been exploring some of the applications of nonlinear dynamics, including chaos theory and fractals, to physiology and medicine [1–9]. The purpose of this presentation is to review briefly selected aspects of this work, with an emphasis on practical applications to bedside medicine and, in particular, to cardiac monitoring.

14.2. CHAOS IN PHYSIOLOGY: HEALTH OR DISEASE?

Traditionally, in physiology, healthy dynamics has been regarded as regular and predictable, whereas disease (including fatal arrhythmias), aging, and

Applied Chaos, Edited by Jong Hyun Kim and John Stringer.
ISBN 0-471-54453-1 © 1992 John Wiley & Sons, Inc.

TABLE 14.1. Some Practical Applications of Chaos Theory to Cardiology

Measuring loss of sinus rhythm interbeat interval complexity prior to cardiac arrest or other complications

Detecting periodic heart rate (sinus rhythm or ectopic beat) dynamics prior to cardiac arrest or other complications

Measuring loss of complexity of cardiovascular dynamics with aging

Measuring loss of heart rate complexity with drug toxicity

Detecting alternans phenomena before electrical or mechanical instability

Detecting cyclic coronary blood flow with impending myocardial infarction

drug toxicity are commonly assumed to produce disorder and even chaos [10]. Over the past decade, we have challenged this conventional notion by proposing the following:

1. The complex variability of healthy dynamics in a variety of physiologic systems has features reminiscent of deterministic chaos.
2. A wide class of disease processes, as well as drug toxicities and aging, may actually *decrease* the amount of deterministic chaos or complexity in physiologic systems (decomplexification).

These postulates have implications both for basic mechanisms in physiology as well as for clinical monitoring, including the problem of anticipating sudden cardiac death (Table 14.1).

What is the evidence that healthy dynamics may be chaotic in the technical sense of the word? A counterintuitive example of what may be physiologic chaos is the timing of the normal heartbeat. Most lay people and, indeed, most clinicians assume that the healthy heart beats with something approaching clocklike regularity under resting, "homeostatic" conditions. This impression is seemingly supported by palpation of one's own pulse or observation of the electrocardiogram (EKG) of a healthy individual—both suggest a high degree of regularity. However, when one analyzes the interbeat interval variations in healthy individuals more carefully, it is apparent that the normal heart rate fluctuates in a highly erratic fashion (Figure 14.1), even at rest. Although these fluctuations are not subjectively perceptible, they are of considerable physiologic importance and reflect the influence of the involuntary autonomic nervous system that regulates the heart rate. Furthermore, the type of variability that one sees under healthy conditions is consistent with nonlinear chaos. First, as noted, the heart rate time series has an irregular appearance (Figure 14.1). Second, spectral analysis of this physiologic variability shows a broadband frequency distribution (Figure 14.2). Third, Lyapunov exponent calculations from long, continuous sections (about

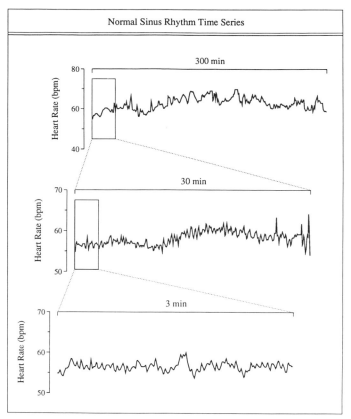

Figure 14.1. Heart rate fluctuations in healthy subjects show a complex type of variability. Furthermore, there are fluctuations across multiple different scales of time. The absence of a characteristic time scale is a *fractal* property of healthy chaotic-appearing dynamics. (Adapted from reference [7].)

90 min) of healthy heart rate data (filtered with singular value decomposition) from our laboratory have yielded a positive exponent [9]. Fourth, preliminary measurements of nonlinear dimensions from other laboratories are also consistent with deterministic chaos [11, 12]. Finally, phase space portraits of healthy heart rate variability reveal complex trajectories that are consistent with a *strange attractor* in a noisy environment (Figure 14.3) [3].

The nonlinear analysis of biologic data sets is still in a fairly early stage, with multiple questions about the optimal way to handle these highly complex heart rate time series yet to be resolved [13]. Such questions relate to the length of the data sets, their stationarity, how if at all the data should be filtered (e.g., with singular value decomposition), how extra heart beats should be handled, and so forth [14]. This is a field of active investigation and there has also already been considerable commercial interest in implement-

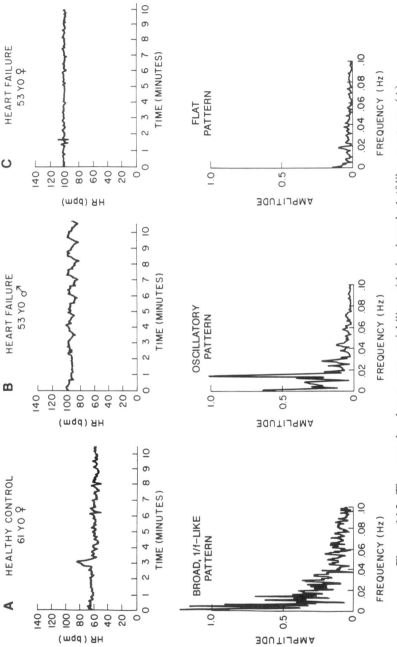

Figure 14.2. The complex heart rate variability with its broad, $1/f$-like spectrum (A) contrasts with the pathologic oscillations (B) or overall loss of variability (C) that may be seen in individuals with certain life-threatening cardiac pathologies. (Adapted from reference [5].)

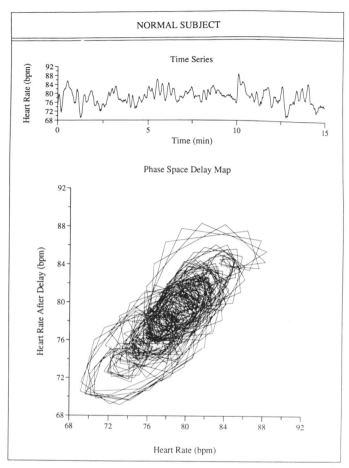

Figure 14.3. The phase space (delay) map provides a geometric representation of a time series. For healthy heart rate variability, the two-dimensional phase space portrait is suggestive of a strange (chaotic) attractor. The delay here is 4 s. The time series has been filtered with singular value decomposition. (Adapted from reference [3].)

ing some of these nonlinear metrics in cardiac monitoring systems. However, caution must be advised here because the appropriate ways to analyze the data remain to be fully determined and the clinical correlates of these nonlinear measurements also need to be better defined. It should be emphasized, however, that traditional statistical measures of variability (e.g., variance or coefficient of variation) cannot be used as reliable surrogates for nonlinear measurements of complexity or chaos [15, 16].

What is the basis of this purported physiologic chaos? The mechanism for apparent chaos in the regulation of the heartbeat has not been defined. As noted previously, the fluctuations of heart rate represent the nonlinear interplay of the sympathetic and parasympathetic branches of the autonomic nervous system of the normal cardiac pacemaker, the sinus node. Efforts are currently underway in our laboratory to devise a deterministic model of heart rate control that will generate chaotic dynamics, as well as other complex and periodic heart rate patterns observed under physiologic and pathologic conditions.

14.3. DISEASE AS DECOMPLEXIFICATION

What happens to physiologic chaotic-like variability under abnormal conditions? According to traditional physiologic theory, *constancy* (homeostasis) [17] is assumed to be the normal state and, therefore, more chaotic behavior might be anticipated under pathologic conditions. However, as we have just seen, it is the normal dynamics of the heartbeat that appears to be most chaotic.* Furthermore, data from our own and other laboratories have demonstrated that there may be a paradoxical loss of chaos (i.e., increase in order) in a variety of different pathologic states. For example, analysis of sinus rhythm heart rate dynamics in patients with severe congestive heart failure or in those just prior to cardiac arrest has revealed a consistent loss of complex variability, sometimes with the abrupt appearance and disappearance of periodic oscillations (Figure 14.2) [4]. These findings, obtained in adults, are consistent with previous reports of heart rate dynamics in the so-called *fetal distress* syndrome, a potentially fatal condition that may be induced by lack of oxygen, by anemia, or by other metabolic problems. It has been appreciated for many years that this syndrome is often characterized by loss of physiologic heart rate variability and sometimes by periodic oscillations in fetal heart rate. We have observed the same patterns in experimental ferrets injected with lethal doses of cocaine. Prior to sudden death from either cardiac arrhythmia or seizure, all the animals showed a striking decline in complex heart rate variability [19]. Loss of heart rate dimensionality has recently been described by Skinner and co-workers [11] in experimental pigs prior to ventricular fibrillation induced by ischemia. Another condition in which there is an apparent loss of physiologic chaos is the aging process. As we grow older, the complexity of our cardiovascular dynamics appears to decrease [15, 20]. Thus the *decomplexification* of physiologic dynamics appears to be a generic response to multiple perturbations associated with pathologic states.

---

*Certain cardiac arrhythmias (e.g., atrial or ventricular fibrillation) may produce erratic cardiac activity, but this pathologic type of variability does *not* appear to be chaotic in the technical sense [3, 18].

14.4. CHAOS IN MEDICINE: IMPLICATIONS AND APPLICATIONS

What are the clinical implications of the loss of complexity or chaos that may occur with disease, drug toxicity, and aging (Table 14.1)? We have proposed that analysis of the nonlinear dynamics of heart rate data and of other physiologic signals, including blood pressure, respiration, hormone fluctuations, and blood-cell variability, may yield diagnostic and prognostic information not accessible by conventional methods [1–8]. For example, in 1992, the most sophisticated commercial analysis programs for heart rate dynamics are still extremely superficial. By and large, the clinical analysis of a 24-hour ambulatory (Holter) electrocardiographic monitor, which may include up to 100,000 heartbeats or more, only provides a summary of the heart rate and range and counts of the number and quantity of various types of extra (abnormal) heartbeats. In addition, there may be some quantitation of heart

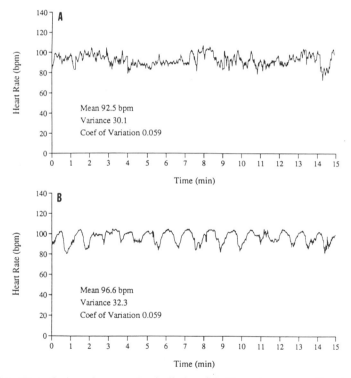

Figure 14.4. Two time series may be indistinguishable using conventional statistical comparisons based on mean and variance, but may represent very different dynamical states. The top panel here shows complex heart rate variability from a healthy individual, the bottom panel shows pathologic heart rate oscillations from a patient with severe cardiac disease. (Adapted from reference [14].)

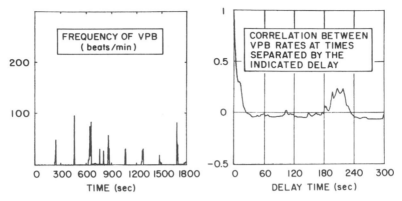

Figure 14.5. Patients at high risk of sudden cardiac death may show complex periodic patterns of ventricular ectopic beat activity. For this case, periodic bursts of ventricular premature beats (VPB) occurred shortly before ventricular fibrillation. The left panel shows the times of VPB bursts in a patient who went into fibrillation at 1855 s. The regularity of the interburst intervals is demonstrated in the right panel. It is seen that a VPB burst is likely to be followed by another one after about 3.5 min. This low-frequency periodicity would not be detected by currently employed clinical analyses of electrocardiographic data. (Adapted from reference [5].)

rate variance. However, other dynamical information contained in the beat-to-beat fluctuations is ignored. As shown in Figure 14.4, two heart rate time series may have essentially identical means and variances but represent very different physiologic processes [14]. Similarly, conventional analysis ignores the often complex and sometimes periodic patterns of abnormal (ectopic) beats that may be precursors of fatal arrhythmias (Figure 14.5) [14, 21, 22].

Currently, research efforts are being devoted to devising new methods of physiologic signal analysis based on nonlinear dynamics and chaos theory. A loss of complexity, as assessed by measurements based on correlation dimension, Kolmogorov entropy, Lyapunov exponent or spectral techniques, will likely provide more sensitive and specific markers of impending cardiac electrical stability, as well as new means of quantitating the aging process [15, 16]. Such measurements may also allow us to detect toxicity due to "recreational" drugs like cocaine [19], as well as from standard therapeutic agents like digitalis or other antiarrhythmic drugs prior to some potentially fatal complication.

Of interest, pathological periodicity and the loss of physiological complexity appear to be markers of hemodynamic as well as electrical instability of the heart. Indeed, periodic dynamics may be important in the pathophysiology of acute myocardial ischemia and infarction. For example, concentrically narrowed coronary arteries in dogs develop cyclic blood flow associated with spontaneous transient coronary thrombosis [23, 24].

The same type of analysis may also apply to other physiologic systems [25–27]. For example, under healthy conditions there are apparently chaotic fluctuations in white blood cell counts from one day to the next [27]. In contrast, in certain cases of white blood cell leukemia, highly periodic oscillations of the blood count may occur [26, 27].

The types of nonlinear analysis described here may also have a broader impact on investigators in a variety of different physiological areas. Until now, physiologists and clinical investigators have tended to ignore or "average out" fluctuations in any given variable such as heart rate or blood pressure. However, a considerable amount of information about the function of the physiologic system may in fact reside in what has previously been considered extrinsic "noise." Nonlinear dynamics promises to alter the way that clinical and physiological investigations are conducted and analyzed, not only with respect to understanding the mechanism of what may be deterministic chaos and not simply noise, but also by providing new understanding of the types of abrupt changes and complex oscillations observed prior to cardiac arrest [3, 4, 11, 14].

ACKNOWLEDGMENTS

This work was supported by grants from the G. Harold and Leila Y. Mathers Charitable Foundation, the National Heart, Lung and Blood Institute, Colin Medical Instruments Corporation, the National Institute on Drug Abuse, and the National Aeronautics and Space Administration.

REFERENCES

1. A. L. Goldberger, V. Bhargava, B. J. West, and A. J. Mandell, On a mechanism of cardiac electrical stability: The fractal hypothesis. *Biophys. J.* **48** 525–528 (1985).

2. B. J. West and A. L. Goldberger, Physiology in fractal dimensions. *American Sci.* **75** 354–365 (1987).

3. A. L. Goldberger and D. R. Rigney, Sudden death is not chaos. In *The Ubiquity of Chaos* (S. Krasner, ed.) pp. 23–34. AAAS Press, Washington, DC, 1990.

4. A. L. Goldberger, D. R. Rigney, J. Mietus, E. M. Antman, and S. Greenwald, Nonlinear dynamics in sudden death syndrome: Heartrate oscillations and bifurcations. *Experientia* **44** 983–987 (1988).

5. A. L. Goldberger and D. R. Rigney, On the non-linear motions of the heart: Fractals, chaos and cardiac dynamics. In *Cell to Cell Signalling: From Experiments*

to Theoretical Models (A. Goldbeter, ed.) pp. 541–550. Academic Press, San Diego, 1989.

6. A. L. Goldberger and B. J. West, Fractals in physiology and medicine. *Yale J. Biol. Med.* **60** 421–435 (1987).

7. A. L. Goldberger, B. J. West, and D. R. Rigney, Chaos and fractals in human physiology. *Scientific American* **262** 42–49 (1990).

8. A. L. Goldberger, Is the normal heartbeat chaotic or homeostatic? *News Physiol. Sci.* **6** 87–91 (1991).

9. D. R. Rigney, J. E. Mietus, and A. L. Goldberger, Is sinus rhythm "chaotic"? Measurement of Lyapunov exponents (Abstract). *Circulation* **82** (Suppl. III) 236 (1990).

10. J. M. Smith and R. J. Cohen, Simple computer model predicts wide range of ventricular dysrhythmias. *Proc. Natl. Acad. Sci. USA* **81** 223–237 (1984).

11. J. E. Skinner, C. Carpeggiani, C. E. Landisman, and K. W. Fulton, Chaotic correlation dimension of the heartbeat is reduced by ischemia, stress or sleep in the pig. *Circ. Res.* **68** 966–976 (1991).

12. A. Destexhe and A. Babloyantz, Is the normal heart a periodic oscillator? *Biol. Cybernet.* **58** 203–11 (1988).

13. P. E. Rapp, T. R. Bashore, I. D. Zimmerman, J. M. Martinerie, A. Albano, and A. I. Mees, Dynamical characterization of brain electrical activity. In *The Ubiquity of Chaos* (S. Krasner, ed.) pp. 10–22. AAAS Press, Washington, DC, 1990.

14. A. L. Goldberger and D. R. Rigney, Nonlinear dynamics at the bedside. In (L. Glass, P. Hunter, and A. MacCulloch, eds.) pp. 583–605. *Theory of Heart.* Springer-Verlag, New York, 1991.

15. D. Kaplan, M. I. Furman, S. M. Pincus, S. M. Ryan, L. A. Lipsitz, and A. L. Goldberger, Aging and the complexity of cardiovascular dynamics. *Biophys. J.* **59** 945–949 (1991).

16. S. M. Pincus, Approximate entropy as a measure of system complexity. *Proc. Natl. Acad. Sci. USA* **88** 2297–2301 (1991).

17. W. B. Cannon, Organization for physiological homeostasis. *Physiol. Rev.* **9** 399–431 (1929).

18. D. T. Kaplan and R. J. Cohen, Is fibrillation chaos? *Circ. Res.* **67** 886–892 (1990).

19. B. S. Stambler, J. P. Morgan, J. Mietus, G. B. Moody, D. R. Rigney, and A. L. Goldberger, Cocaine alters heart rate dynamics in conscious ferrets. *Yale J. Biol. Med.* **64** 143–153 (1991).

20. L. A. Lipsitz and A. L. Goldberger, Loss of "complexity" and aging. Potential applications of fractals and chaos theory to senescence. *J. Amer. Med. Assoc.* **267** 1806–1809 (1992).

21. L. Glass, A. L. Goldberger, M. Courtemanche, and A. Shrier, Nonlinear dynamics, chaos and complex cardiac arrhythmias. *Philos. Trans. Roy. Soc. Ser. A* **413** 9–26 (1987).

22. M. Courtemanche, L. Glass, and M. D. Rosengarten, and A. L. Goldberger, Beyond pure parasystole: Promises and problems in modelling complex arrhythmias. *Am. J. Physiol* **257** (*Heart Circ.* **26**) H693–706 (1989).

23. J. D. Folts, E. B. Crowell, Jr., and G. G. Rowe, Platelet aggregation in partially obstructed vessels and its elimination with aspirin. *Circulation* **54** 365–370 (1976).

24. J. H. Ashton, C. B. Benedict, C. Fitzgerald, S. Raheja, A. Taylor, W. B. Campbell, L. M. Buja, and J. T. Willerson, Serotonin as a mediator of cyclic flow variations in stenosed canine coronary arteries. *Circulation* **73** 572–578 (1986).

25. W. J. Freeman, The physiology of perception. *Scientific American* **264** 78–85 (1991).

26. M. C. Mackey and L. Glass, Oscillation and chaos in physiological control systems. *Science* **197** 287–289 (1977).

27. A. L. Goldberger, K. Kobalter, and V. Bhargava, $1/f$-Like scaling in normal neutrophil dynamics: Implications for hematologic monitoring. *IEEE Trans. Biomed. Eng.* **33** 874–876 (1986).

PART IV

CHAOTIC TIME SERIES AND FORECASTING

CHAPTER 15

NONLINEAR MODELING OF CHAOTIC TIME SERIES: THEORY AND APPLICATIONS

MARTIN CASDAGLI,* DEIRDRE DES JARDINS,[†] STEPHEN EUBANK,*[‡]
J. DOYNE FARMER,[§]JOHN GIBSON,* JAMES THEILER[‡]
Theoretical Division
Los Alamos National Laboratory
Los Alamos, NM 87545

NORMAN HUNTER
WX Division
Los Alamos National Laboratory
Los Alamos, NM 87545

We review recent developments in the modeling and prediction of nonlinear time series. In some cases, apparent randomness in time series may be due to chaotic behavior of a nonlinear but deterministic system. In such cases it is possible to exploit the determinism to make short-term forecasts that are much more accurate than one could make from a linear stochastic model. This is done by first reconstructing a state space and then using nonlinear function approximation methods to create a dynamical model. Nonlinear models are valuable not only as short-term forecasters, but also as diagnostic tools for identifying and quantifying low-dimensional chaotic behavior.

*Also, Santa Fe Institute, 1660 Old Pecos Trail, Santa Fe, NM 87501.
[†]Also, Center for Nonlinear Studies, and Advanced Computing Laboratory, Los Alamos National Laboratory.
[‡]Also, Center for Nonlinear Studies, Los Alamos National Laboratory.
[§]Present address: Prediction Co., 234 Griffin Avenue, Santa Fe, NM 87501.

Applied Chaos, Edited by Jong Hyun Kim and John Stringer.
ISBN 0-471-54453-1   © 1992 John Wiley & Sons, Inc.

During the past few years, methods for nonlinear modeling have developed rapidly and have already led to several applications where nonlinear models motivated by chaotic dynamics provide superior predictions to linear models. These applications include prediction and fluid flows, sunspots, mechanical vibrations, ice ages, measles epidemics, and human speech.

15.1. INTRODUCTION

Chaos is the irregular behavior of simple deterministic equations. Irregular fluctuations are ubiquitous in both natural and artificial systems. Chaos is an appealing notion to a physicist confronted with a dynamical system that exhibits aperiodic fluctuations, because it implies that these fluctuations might be explained with relatively simple physics, in terms of only a few equations of motion. If it is possible to model these apparently complicated variations with simple deterministic equations, then it becomes possible to predict future fluctuations, at least in the short term, and perhaps even to control them.

Although the mathematical properties of nonlinear equations have been studied since the time of Poincaré [108], the physical implications of chaos seem not to have been widely appreciated until the numerical work of Lorenz [90], which provided researchers with a simple and specific example of chaos. For a technical introduction, see, for instance, Crutchfield et al. [36], Eckmann and Ruelle [42], and Eubank and Farmer [45].

Traditionally, the modeling of a dynamical system proceeds along one of two lines. In one approach, deterministic equations of motion are derived from first principles, initial conditions are measured, and the equations of motion are integrated forward in time. Alternatively, when a first-principles model is unavailable or intractable or when initial conditions are not accessible, then the dynamics can be modeled as a random process, using nondeterministic and typically linear laws of motion.

Until recently, the notions of determinism and randomness were seen as opposites and were studied as separate subjects with little or no overlap. Complicated phenomena were assumed to result from complicated physics among many degrees of freedom, and thus they were analyzed as random processes. Simple dynamical systems were assumed to produce simple phenomena, so only simple phenomena were modeled deterministically.

Chaos provides a link between deterministic systems and random processes, with both good and bad implications for the prediction problem. In a deterministic system, chaotic dynamics can amplify small differences, which in the long run produces effectively unpredictable behavior. On the other hand, chaos implies that not all random-looking behavior is the product of complicated physics. Under the intoxicating influence of nonlinearity, only a few degrees of freedom are necessary to generate chaotic motion. In this case, it is possible to model the behavior deterministically and to make short-term predictions that are far better than those that would be obtained from a linear stochastic model. Chaos is thus a double-edged sword; it implies that even approximate long-term predictions may be impossible, but that very accurate short-term predictions may be possible.

This paper focuses on situations in which we cannot find a good model from first principles. Instead we must try to find a model directly from the data. We are interested in time series that arise from observations of a deterministic dynamical system, such as a set of differential equations. Of course, the dynamics is never strictly deterministic, due to *dynamical noise*, which perturbs the states of the system; nor are the measurements completely accurate, due to *observational noise*. We are able to exploit most successfully those cases where the dynamical and observational noise are reasonably small and where much of the apparent randomness is caused by low-dimensional chaotic behavior.

An important consideration is the dimension of the dynamics, which indicates the number of irreducible degrees of freedom. We have noted that complex aperiodic behavior can result from deterministic physical systems with only a few degrees of freedom. However, it is also the case that a dissipative dynamical system that has many nominal degrees of freedom (such as a fluid with 10^{23} molecules, or a power grid with 10^6 customers), may settle down, after an initial transient, to motion in which only a few degrees of freedom are active. Dimension counts the number of degrees of freedom necessary to describe this motion and thus quantifies the difficulty of modeling the system's behavior. The methods that we describe here are most effective when, at some level of approximation, the dimension is moderately low.

Very often one observes a time series with fewer variables than are needed to describe the dynamical system fully. Indeed, the time series in many cases consists only of a sequence of scalar values. Building a dynamical model directly from the data involves two steps:

1. *State space reconstruction.* A state $s(t)$ is an information set, typically a real vector, which fully describes the system at a fixed instant in time t. If it is known with complete accuracy and if the system is strictly deterministic, then the state contains sufficient information to determine the future of the system. The goal of state space reconstruction is to use the immediate past* behavior of the time series to reconstruct the current state of the system, at least to a level of accuracy permitted by noise.

2. *Nonlinear function approximation.* The dynamics is a function f that maps the current state $s(t)$ to a future state $s(t + T)$. An approxima-

---

*For purposes of noise reduction it is also useful to consider the future evolution of the state; we will not address this issue here, but the reader is referred to references [49], [50], and [78] for further discussion.

tion to the dynamics can be found by fitting a nonlinear function to the graph of all pairs of the form $(s(t), s(t + T))$. Note that to model chaotic behavior, the functional form of f *must* be nonlinear.

The field of nonlinear time series analysis is an old one, with several classic contributions [57, 113, 155, 159], but there has been a burst of new work in the last few years [28, 35, 37, 48, 49, 62, 85, 96], in part because of the new aspect introduced by deterministic chaos and in part due to the availability of significant computational power.

In this paper, we review some recent developments. In the first half, we discuss the theory of nonlinear prediction of chaotic time series, which we break down into the problems of state space reconstruction, nonlinear function approximation, statistical characterization, and treatment of input–output systems. We conclude the theoretical discussion with a comparison of this approach with that of nonlinear time series analysis based on the theory of random processes. In the second half of the paper, we discuss several applications including prediction of fluid flows, sunspots, mechanical vibrations, ice ages, measles epidemics, and human speech.

15.2. STATE SPACE RECONSTRUCTION

In this section we review currently used techniques for state space reconstruction, and outline theoretical results recently obtained by some of us [30] concerning how observational noise affects the reconstruction problem.

15.2.1. Review of Currently Used Methods

Typically, the measurements making up a time series $\{x(t_i)\}$, $i = 1, \ldots, N$, are of lower dimension than the dynamics that generates them. For example, in this review, we assume the time series is generated by a dynamical system,

$$\frac{ds}{dt} = F(s(t)) + \eta(t)$$

$$x(t) = h(s(t)) + \xi(t) \tag{15.1}$$

where s is a d-dimensional state, h is a scalar-valued measurement function, ξ denotes observational noise, and η denotes dynamic noise. In most real problems, the only observable is $\{x(t_i)\}$; s, F, η, h, and ξ must be "reconstructed" from the time series.

To construct a dynamical model for the time series, we must first reconstruct a state space. The past behavior of the time series contains information about the present state. If for convenience the sampling time τ is assumed to be uniform, this information can be represented as a *delay vector* of dimension m,

$$\mathbf{x}(t) = (x(t), \ldots, x(t - (m - 1)\tau))^{\dagger} \qquad (15.2)$$

where m is called the *embedding dimension*, and the dagger ($\dagger$) denotes the transpose (by convention, all states are taken to be column vectors). The use of the past history to construct a state is a standard practice in time series analysis and dates back at least as far as the work of Yule [163]. Work by dynamical systems theorists, including Packard et al. [106], Ruelle [116], and Takens [144], show that invariants such as fractal dimension and Lyapunov exponents are preserved and can, in principle, be estimated from the time series (see Section 15.4). Takens considered the global properties of the map that takes the original state $s(t)$ to the delay vector $\mathbf{x}(t)$ and proved that, in the absence of noise, if $m \geq 2d + 1$, then this map generically forms an *embedding*; that is, the reconstructed state space is diffeomorphic to the original state space [144]. (This result has been extended recently by Sauer, Yorke, and Casdagli [122].) It also follows that there exists a smooth map P: $\Re^m \to \Re$ satisfying

$$x(t + \tau) = P(x(t), x(t - \tau), \ldots, x(t - (m - 1)\tau)) \qquad (15.3)$$

The techniques described in Section 15.3 can then be used to estimate P and to make predictions.

Delay vectors are currently the most widely used choice for state space reconstruction. Unfortunately, in order to use them it is necessary to choose the delay parameter τ. Although Takens' theorem suggests that this choice is not important, in practice—because of noise contamination and estimation problems—it is crucial to choose a good value for τ. If τ is too small, each coordinate is almost the same, and the trajectories of the reconstructed space are squeezed along the identity line; if τ is too large, in the presence of chaos and noise the dynamics at one time become effectively causally disconnected from the dynamics at a later time, so that even simple geometric objects look extremely complicated. Most of the research on the state space reconstruction problem has centered on theoretical approaches to choosing τ and m in an optimal fashion for delay coordinates. The proposals for doing this include information-theoretic [2, 52, 54] and geometrical [33, 89, 90] properties.

The other method of state space reconstruction in common use is the principal components technique. This is a standard procedure in signal processing and was first applied to chaotic dynamical systems by Broomhead and King [24]. The simplest way to implement their procedure is to compute a covariance matrix $C_{ij} = \langle x(t - i\tau)x(t - j\tau)\rangle_t$, where $|i - j| < m$, and then compute its eigenvalues, where $\langle \ \rangle_t$ denotes a time average. The eigenvectors of C_{ij} define a new coordinate system, which is a rotation of the original delay coordinate system. The eigenvalues are the average root-mean-square projection of the m-dimensional delay coordinate time series onto the eigenvectors. Ordering them according to size, the first eigenvector has the maximum possible projection, the second has the largest possible projection for any fixed vector orthogonal to the first, and so on. This procedure can be useful in the presence of noise; discarding eigenvalues whose size is below the noise level can reduce noise. The number of singular values above the noise level has been used as an estimate for the appropriate embedding dimension of the data [24], but there are pitfalls with this idea [59, 97]. In some cases this technique compresses information in a delay vector down to a smaller number of dimensions, which can be advantageous for model fitting.

Filtering is another procedure that is often used in state space reconstruction, and can be used in combination with any reconstruction technique. Results of Badii et al. [6] show that some types of filtering can increase the dimension of a time series and should therefore be undertaken with caution. However, Mitschke has shown numerically that this effect can be mitigated if acausal filters are used [98], and Scargle [123] has found circumstances when filtering can be beneficial. Gibson et al. [59] recently showed that principal component coordinates are closely related to (appropriately low-pass filtered) derivative coordinates, considered earlier by Packard et al. [106] and Takens [144]. Furthermore, for small delay times, they showed that the principal component basis reduces to Legendre polynomials, which are convenient because they do not require recomputation for each problem.

At this point there is no clear statement as to which of these methods is superior. Fraser has demonstrated situations in which delay coordinates are superior to principal components [53]. However, we have observed examples where the opposite is true. The situation at this point is inconclusive, and it is not clear what causes one coordinate system to be better than another. One of the central motives for defining noise amplification in Section 15.2.2 is to compare different methods of state space reconstruction. This gives guidance for optimizing the parameters of a particular method or for comparing two different methods.

Principal component, derivative, and delay coordinates, and filtered versions thereof, are all related to each other by linear transformations. However, the transformation from delay coordinates to the original coordinates is typically *non*linear. As Fraser has demonstrated [53], nonlinear coordinate transformations can be greatly superior. The method of local principal components proposed in Section 15.2.2 is a nonlinear coordinate transformation, which is optimal in a certain theoretical sense.

15.2.2. Noise Amplification of Reconstructions

We now outline a recent theoretical approach that quantifies how observational noise gets amplified in the process of state space reconstruction. We

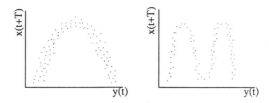

Figure 15.1. Two hypothetical scenarios for prediction in a one-dimensional state space. The horizontal axis is the reconstructed state $y(t)$, and the vertical axis is the future value of the time series, $x(t + T)$: (a) coordinate system with high noise amplification; (b) coordinate system with low noise amplification. This is evident from the "thickness" of the distribution of points at any given y. However, because the functional form of (b) appears "more complicated," with a limited amount of data (b) might result in larger estimation error than (a).

also discuss a method for nonlinear reconstruction, which produces coordinates that minimize noise amplification, while making the dimension of the state space as small as possible. For details see reference [30].

Let $y(t) = \Psi(\mathbf{x}(t))$ be a reconstructed state at time t. We quantify the quality of the reconstruction Ψ according to its *predictive value* as follows. Select a prediction time T, and imagine plotting the time series data according to Figure 15.1, where the horizontal axis should be thought of as being m-dimensional. Then if there are enough points,* the predictability of $x(t + T)$ is approximately equal to the "thickness" of the graph in the vertical direction. Now suppose that the observational noise has variance ε^2, which is small. Then the "thickness" is expected to be proportional to ε, with a constant of proportionality that depends *solely* on *deterministic quantities*, such as the dynamics f, the observation function h, and the reconstruction map Ψ. This motivates us to define the noise amplification $\sigma(T, t)$ of a reconstruction by

$$\sigma(T, t) = \lim_{\varepsilon \to 0} \frac{1}{\varepsilon} \sqrt{\text{Var}(x(t + T)|y(t))} \qquad (15.4)$$

where $\text{Var}(x(t + T)|y(t))$ is the variance of $x(t + T)$ given $y(t)$. The circumstances in which this limit exists are investigated in reference [30]. Taking the limit as the noise goes to zero is quite different from simply *setting* the noise to zero, as was effectively done by Takens. When the noise is set to zero, all reconstructions that are embeddings are equivalent. In the *limit* as

---

*If there are fewer points, estimation error will dominate. In this case, the following analysis gives a lower bound to predictability. The lower bound is only attainable in the limit of a large number of points (or if an explicit functional form is known for f and h, as assumed in reference [16]).

the noise goes to zero, however, two embeddings may have quite different noise amplifications, as illustrated in Figure 15.1.

In the case of independent identically distributed (IID) Gaussian observational noise, it is possible to derive a formula for the noise amplification. In the case of delay coordinates, consider the action of "Takens' embedding map" $\Phi: \Re^d \to \Re^m$ defined by

$$\Phi(s) = \left(h(s), h(f^{-\tau}(s)), \dots, h(f^{-(m-1)\tau}(s)) \right)^{\dagger} \qquad (15.5)$$

Here f^τ denotes the time τ map of the differential equation (15.1). A noisy reconstructed state $\mathbf{x}(t)$ induces a probability density function

$$p(s|\mathbf{x}) = Ap(s)\exp\left[-\frac{1}{2\varepsilon^2} \|\mathbf{x}(t) - \Phi(s)\|^2 \right] \qquad (15.6)$$

where A is a normalization constant; $p(s)$ represents prior information, for example, knowledge of the invariant measure on an attractor. For convenience, in this review we simply assume that $p(s) = $ constant. In general, $p(s|\mathbf{x})$ is a very complicated function of s, due to the nonlinear function Φ. However, if Φ is an embedding, in the limit as ε tends to zero, $p(s|\mathbf{x})$ tends to a Gaussian of covariance matrix $\varepsilon^2 \Sigma$, where the *distortion matrix** Σ is given by

$$\Sigma = \left(D\Phi^\dagger D\Phi \right)^{-1} \qquad (15.7)$$

where the derivative $D\Phi$ is evaluated at $s(t)$. By transforming this Gaussian by the linear map $Dh\, Df^T$ and convolving with observational noise, it follows that the noise amplification is given by

$$\sigma^2(T, t) = 1 + DhDf^T \Sigma \left(Df^T \right)^\dagger Dh^\dagger \qquad (15.8)$$

The noise amplification depends on the reconstruction technique solely through the distortion matrix Σ. We therefore focus attention on Σ, because it is independent of T. From (15.7) it is seen that the larger the singular values of $D\Phi$, the smaller Σ and, consequently, the noise amplification. The more Φ stretches, the more concentrated $p(s|\mathbf{x})$ is forced to be.

Finally, we consider how to minimize the noise amplification over general reconstructions $\Psi: \Re^m \to \Re^{m'}$, where m and m' are held fixed with $m' < m$, and m' chosen large enough to provide an embedding. This is a question of how optimally to compress information into m' variables. In the case of IID

---

*The term "distortion" was originally used for a closely related quantity defined by Fraser [53].

Gaussian noise, (15.7) can be generalized to

$$\Sigma(t) = \left(D\Phi^{\dagger}D\Psi^{\dagger}(D\Psi D\Psi^{\dagger})^{-1}D\Psi D\Phi \right)^{-1} \tag{15.9}$$

Now, the eigenvalues of $\Sigma(t)$ can be simultaneously minimized by choosing $D\Psi(t) = U^{\dagger}$, where U is obtained from the singular-value decomposition (SVD) of $D\Phi$ evaluated at $s(t)$, that is, $D\Phi + UWV^{\dagger}$. Because $s(t)$ varies, Ψ is a globally nonlinear reconstruction. Geometrically this is an intuitive result, because the matrix $U^{\dagger}$ projects noisy delay vectors back onto the tangent space to $\Phi(\Re^d)$. This is a practical result, because it is possible to estimate the tangent space directly from a time series, without knowing an explicit form for Φ (see [23]). We refer to this nonlinear reconstruction technique as *local principal components*. It already has been used in modeling by others for different reasons [22, 156]. The foregoing analysis provides a solid theoretical reason for using the technique, although numerous complications have been ignored, such as the effects of estimation error, correlated noise, and dynamical noise.

15.3. NONLINEAR FUNCTION APPROXIMATION

The goal of forecasting is to predict the future behavior of a time series from a set of known past values. We will call the available data the *historical record*, or simply the *data set*, to distinguish it from other future or past values of the time series.

Once we have chosen a method for state space reconstruction, the data set can be mapped into the state space. If we plot states at time t versus states at time $t + T$ for different values of t, they are related according to $s(t + T) = f^T(s(t)) + \eta(t)$, as shown in Figure 15.1. We can construct a model for the dynamics by approximating f, that is, by finding a function $\hat{f} \approx f$. The approximation $\hat{f}$ also can be used to predict future states, which in turn can be projected back onto the time series to predict its future behavior.*

The use of a state space representation is most effective when the dynamics is stationary or can be mapped onto something that is stationary. Forecasting involves *extrapolation* in time, in the sense that one uses data from one domain (the past) to extrapolate to the behavior of data in a disjoint domain (the future). Extrapolation is inherently a difficult problem. State space reconstruction makes it possible to convert extrapolation of the time series into a problem of interpolation in the state space.

The most commonly used approach of time series forecasting assumes that the dynamics $\hat{f}$ is linear [15]. This has several advantages. Linear functions

---

*In general there is noise present, so that for optimal prediction the approximation $\hat{f}$ should not fit the data set perfectly.

have a unique representation, and they are relatively easy to approximate. However, linear dynamical systems are exceptional—virtually any real system contains nonlinearities. Furthermore, linear dynamical systems have a highly limited repertoire of dynamical behavior. In particular, linear dynamical systems cannot produce limit cycles or chaotic dynamics. To model chaotic dynamics, we are forced to consider nonlinear functions.

The importance of using nonlinear functions cannot be overemphasized. Linear modeling techniques are sometimes motivated by the Wold decomposition theorem, which states that a random process can be decomposed into the sum of an autoregressive process and a noise process whose values are uncorrelated. This uncorrelated process is not, however, necessarily unpredictable. For example, a simple nonlinear mapping such as the logistic map $x_{t+1} = 4x_t(1 - x_t)$ can produce time series for which the autocorrelation function $\langle x_t x_{t+\tau} \rangle - \langle x_t \rangle^2 = 0$ unless $\tau = 0$. The Wold decomposition for this case represents the entire process as unpredictable noise. In contrast, by exploiting nonlinearity we can express this as a purely deterministic system.

Approximating a function requires two choices. We must choose a representation, and we must choose a method of selecting the parameters of the representation. For example, we might choose multidimensional polynomials of a given order as a representation, and minimization of the least-mean-squares distance $\hat{s}(t + T) - s(t + T)$ as a criterion for selecting parameters [where $\hat{s}(t + T) = \hat{f}^T(s(t))$]. There is an infinite variety of possible representations and, in the absence of any prior knowledge, no immediate reason to prefer one representation over another. However, there do seem to be general properties of some representations that make them superior to others across a wide class of practical problems. In many cases the selection of a particular function representation is based purely on expedience—a representation may be chosen because it allows a fast algorithm for fitting parameters or because it happens to work well on a given problem.

It is beyond the scope of this problem to review the field of function approximation, but we will describe a few aspects that we have found useful, first discussing representations and then discussing generalization and parameter selection.

15.3.1. Nonlinear Function Representations

15.3.1.1. Global Function Representations. Any complete set of special functions can be used to represent nonlinear functions. The following is a list of functions that have been employed in approximation problems.

- *Polynomials* are of the form

$$\hat{f}(s) = \sum_{a_{i_1 \dots i_d}} s_1^{i_1} \cdots s_d^{i_d} \tag{15.10}$$

They have the advantage that the parameters a can be determined by a linear least squares algorithm, which is fast and gives a unique solution. They have the disadvantage that they blow up as $s \to \infty$ and consequently do not extrapolate well outside the domain of the data set. Furthermore, they behave poorly under iteration.

- *Rational polynomials* offer some improvements—by taking ratios of polynomials of the same order, $\hat{f}(s)$ approaches a constant as $s \to \infty$.

- *Wavelets* are a localized generalization of Fourier series [140]. Their most immediate use is as a means of signal decomposition and, hence, state space representation. They also present an interesting possibility as a function representation within the state space.

- *Neural nets* are currently very popular [34, 47, 119]. There are many different varieties of neural nets, and it is beyond the scope of this review to discuss them all. Furthermore, for cultural reasons it is often difficult to say precisely what deserves to be called a neural network and what is simply a nonlinear function approximator. Perhaps the most popular varieties amount to a function representation consisting of sums and compositions of sigmoid functions. The parameters are called *weights* and *thresholds*, and the parameter-fitting algorithms are called *learning algorithms*. A popular example is back-propagation, which is a nonlinear least squares fitting algorithm.

   Such neural nets have been used by Lapedes and Farber [85, 86], Weigend, Huberman, and Rumelhart [158], and others in time series forecasting problems involving chaos. These neural networks have the disadvantage, when compared to some other methods of function approximation, that for most algorithms parameter fitting is slow; it is also not clear why the sigmoid function should have a special role in function approximation. However, in some problems they seem to do a good job of function approximation.

- *Radial basis functions* are of the form

$$\hat{f}(s) = \sum_i a_i \phi(\|s - s_i\|) \tag{15.11}$$

where ϕ is an arbitrary function, s_i is the value of s associated with the ith data point, and $\|s - s_i\|$ is the distance from s to the ith data point [111]. This functional form has the advantage that the least squares solution of a_i is a linear problem. When used as before, the number of free parameters is equal to the number of data points, giving an interpolation rather than an approximation, because $f(s_i) = \hat{f}(s_i)$. However, it is also possible to generalize this for approximation, by choosing the s_i so that they are not necessarily centered on data points [100, 107]. Radial basis functions or variants thereof are widely used in the fields of computer graphics and vision [107] and have also been used by several authors in time series prediction [25, 28, 100].

- *Adaptive expression trees*, used in conjunction with the hierarchical genetic algorithm provide an interesting new approach to function approximation, which has not been explored in the context of time series forecasting [79].

15.3.1.2. Local Function Approximation. Local function approximation extends the methods listed in Section 15.3.1.1, by imposing a metric and explicitly basing the contributions of the points in a data set on their proximity to s. When a finite cutoff is used, local methods automatically implement a divide-and-conquer strategy. Parameter selection can be based solely on nearby points. There are fast algorithms for finding nearest neighbors efficiently in moderately low dimensions [11, 12]. For functions with complicated behavior, local methods typically give more accurate approximations than global methods.

An added benefit of local approximation is that there are scaling laws for the a priori accuracy of approximation. In the limit where there are large numbers of data points, so that the spacing ε between data points is small and the noise is small, the approximation error $E(s)$ scales according to $f^{(q)}(s)\varepsilon^q$. The value q is the *order of approximation*, which depends on the method of approximation and may also depend on the data set; $f^{(q)}$ is the qth derivative of f. It obviously is desirable to make q as large as possible, but with present methods it is often not possible to make it larger than 2. For a dynamical system with largest Lyapunov exponent $\lambda_{\max}$ and information dimension D, in the limit as first $N \to \infty$ and then $T \to \infty$, for local iterated approximation* the errors scale as [45, 49]

$$\langle \log |E| \rangle \sim -\frac{q}{D} \log N + \lambda_{\max} T \tag{15.12}$$

There are several ways to do local approximation:

1. *Kernel density estimation* is a method for estimating probability density functions from continuous data. The basic idea is to assign each point an "influence function" or *kernel* that decays with distance, giving an estimator of the form $p(s) = \Sigma_i \phi(\|s - s_i\|)$. The kernel ϕ may be localized, for example, a step function, or it may be a monotonically decreasing function such as an exponential. There are many other variants as well—for a review

---

*For iterated approximation, to make a forecast for time T a map $\hat{f}^{t_c}$ is constructed for $t_c \ll T$ and then iterated to produce forecasts for time T. This method takes advantage of the recursive structure of a dynamical system, and for $q > 1$, as $N \to \infty$ and $T \to \infty$, iterated forecasting is superior to direct forecasting, whose scaling law is the same as (15.12) except that there is a factor of q multiplying the last term [49].

see Silverman [131]. Kernel density estimation can also be used for function approximation, by computing the expected value of $s(t + T)$ according to

$$\langle s(t + T) \rangle = \int s(t + T) p(s(t + T)|s(t)) \, ds(t + T) \qquad (15.13)$$

Kernel density estimation is used to compute p. By careful choice of the kernel function it is sometimes possible to achieve $q = 2$ with this method, but in general for kernel density estimation the order of approximation $q = 1$. Nonetheless, kernel density estimation has the advantage of being quite stable, and in noisy situations it can be an effective method of approximation.

Kernel density estimation is obviously related to radial basis functions. The difference is that for radial basis functions the contribution of the kernel from each point depends on s and is computed for each s based on least squares, whereas for kernel density estimation the kernels are usually fixed or are changed in a uniform manner; for example, for an exponential kernel the decay parameter may be adjusted. Note that when ϕ is a decreasing function, radial basis functions are essentially a local method. However, in many situations good results are obtained when ϕ is chosen to be *increasing*, so that the representations are highly nonlocal [111].

2. *Local fitting* can be done with any global representation, by fitting the parameters based on points within a given region. This is analogous to using splines for interpolation. The basic idea is illustrated in Figure 15.2. This approach allows for considerable flexibility in building a globally nonlinear model while fitting only a few parameters in each local patch. A commonly used example is fitting a linear map to the nearest k points [48, 49]. For low noise levels and N sufficiently large, it is usually possible to achieve second-order approximation, even for scattered data in higher dimensions. In principle, it is possible to achieve higher orders of approximation by fitting higher-order polynomial maps; in practice, this may or may not give an improvement, depending on the context. Local fitting has the advantage of

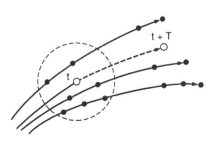

Figure 15.2. Local forecasting for a dynamical system. The current state $s(t)$ and its unknown future value $s(t + T)$ are represented by open circles. The black dots are previous states $s(t')$ for $t' < t$. To make a prediction, we fit a function between the previous states and the states they evolve into a time T later. The prediction is given by the value of the function at $x(t)$. Local modeling techniques use only those previous states that are nearby, for instance, the ones within the circle.

being quick and accurate, but the disadvantage that the approximations are discontinuous. Continuity may be guaranteed by choosing neighboring points for the local approximation so that they form appropriate simplices [96] or by weighted local fitting. In some cases it may be desirable to select parameters using more robust criteria than least squares [66].

Local function approximation is similar in nature to Tong's threshold autoregressive models [153]. The main difference has to do with the scale and mode of use. Tong typically constructs two or three different linear models depending on the state of the system; in contrast, researchers in dynamical systems frequently build a different model for *each* state.

3. *Weighted local fitting* uses a function representation of the form

$$\hat{f}(s) = \sum_i \psi(s - s_i)\phi(\|s - s_i\|) \tag{15.14}$$

where the free parameters are in the function ψ. For example, ψ might be a linear polynomial and ϕ a Gaussian; this case has been studied by Jones and co-workers [76, 95] and Stokbro and Umberger [136]. Weighted local fitting can be viewed as a combination of simple local fitting and kernel density estimation, or alternatively as a generalization of radial basis functions [compare (15.14) to (15.11)]. Simple local fitting is the special case in which ϕ is chosen to be a step function. Note that it is not necessary to restrict the s_i to be data points; for generalization purposes they may be viewed as "centers" of the approximation scheme and may be chosen so that the number of different s_i is less than N. This may be desirable for improved generalization or for data compression. However, it also introduces extra free parameters and an additional optimization problem—the wrong choice of the s_i may give poor generalization. Weighted local linear maps have the advantage that by the appropriate choice of ϕ the approximation $\hat{f}$ is guaranteed to be continuous.

15.3.1.3. Compositions and Combinations. *Compositions* of function representations can be used to extend the power of any of these methods, by expanding $\hat{f}$ as $\hat{f} = f_1 \circ f_2 \circ \cdots$. A familiar example is the use of hidden layers in neural nets. Function composition is one of the important ideas underlying iterated function systems [9] and the group method of data handling (GMDH) [48]. Function composition can take advantage of recursive structure—for example, function composition in *time* is the underlying reason for the superiority of iterated versus direct approximation [49]. In some loose sense, composition significantly increases the complexity of the space of functions that can be represented with a given number of free parameters. However, in general it also significantly increases the complexity of the problem of fitting parameters.

On the other hand, there is some evidence that in simple problems, simple parameter-fitting algorithms such as that employed by the GMDH method, which are quite fast, can still give improved generalization [40, 46, 69, 157, 161]. The basic idea behind these methods is first to optimize the parameters for a single-layer representation to produce an approximation $\hat{f}_1$. This produces a series of errors $E(t) = \hat{s}(t + T) - s(t + T)$. This can then be used to make a map $\hat{f}_2$ (possibly with another representation) from $\{E(t)\}$ to $s(t + T)$. Forecasts can then be made using $\hat{f} = \hat{f}_1 \circ \hat{f}_2$. This can be repeated more than once, until no further improvement is observed [40, 69].

Combinations of different methods can also be used to improve performance. This can be done by trying several different representations $\hat{f}_1(i)$ and using each of them to generate a series of errors $\{E^i(t)\}$. These can then be combined, using $\{E^i(t)\}$ as an input to another layer $\hat{f}_2$, just as described for compositions, except that there are now several input series. This method has been called stacked generalization by Wolpert [161]. Another interesting alternative is to decimate the input, for example, by randomly deleting points from the data set, use each data set to construct a different approximation $\hat{f}_1(i)$, and then combine them with another layer $\hat{f}_2$, as done by He and Lapedes [68].

15.3.2. Generalization and Model Selection

The function representations of the previous section are just a few selections from an infinite menu, and for each representation there are an infinite number of possible models, depending on the choice of parameter values. To choose from among the items on this vast list of possibilities, one needs some criterion for selecting the best model.

A criterion that naively comes to mind is to choose the model with the best fit to the data. This amounts to choosing $\hat{f}$ so as to minimize

$$\hat{\sigma}^2 = \sum_{i=1}^{N} \left[x(t_i) - h\left(\hat{f}\left(s(t_i)\right)\right)\right]^2 \tag{15.15}$$

There is, however, an obvious flaw: For any function representation that forms a complete set, by using a sufficient number of parameters it is (by definition) always possible to fit the data perfectly. For the purposes of making predictions this may be a very poor choice, however. For example, suppose the data consist of a finite number of samples from a smooth function whose derivatives are all small. A polynomial whose order is equal to the number of data points will always fit perfectly, but will typically oscillate wildly between samples.

In prediction problems the goal is to *generalize* from the data we know to the data we do not know. A polynomial fit that oscillates wildly between data points that appear to lie on a smooth curve will *probably* be a terrible choice as a prediction of the next data point that was not used in the fit; a polynomial with fewer free parameters is probably a better choice. How many parameters provide the optimum? A generalization criterion attempts to rank the possible choices according to the probability that they will yield a good guess. For a given representation, such as a polynomial, it should provide the optimal number of free parameters and a procedure for choosing them. It should also provide a criterion for choosing between different function representations.

One class of generalization criteria is motivated by Occam's razor, that is, the principle that the best model is the simplest. These criteria consist of scalar functions with two terms—the first measures goodness of fit, and the other penalizes "larger" models, for example, by counting the number of free parameters. The first criterion of this type was based on maximum likelihood and was originally proposed by Akaike [4]. For a linear model it is roughly

$$\text{AIC} = N \log \hat{\sigma}^2 + 2M \qquad (15.16)$$

where M is the number of free parameters. Another criterion, proposed by Rissanen, is based on the minimum description length principle, that is, compressing the description of the data by a given number of bits of information is equivalent to improving the ability to predict by the same number of bits [114, 115]. In some cases the minimum description length criterion has been shown to be equivalent to another criterion due to Schwartz, based on a Bayesian analysis [128, 115]. For a linear model it is roughly

$$\text{MDL} = N \log \hat{\sigma}^2 + M \log N \qquad (15.17)$$

Akaike's criterion tends to overestimate M, and it is our impression that the MDL criterion is generally more reliable [51, 115].

These criteria are quite useful for comparing models of varying sizes within a given global function representation. There is some arbitrariness, however, in the measure of the "size" of two models, which makes comparisons between different function representations more tenuous. (Some of the problems with using Occam's razor in this context have been discussed by Wolpert [162].) Furthermore, using these criteria with local approximation methods is not straightforward.

Another class of generalization criteria is based on the principle of cross validation [43, 137–139]. The basic concept is to hold back a "sample" of the data, fit several models to the sample, and then choose the model that makes the best predictions out of sample. The most common procedure is to drop the data points one at a time. A more straightforward approach in the case of time series is to break the data into two temporally distinct samples: a "past," which is used for training, and a "future," which is used for testing. The question then arises as to how much of the data to use to build the model and how much to use for testing. We have recently shown that for local approximation, in the limit where the scaling law of (15.12) holds, the splitting of the data that maximizes the statistical significance of the cross validation depends only on the length of the data set [151]. In the large data limit, it turns out that the optimal split favors testing over training.

Generalization criteria related to minimum description length principle, which trade goodness of fit off against the number of free parameters, have the advantages that they are easy to use and that they make use of all the available data in fitting the model. Cross validation has the disadvantage of being more time consuming, but the advantage of being more easily applied to local approximation methods, where counting the number of free parameters and therefore estimating the "size" of the model is not straightforward. Stone has shown that cross validation and Akaike's generalization criterion are asymptotically equivalent [138]. However, there are still many questions concerning generalization criteria that remain to be resolved.

15.4. NONLINEAR STATISTICS

Because chaos is inherently nonlinear, conventional linear statistical measures such as autocorrelation or Fourier power spectra are inadequate to describe it. Both chaotic and random time series have broadband spectra, for example, so that distinguishing them requires some other nonlinear statistical average. The short-term error of a nonlinear predictor is one such statistic—it is small for chaotic systems and large for random processes—but there are other nonlinear statistics that describe different aspects of chaos. The Lyapunov exponents quantify the rate of divergence of nearby trajectories, and the dimension indicates the number of degrees of freedom in the dynamical system. Also useful, although more qualitative in nature, are the recurrence plots of Eckmann, Kamporst, and Ruelle [41], which give a graphical indication of the kind of correlations that exist over the whole time series. These are statistics that, unlike the power spectrum, are sensitive to *nonlinear* correlations in a time series.

In this section we briefly discuss dimension estimation and several tests for nonlinear structure that have evolved from dimension algorithms. Although we do not wish to imply that Lyapunov exponent estimation is any less worthwhile, we will nonetheless defer discussion of this topic to references [1], [26], [42], [121], and [160].

Dimension is of particular interest to us as modelers, because it is low-dimensional determinism (not chaos, per se) which is the primary prerequisite for the modeling procedures of Sections 15.2 and 15.3 to apply. It is true that there is a qualitative difference in the time scale over which the trajectory of a chaotic versus a nonchaotic system can be usefully predicted, but unless the system is low-dimensional, prediction will be impossible in either case.

15.4.1. Dimension Estimation

The development of algorithms for estimating the dimension of an attractor directly from a time series has been an active field of research over the last decade. The objective of these algorithms is to estimate the fractal dimension of a hypothesized strange attractor in a reconstructed state space. If the time series is deterministic and of finite dimension, the estimated dimension of the reconstructed attractor should converge to the dimension of the strange attractor as the embedding dimension is increased. If the time series is random, the estimated dimension should be equal to the embedding dimension. Unfortunately, these statements hold only in the limit of arbitrarily long noise-free stationary time series; in some cases, the convergence can be excruciatingly slow [109].

Historically, the first numerical algorithms were based on a "box-counting" principle [120], although this was found to be impractical for a number of reasons [65]. The correlation dimension developed by Grassberger and Procaccia [63, 64] and independently by Takens [145] considers the statistics of distances between pairs of points, and this remains the most popular way to compute dimension.

Briefly, this method defines a correlation integral $C(m, N, r)$ that is an average of the pointwise mass functions $B(\mathbf{x}; m, N, r)$ at each point $\mathbf{x}$ in the reconstructed state space. Here, m is the dimension of the embedding space, N is the number of points, and $B(\mathbf{x}; m, N, r)$ is the fraction of points (not including $\mathbf{x}$ itself) within a distance r of the point $\mathbf{x}$. The asymptotic scaling of $C(m, N, r) \sim r^{\nu}$ for small r defines the correlation dimension ν.

The pointwise mass functions $B(\mathbf{x}; m, N, r)$ are kernel-density estimators for the natural measure of the strange attractor at points x_i on the attractor and as such more accurately characterize the attractor than the crude histograms that box counting provides. A further advantage is that the correlation integral scales as r^{ν} down to distances on the order of the smallest distance between any pair of points; this range is significantly greater than box counting permits.

Other approaches for estimating dimension include using the statistics of kth nearest neighbors [8],* using Lyapunov exponents to estimate a "Lyapunov dimension," which is related to the actual geometric dimension [77], and using m-dimensional hyperplanes to confine the data in local regions [56, 23, 33]. It also has been suggested that prediction itself can provide a robust test for low-dimensionality [49, 143]: Basically, the smallest embedding dimension for which good predictions can be made is a reliable upper bound on the number of degrees of freedom in the time series.

A number of authors have attempted to identify and quantify specific sources of error in dimension estimates arising from finite-data effects

---

*These are often called "fixed-mass" algorithms because they use balls not of a fixed size, as the correlation dimension does, but of variable size to accommodate a fixed number of nearest neighbors.

[132, 117], statistical precision [21, 39, 149], geometrical effects [32, 103], autocorrelation [146], linear filtering [6], discretization (as by an analog-to-digital converter) and additive noise contamination [10, 99, 134], and lacunarity in the attractor [7, 133, 135, 147]. Also, Osborne and Provenzale [104] observed that standard dimension algorithms may erroneously identify random time series with $1/f^\alpha$ power spectra as low-dimensional, even in the limit of a large number of data points. However, it has been argued [150] that a good dimension estimator need not be fooled by such time series. General discussions of practical issues can be found in references [5], [27], [58], [94], and [148]. A rich source of discussion on a variety of related topics can also be found in the following conference proceedings: references [3], [31], and [93].

From these papers emerges the general impression that reliable and precise estimation of fractal dimension from a time series is possible for sufficiently "nice" data, but that pitfalls abound. Estimates of low dimensions from even large data sets must be treated with caution. The statistical technique we describe next is an example of the kind of cautious approach that we feel is necessary for the reliable use of dimension in nonlinear signal analysis.

15.4.2. Testing for Nonlinear Structure

Although the dimension was originally proposed as a way to distinguish nonlinear deterministic systems from linear stochastic systems ("is it chaos or is it noise?"), a potentially more valuable use for the dimension algorithm is as a statistical tool for distinguishing nonlinear stochastic systems from linear stochastic systems. This is of particular interest to forecasters, who would like to know whether a given time series contains the kind of nonlinear structure that might be exploited by a nonlinear prediction algorithm. It may not really be crucial that the system is completely deterministic, because an approximate forecast can still be useful. At issue is whether a nonlinear forecasting algorithm would provide significantly better predictions than a linear forecasting algorithm. The goal is to reject the null hypothesis that the original time series is just linearly correlated noise. Rejecting a null hypothesis is less ambitious than estimating a dimension, but it is also a statistically more well-posed problem and, in general, can be done more reliably.

There are a number tests for nonlinearity that have been developed by statisticians; see Tong [154] for some examples and Section 15.6 for a discussion of how these tests were developed from a different point of view (and for different data sets) from those that have arisen with the modern study of nonlinear dynamical systems. Recently, a number of tests for nonlinearity have been developed, primarily by the economics community [18–21], which adapt the statistics that were developed for characterizing chaotic systems into tests for nonlinearity.

One of the first of these was developed by Brock, Dechert, and Scheinkman and is known as the BDS test [18]. This test adapts the correlation integral of

Grassberger and Procaccia into a test against IID noise. The BDS statistic is defined by

$$\text{BDS}(m, N, r) = \sqrt{N}\left[C(m, N, r) - C(1, N, r)^m\right] \qquad (15.18)$$

If the time series is IID, then the BDS statistic converges (for large N) to a normal distribution with mean 0 and variance $V(m, r)$ that is independent of N. If there is any correlation in the data at all, linear or nonlinear, then $C(m, N, r) > C(1, N, r)^m$ and the statistic diverges as $\sqrt{N}$. Although the test does not distinguish linear from nonlinear determinism, it can (at least in principle) be modified to do so by "subtracting out" the linear component* and applying the BDS test to the residuals.

An approach we advocate is to start directly with a null hypothesis that the time series is linearly correlated noise. Given a raw data set, create an ensemble (in practice, 10 or 20) of "surrogate" data sets that are stochastic except that they contain the same linear correlations as the original raw time series. A practical way to do this is to take a Fourier transform of the raw data, randomize the phases (while keeping the magnitudes intact), and then invert the Fourier transform. The resulting time series will have the same power spectrum as the raw data set, but will in all other respects be random. Now compute the dimension (or any other nonlinear statistic, such as Lyapunov exponent or nonlinear forecasting error) for the raw time series and call the value obtained v_r. Compute the dimension for each of the surrogate time series; call these $\{v_i\}$, and let v_s and σ_s be the mean and standard deviation of the set $\{v_i\}$. If the difference between the raw and the surrogate dimension is significantly larger than the standard deviation of surrogate dimensions, that is, if $|v_s - v_r| \gg \sigma_s$, then one can be confident that there is significant nonlinear structure in the time series that is not captured in the linear stochastic surrogate data sets.

This approach is related to a technique known as bootstrapping, a robust statistical tool that, with the advent of inexpensive computing, is increasingly favored by statisticians [43]. We have found evidence that bootstrapping with surrogate data sets can often provide a more powerful test than simply looking at residuals. In Figure 15.3, for example, we show in the case of a time series obtained from the Hénon attractor that residuals exhibit less nonlinear significance than the original raw data. Townshend [157], on the other hand, has found that judicious use of the linear residuals can be helpful in nonlinear speech processing (see Section 15.7.6).

A simplified version of this test was applied by Grassberger [61] to reject an earlier claim of low-dimensionality in a climatic time series. The systematic approach advocated here is also described by Theiler [147, 148] for the case of dimension, but a more complete discussion along with many numeri-

---

*This is done by fitting a linear autoregressive (AR) model to the data.

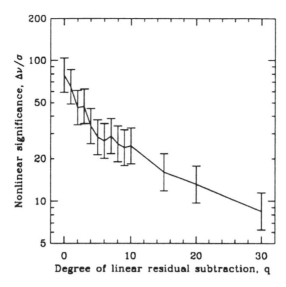

Figure 15.3. Nonlinear significance, measured as $|\nu_s - \nu_r|/\sigma_s$, where ν_s and σ_s are the mean and standard deviation, respectively, of the dimensions computed by the correlation algorithm for 15 "surrogate" data samples, and ν_r is the dimension computed for the raw data set. For $q = 0$, the raw data set is the time series derived from the Hénon attractor, and for $q > 0$ is the qth-order autoregressive (AR) residual. It is evident that attempting to "subtract out" the linear component only decreases the power of the test to discriminate between linear and nonlinear determinism.

cal experiments will be described in a forthcoming paper [151]. Ellner [44] also has emphasized the importance of testing against a plausible alternative (or null hypothesis) and has noted that there is some art in the choice of this alternative.

An exploratory test for nonlinearity has been investigated by Casdagli [29]; here the error of a local linear predictor is plotted against a smoothing parameter. Maximum smoothing corresponds to linear stochastic prediction, and minimum smoothing corresponds to perfect determinism. If the prediction is improved by any amount of smoothing less than maximum smoothing, it is taken as an indication of nonlinearity.

15.5. INPUT – OUTPUT SYSTEMS

In this section, we review a dynamical systems approach to the nonlinear modeling of input–output systems, developed by Casdagli and Hunter [30, 73].

Figure 15.4. Conceptual model of a single-input–single-output system.

15.5.1. Motivation

The formulation of autonomous models from time series using delay coordinates, as in Section 15.2, is useful as a means of obtaining states from a single time series. For many applications, however, other model forms may be more appropriate. In many instances of time series analysis a system may be modeled using both an *input* time series $u(t)$ and an *output* time series $x(t)$ as shown conceptually in Figure 15.4. We assume this system is described by the dynamical system

$$\frac{ds}{dt} = F(s(t), u(t))$$

$$x(t) = h(s(t)) \tag{15.19}$$

where $s(t)$ denotes a d-dimensional state, and $h: \Re^d \to \Re$ is a scalar measurement function. For simplicity we have ignored noise, although in general noise should also be included.

The input–output formulation of (15.19) is appropriate when describing numerous mechanical, chemical, environmental, biological, climatic, and economic systems. We will illustrate specific applications to mechanical oscillators and to climatic time series data in Section 15.7. The input time series can take a variety of forms. If it is periodic, a Poincaré section reduces the system to an autonomous mapping problem. In many applications the input time series is essentially random. This is the typical case in vibration tests performed on mechanical systems. Then, if only the output time series were observed, it would also be a random time series, with possibly some nonlinear structure. However, if both the input and output time series are observed, we describe in Section 15.5.2 how it may be possible to build a purely deterministic model for the *pair* of time series, which consequently gives rise to more-accurate short-term forecasts.

There are many forms other than periodic or random inputs that arise in applications. For example, the input may be quasiperiodic (see Section 15.7.4), or chaotic. Hübler has proposed to use inputs that are chaotic transients to obtain optimal transfer of energy between the input and output [70]. In the case that the input is a stationary (i.e., ergodic) time series, it is possible to define a notion of chaos for input–output systems that describes the sensitivity to perturbations in the initial condition $s(0)$, assuming that the entire input time series is held constant [30, 73] This definition of chaos is useful in understanding when long-term prediction of the output time series is possible, assuming the input time series is constantly monitored or controlled.

15.5.2. State Space Reconstruction and Modeling

In a manner analogous to that used for the autonomous case, we may attempt to model a driven system based on a delay coordinate reconstruction using both input and output values, according to

$$x(t) = P\big(x(t-\tau), x(t-2\tau), \ldots, x(t-m\tau),$$
$$u(t), u(t-\tau), \ldots, u(t-(l-1)\tau)\big) \qquad (15.20)$$

The reconstruction now explicitly includes l delays of the input in addition to m delays of the output. It can be shown that, if the inputs occur at discrete* time intervals spaced τ apart, then an extension of Takens' embedding theorem holds [30]. It follows that generically there exists a differentiable function P satisfying (15.20) if $m \geq 2d + 1$ and $l \geq 2d + 1$.

Just as in the autonomous case, a number of functional forms have been suggested in the formulation of input–output models in (15.20). These include polynomials [14], radial basis functions [72], local or threshold polynomial models [82], local linear models [72], and local splines [55]. In addition to the standard linear model forms, we have experimented with polynomials, radial basis functions, and local linear models. Of these forms, our best results have been obtained with local linear models. In our experience, global polynomial model forms have limited utility for iterated time series predictions because they tend to become unstable under repeated iteration.

Optimal reconstruction techniques for autonomous models have been investigated extensively. Such extensive investigation of reconstruction has not, to our knowledge, been applied to models of driven systems. One exception is canonical variate analysis, which has also been proposed as a means of optimal reconstruction for driven systems [87].

15.6. COMPARISON TO TRADITIONAL STOCHASTIC NONLINEAR MODELING

Linear stochastic modeling has a long history, going back at least to Yule [163] and more recently, Box and Jenkins [15]. Over the last 10 years, nonlinear stochastic modeling has become a growing discipline; for a review see the book by Tong [154]. The stochastic nonlinear modeling approach to

---

*If the inputs occur continuously in time, then (15.20) only holds approximately, with an accuracy determined by how much the observations $u(t)$, $u(t-\tau), \ldots, u(t-(l-1)\tau)$ constrain u over the entire interval $[t, t-(l-1)\tau]$.

time series analysis has, at first sight, much in common with deterministic nonlinear modeling. Rather than describe the stochastic approach, we will instead list the major contrasts to the deterministic approach.

1. *Stochastic versus Deterministic modeling*. Most statisticians believe that naturally occurring time series that are irregular have a substantial stochastic component, in addition to any nonlinearity that may be present. The alternative view is that most of the irregular component is due to chaos, so that the process can be made largely deterministic by fitting the correct model (assuming the chaos is of low-enough dimension). Because statisticians believe that most of the irregular behavior is stochastic, they tend to use far fewer parameters than one would with the alternative point of view. This avoids the problem of overfitting and is certainly a sensible approach if the irregularity in the time series is predominantly due to stochastic or high-dimensional effects. However, this approach can give much less accurate forecasts if there is only a small stochastic component and the behavior is of low-enough dimension.* Evidently there is much unexplored middle ground between the two approaches, which is presently beginning to be investigated by both communities.

2. *Data Sets*. The data sets analyzed by statisticians have typically been more limited in size and in scope. With a large degree of stochastic behavior, models eventually saturate, so that increasing the number of data points only gives a small increase in the quality of the model. As evidenced by the scaling laws [see (15.12)] nonlinear models for chaotic behavior get increasingly better as the number of points is increased. Thus there is a large incentive to use longer data sets. For many physical systems, it is not difficult to obtain long data sets (for instance, see Sections 15.7.1 and 15.7.3). Recently, a variety of new time series have been analyzed with nonlinear methods, many of them motivated by the search for chaotic behavior.

3. *State Space Reconstruction*. This has not been considered in detail by statisticians, who often use raw delay coordinates, with the lag time τ chosen at the sampling rate. However, as described in Section 15.2, there can be practical advantages to other methods of state space reconstruction. Also there are very interesting theoretical questions that present themselves clearly in the low-noise limit, which have naturally been bypassed by statisticians because they have been less interested in this limit.

---

*Many pseudorandom number generators rely on low-dimensional chaos to appear random. However, in this case deterministic modeling fails to give good short-term predictions, due to the highly discontinuous nature of the underlying dynamical system. This is not expected to be the case in naturally occurring systems.

15.7. APPLICATIONS

15.7.1. Fluid Flows

The original Landau scenario [83, 84] for the onset of fluid turbulence held that the irregular motion was due to an increasing number of linearly unstable modes. This notion was challenged by Ruelle and Takens [118], who argued that as few as three unstable modes, interacting nonlinearly, could describe the chaotic motion of the fluid. This picture was confirmed by Gollub and Swinney [60], who observed a low-dimensional strange attractor in an experiment with rotating fluid between two concentric cylinders.

Near-transition fluid flows are thus a natural application for nonlinear modeling. Data from an experiment involving convective flow between two flat plates provided one of the first real tests for the techniques of nonlinear prediction described here; in this case, predictions were obtained that were 40 times better than those of standard linear methods [48].

15.7.2. Sunspots

Since its discovery in 1700, the sunspot cycle has attracted perhaps more attention than any other time series, due to its interesting mixture of regularity and irregularity (see Figure 15.5). The conventional measure of sunspot activity is the Wolfer number, a quantity that reflects both the number of sunspot groups and the number of individual sunspots, while compensating for varying atmospheric conditions that affect observability. Maxima in activity appear to occur approximately every 11 years, but the magnitudes of the maxima are irregular. Sunspot activity is an indicator of general solar activity, including output of shortwave radiation and energetic particles. These phenomena strongly influence conditions in the upper atmosphere and interplanetary medium and, consequently, have a direct bearing on the lifetimes of satellites in low earth orbit and the frequency of dangerous bursts of radiation likely to be encountered by astronauts. Reliable advance knowledge of sunspot activity is therefore extremely important.

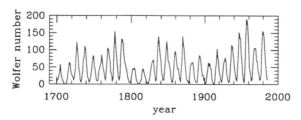

Figure 15.5. Yearly sunspot number for the last three centuries. Note the 11-year cycle with varying amplitude. Also note the difference between the rising and falling slopes—a time asymmetry that cannot be captured by a linear model.

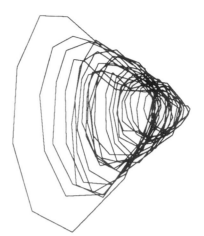

Figure 15.6. A three-dimensional phase plot of the sunspot series in coordinates of principal components. This view suggests structure similar to that of the Rössler attractor.

Stochastic linear models were developed by Yule [163]. These were found to be inadequate due to the strong time asymmetry in the time series, which cannot be captured by a linear Gaussian model. Stochastic nonlinear models were developed by Tong and Lim [160] to model this aspect of the time series, essentially by a noisy limit cycle, and better forecasts were obtained. Not surprisingly, predictions obtained from these statistical methods are superior to those obtained from first-principles models, due to the difficulty in accurately modeling the physics of the interior of the Sun. More recently, the sunspot time series has been modeled by several others, using models that are motivated by the notion that the time series may be chaotic [38, 81, 158]. Mundt, Maguire, and Chase [102] have done a particularly detailed study using a variety of techniques.

Further evidence of nonlinearity can be seen in a three-dimensional phase portrait, shown in Figure 15.6. The basic shape of the phase portrait is conical. The 11-year period is associated with uniform rotation about the cone's axis. The amplitude of the cycle is related to the radius of the rotation. The tendency for some trajectories to break off the cone and shoot toward the vertex suggests that the dynamics is nonlinear. In fact, the geometry of this attractor is reminiscent of Rössler's funnel.

We have modeled the sunspot activity by fitting deterministic, local linear models. This is to explore the possibility that the irregularities in the sunspots are a manifestation of low-dimensional chaos. We have been careful to compare our results to those of the stochastic models mentioned previously. We fit a model to the 232 Wolfer sunspot numbers in the years 1716–1948. Forecasts were made one year ahead on the remaining data. In Figure 15.7 the error index is plotted against the number of neighbors used in a three-dimensional local linear model, with time lag two years. In this case a local linear model with 220 neighbors is essentially equivalent to a stochastic linear model. A medium number of nearest neighbors indicates a stochastic

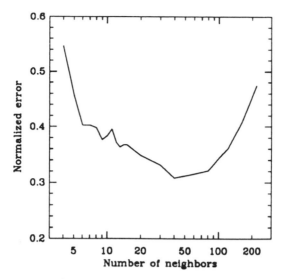

Figure 15.7. Average error for one-step-ahead local linear predictions versus the number of neighbors used for function approximation. The right-hand edge, at 220 neighbors, is equivalent to a linear stochastic model. The embedding dimension was $m = 3$.

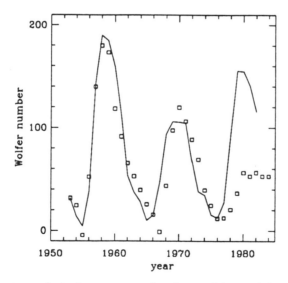

Figure 15.8. Forecasts (□) of sunspot number iterated forward from 1950, and the true time series. Prediction accuracy degrades for long predictions; hence the forecasts are better around 1960 than they are 20 years later in 1980. Note that this nonlinear model reproduces the time asymmetry of the data.

nonlinear model (Tong and Lim used essentially about 100 neighbors). A small number of nearest neighbors indicates a deterministic model. The optimal choice was found to be 40 nearest neighbors, which gives about a 50% improvement over linear models and a slight improvement over Tong's model. It is not clear what statistical significance to attach to this result, because predictions were only compared over two cycles.

The model was also iterated to obtain multistep forecasts, as shown in Figure 15.8. The boxes represent predictions, the solid lines observed data. Note that the rising and descending slopes of the forecasts are different, a quality of sunspot cycles that global linear forecasts are unable to reproduce. For reasons we do not understand, to obtain optimal results in Figure 15.8 required only 10 neighbors.

15.7.3. Mechanical Vibrations

Mechanical vibrations provide an ideal testbed for nonlinear input–output modeling approaches that were discussed in Section 15.5. The input is a force or acceleration, which may be a single pulse (like an earthquake), an irregular oscillation (like ocean waves), or a continuous random signal. . . typical experimental situation in structural mechanics involves vibrating a structure (perhaps a space station or a model of a skyscraper) with a known input and measuring the response. One would like to have the ability to predict the detailed response of the system from a known input and also to be able to predict long-term statistical averages, such as the average—or maximum—amplitude of the response (is it enough to make the structure break?) or its power spectrum.

In this section, we discuss input–output modeling for two experimental situations: the first is an analog Duffing oscillator that describes in a generic way the kind of mechanical systems that are generally studied; the second is a mechanical beam moving in a double potential well. Additional examples, including applications to a heat exchanger system [14], ship rolling [82], and a van der Pol oscillator [82] have been reported in the literature.

15.7.3.1. Analog Duffing Oscillator. As a generic example of a system driven by an input acceleration, we consider the following Duffing-like* equation:

$$y'' + 2\zeta\omega_n(y' - y_0') + \omega_n^2(y - y_0)$$

$$+ \alpha\omega_n^2(y - y_0)^2 + \beta\omega_n^2(y - y_0)|y - y_0| = 0 \qquad (15.21)$$

---

*The Duffing oscillator is usually written as a differential equation with a cubic term. Here, to simplify the analog circuitry, we use $y|y|$ instead of y^3, which although not cubic does preserve the antisymmetric nature of the cubic nonlinearity.

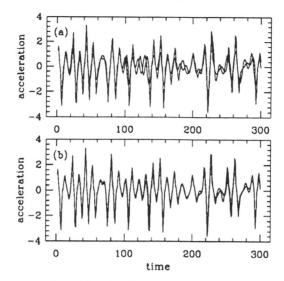

Figure 15.9. Response time histories for a strongly nonlinear Duffing oscillator: (*a*) linear model; (*b*) nonlinear model. In both graphs, the measured and predicted responses are superimposed. Both models provide reasonable approximation to the response even for long times, although the nonlinear model is clearly superior.

The input is the acceleration* y_0'' and the response is y. Here, ω_n is the natural frequency of the linear restoring force, ζ is the damping coefficient, and α (resp., β) is the coefficient of the antisymmetric (resp., symmetric) nonlinear restoring force. The behavior exhibited by a driven Duffing system is characteristic of many of the structural nonlinearities encountered in practice.

The equation is simulated on an analog computer, with parameter values $\alpha = 3500$, $\beta = 3500$, $\zeta = 0.04$, and $\omega_n = 2\pi(11.5)$, so that the natural frequency in the absence of nonlinearity is 11.5 Hz. The input acceleration y_0'' is Gaussian random noise band limited in the 5- to 15-Hz range, with a drive level of 0.405 V. Substantial second- and third-harmonic responses alert us to the presence of a significantly nonlinear system; in fact, the magnitude of the nonlinear terms substantially exceeds that of the linear terms.

The system input and response are digitized and are modeled according to the methods described in Section 15.5. Figure 15.9 compares the iterated predictions and measured responses of linear and nonlinear time series models for this case of the strongly nonlinear driven Duffing oscillator. The superiority of the nonlinear (local linear) model is evident. For this strongly nonlinear system the one-step prediction error (rms error normalized by the

---

*Note, although the term y_0'' does not explicitly appear in (15.21), integration of y_0'' determines y_0' and y_0, and these do appear.

response) for the linear model is 0.023 and for the nonlinear model is 0.014. The average iterated prediction error is computed over a range of 300 samples (about 2.5 s) and found to be 0.62 for the linear model and 0.32 for the nonlinear (in this case, a local linear) model.

For autonomous chaotic systems, even approximate long-term prediction is usually impossible; however, for input–output systems, depending on how directly the response depends on the input, iterated predictions can sometimes give quite reasonable results. The example in the next section is a case for which long-term prediction of the system response is not possible.

15.7.3.2. Beam Moving in a Double Potential Well. As our next illustration of the application of nonlinear time series analysis to driven systems, we choose a system that is clearly exhibiting chaotic behavior, the "chaos beam" described by Moon [101]. In this example, discussed in detail by Hunter [73], an experimental oscillator was built that exhibited chaotic transitions between two stable states. Measured acceleration and strain data from this system constitute the input and output time series.

Typical predicted and measured response time series for this system are shown in Figure 15.10. Unlike the case in the previous section, chaotic behavior of the system leads the predicted and measured response to rapidly diverge. The iterated prediction error is essentially unity.

However, for this system, long-term iterated forecast error provides a poor measure of model validity. Even for a very good model, small errors will grow

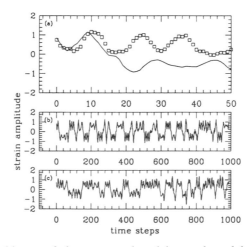

Figure 5.10. Time history of the measured and iterated model predictions for the chaos beam: (*a*) iterated predictions (□), which quickly diverge from the actual measured time history (—); (*b*) long-time behavior of the actual time series; (*c*) long-time behavior of the iterated predictions. Note that although the system does not permit long-term prediction of the time series itself, qualitative aspects of the long-time behavior are preserved.

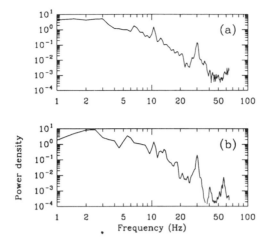

Figure 15.11. (*a*) Power spectrum of the measured time history of the chaos beam; (*b*) power spectrum of the iterated model. Note that the model is able to predict the peaks at 6.5, 10.5, and 30 Hz, as well as being able to describe the general shape and magnitude of the power spectrum.

exponentially in time. For example, the model will be incapable of predicting which well the beam will be in at some time far in the future. However, part of what is desired from an input–output model is knowledge of how the response will depend on the input in terms of various statistical averages. Figure 15.11 shows the power spectra of iterated and measured response time series. Although the model is not good at making long-term forecasts of the specific time history response, it does successfully predict the power spectrum.

15.7.4. Ice Ages

As another example of the nonlinear modeling of input–output time series, we consider an application involving long-term climatic data, specifically the long-term, rather complex variation of global ice volume. The modeling of a geologically derived time series presents a difficult problem, because geologic time series are often noisy, and event timing may be uncertain.

The total solar energy incident on the upper atmosphere over a 24-hour period varies with time of day, latitude, and season. Changes in the tilt of the Earth's axis (obliquity), of the season in which the Northern hemisphere is tilted most away from the Sun (precession), and in the shape of the Earth's orbit (eccentricity) change the effective solar energy received in a given month at a given location. Over millennia the energy received from the Sun, termed solar insolation, varies about 20% at latitudes above 45° North. Figure 15.12*a* shows the time series of solar insolation at 60° North latitude,

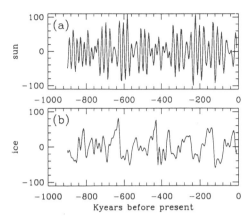

Figure 15.12. (*a*) Relative July solar insolation at 60° north latitude; (*b*) relative global ice volume based on oxygen isotope analysis of ocean core V28-239.

as calculated from celestial mechanics [13]. Figure 15.12*b* shows, for comparison, an estimate of the variation in global ice volume derived from an oxygen isotope analysis of bottom-dwelling organisms in ocean sediment cores.

In 1941, Milkanovich postulated that the summer insolation at relatively high northern latitudes was a determining factor in the evolution of the ice ages. Winters at these latitudes are always cold enough for snow to accumulate. A critical factor in the development of ice sheets is summer melting, as consistent incomplete summer melting leads to ice sheet growth. Recently Imbrie and co-workers [75, 74] showed that common frequency components with periods characteristic of precession, whose period is 21,000 years, and obliquity, whose period is 41,000 years, are strongly present in the ice volume time series illustrated in Figure 15.12*b*. They also pointed out that the dominant component in the ice volume time series, whose period is 95,000 years, is absent from the solar insolation time series.

Although components with a 95,000-year period are absent from the insolation time series, they are present in the envelope of the series. A dominant variation in eccentricity, with a period of about 100,000 years, modulates the insolation signal, causing regions of low variance to occur about every 100,000 years. It is our contention that regions of low variance in the insolation signal correspond to glacial periods.

Numerous climatic models have been proposed to explain the complex variation in ice volume and to link the variation in ice volume to that of solar insolation. These include models by Imbrie and Imbrie [74], Suarez and Held [141], and Pollard [110]. These models generally use solar insolation as an input and invoke nonlinear processes involved in the growth and decay of ice sheets to derive nonlinear differential equations governing the process. The models have met with varying degrees of success. In particular, the model by

Imbrie and Imbrie, based on a first-order nonlinear differential equation, gives a reasonable simulation of the two most recent ice ages.

Using our modeling process, we apply the methods of Section 15.5.2 to the insolation and ice volume time series. The insolation time series is used directly. A low-pass filter is applied to the ice volume time series to remove noisy signals whose period is less than 14,000 years. In addition, there are uncertainties in the time scale for the ice volume data. The fundamental time scale for ice volume is derived by assuming that depth is proportional to time. Corrections for magnetic reversal boundaries detected in the ocean cores are made. In addition some "tuning" of the time scale has been done by phase locking components of eccentricity, precession, or obliquity with corresponding components of the ice volume time series. We have studied data corrected for stage boundaries, for eccentricity, and also used a correction developed by Hays [67] which is based on phase locking corresponding components of obliquity and precession in the ice volume and insolation time series.

A systematic search of possible nonlinear and linear models was made for data from two ocean cores, V28-238 and V28-239. A cross-validation procedure based on successively selecting segments of the ice volume time series for prediction, while training on the remainder of the series, was used. Iterated predictions over a period of about 90,000 years were made, sufficient to predict about one ice age into the future. These results were compared to those obtained using simulated insolation and ice volume time series whose power spectra were identical to those of the actual time series but with randomized phases. Nonlinear models (in a few cases, linear time series models) gave results whose average iterated prediction errors were between 3 and 5 σ below those obtained for the simulated time series. The best models were nonlinear. Twenty lagged values of solar insolation (about 60,000 years) were used for the model input and between 4 and 20 lagged ice volume values (12,000 to 20,000 years) as lagged response values. Local singular-value decomposition, as discussed in Section 15.5.2, indicated that the solar-insolation–ice-volume system has approximately three significant state variables.

Because solar insolation data are available for the future, nonlinear forecasting techniques can be applied to predict the next ice age. For this prediction eccentricity-tuned data were used from core V28-239. Except for the last 60,000 years, the entire past insolation and ice volume data sets were used for training. In Figure 15.13, measured and predicted ice volume for the last 150,000 years are compared. Continued iteration of the model is then used to predict a future peak in global ice volume, occurring, as shown in Figure 15.13, in about 40,000 years.

15.7.5. Measles

Sugihara and May [143] recently analyzed data from measles and chicken pox epidemics in New York City. They used monthly measles data from 1928

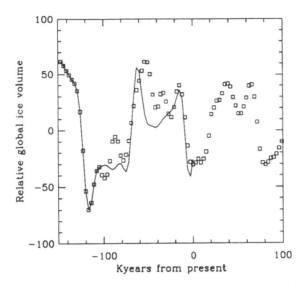

Figure 15.13. Prediction of the next ice age, assuming no human-induced effects. Shown are the ice age data up to the present (—) and the iterated predictions (□), starting from about 150,000 years in the past (so that comparisons with the available data can be made) and continuing into the future.

until 1963, when immunization began to alter the dynamics of the data series, and monthly chicken pox data from 1949 to 1972. Both data sets show strong periodic annual cycles, with additional fluctuations. There has been some controversy in the early nonlinear dynamics literature as to whether there is a low-dimensional chaotic attractor in the measles data or whether it is better represented by a limit cycle with noise. Schaffer and others [124–126] analyzed the data using calculations of dimension and Lyapunov exponents. Their results consistently indicated the presence of a chaotic attractor with a dimension between 2 and 3. They also produced pictures of a reconstructed attractor that were strongly reminiscent of the Rössler attractor. However, a linear analysis by Schwartz [129] seemed to indicate that the data were merely a limit cycle with additive white noise. Neither hypothesis could be completely rejected. Even though it was known that chaotic time series can fool all the conventional linear tests for white noise, the calculation of single-number statistics such as the dimension appeared somewhat suspect, especially given the small size of the data set (432 points). Although the pictures of the reconstructed attractor were strongly suggestive, they could not be subjected to any decisive statistical tests. Using a nonlinear predictive model based on dividing the reconstructed data into simplices and using a kernel-density estimator, Sugihara and May developed a test to distinguish between the two cases. They divided the data set into two parts, constructing a simple nonlinear model from the first half and then running iterated forecasts over the second half, comparing predicted with actual values. The signature of a chaotic time series, they argued, is that the prediction accuracy falls off with increasing prediction time (at a rate proportional to the largest Lyapunov exponent), whereas for a noisy limit cycle, the prediction accuracy should be roughly independent of the prediction time.

The results can be further corroborated by comparing the accuracy of predictions obtained from the nonlinear model with predictions obtained from a linear model. If the data are chaotic, the nonlinear predictions should be significantly better.

Their paper showed the results of careful analyses of both artificially constructed test data sets and of the real-world measles and chicken pox data. Their method worked well at distinguishing between chaos and limit cycles with additive noise in the artificially constructed data sets. The measles data showed a strong falloff in prediction accuracy with increasing prediction interval, and the calculation of the optimal embedding dimension of between 5 and 7 was in agreement with Schaffer's calculation of a Grassberger–Procaccia dimension of between 2 and 3. The nonlinear forecasts also appeared to be significantly better (with a correlation between predicted and observed values of 0.85) than the linear autoregressive forecasts (with a correlation of 0.72). The chicken pox data, on the other hand, appeared to be a noisy limit cycle. The prediction accuracy was roughly independent of the prediction interval, and the accuracy of the nonlinear model (correlation of predicted with actual values of 0.82) was slightly worse than the accuracy of the linear autoregressive model (correlation of 0.84).

There is some a priori biological justification of the difference between the measles and chicken pox data. The rapidly reproducing and highly infectious nature of the measles virus tends to produce long-lasting interepidemic oscillations, and this is probably responsible for the lively dynamics observed. Chicken pox, on the other hand, has a much lower basic reproductive rate, so it may be less likely to show any regularities in the dynamics other than basic seasonal ones associated with schools opening and closing. Also, any fine details in the dynamics of the chicken pox data may be overwhelmed by sampling error, because the reporting of cases was not compulsory for the time period in question.

These findings for New York City measles were later confirmed by similar results in the United Kingdom, where it was found that nonlinearities were difficult to detect in highly aggregated data for the country as a whole, but on a city-by-city basis, the evidence for chaos was stronger [142].

On the other hand, Ellner [44] has criticized this approach and emphasized the need for comparison of raw data against plausible alternatives. In particular, he has exhibited stochastic models that show the same features as the presumably chaotic measles data. He also notes that the results can be sensitive to how the data are transformed; for instance, a time series of the logarithm of number of measles cases shows no decrease in prediction error versus prediction time, suggesting that the data follows a limit cycle with noise.

15.7.6. Speech

Although long-term prediction of speech is an absurdity, the problem of short-term prediction is strongly related to the commercially important problem of coding and transmission. Any predictive description of speech will reduce the range of possible future acoustic waveforms, given some past knowledge of the waveform. This reduces the effective entropy of the speech signal, so that it may be encoded and transmitted more efficiently. Several authors have analyzed speech data using techniques motivated by chaotic dynamics (see references [80] and [152]). In this section we review work of Townshend, who recently looked into the problem of short-term prediction of speech using nonlinear models [156, 157].

Linear models have been used successfully to predict speech for years. The basic idea is that speech is the product of a source, the glottis, exciting a set of linear passive resonators, the vocal tract. Linear predictive coding models these resonances by using an all-pole filter, that is, by finding the parameters a_i in the relation

$$x(t) = \sum_{i=1}^{m} a_i x(t - i) + e(t) \tag{15.22}$$

which minimize the root-mean-square average of the residuals $e(t)$. Because the resonances are not fixed, but change slowly, the predictor parameters are usually updated at a commensurate rate, typically every 10–20 ms. To be useful in coding, only a limited number of parameters can be fit. Because most of the prediction gain is obtained with the first 12 parameters, these are usually the only ones used.

For comparison, Townshend used a local linear predictor based on global and local principal value reconstructions (see Section 15.2.2). He then tested the two predictors on data from a single speaker with a sampling rate of 8 kHz. A 30-s section of speech was used to fit the nonlinear model, then it was tested on a 2-s sentence. The effectiveness of the method was measured in terms of the difference between the energy of the original signal and the energy of the residual after prediction. For the nonlinear predictor, this prediction gain was 9.2 dB (in other words, the residual amplitude was 35% of the signal amplitude). The LPC model actually did significantly better. Even a fixed LPC model, where the parameters were set with the 30-s segment and then tested, unchanged, on the 2-s segment, did significantly better with a prediction gain of 11.1 dB (or 28% residual to signal amplitude). So it can be concluded that the dominant predictable component of speech is due to linear effects.

Townshend then attempted to use the two predictors in tandem, applying the nonlinear predictor to the residual of the linear predictor, and obtained a prediction gain of 2.8 dB over the linear predictor alone. This result indicates that speech has significant nonlinear components, and it may ultimately prove to be of commercial use for speech encoding or for noise reduction.

15.7.7. Other Applications

There is a growing literature on the search for nonlinearity in economic data. Although the results are often inconclusive (see references [19]–[21], [112], and [127]), some recent results which focus on determining special regions of the state space where significant nonlinearity resides have been reported by Packard and co-workers [105] and LeBaron [88].

Lorenz [92] attempted spatiotemporal modeling of the weather by searching for close neighbors in cloud patterns; the attempt failed, presumably due to the high dimensionality of the weather. Nonlinear prediction was recently used by Serre and co-workers to test for the presence of a chaotic attractor in the observed output of a variable white dwarf star [130]. The time series consisted of observations for a single night, comprising 28 periods of the star. The results were suggestive of low-dimensional chaos, in that they were able to get very good prediction results using a local linear predictor. However, a more comprehensive analysis is still needed, including quantitative comparison to a linear model, and plots of prediction accuracy versus size of neighborhood, as we did for the sunspot series. Hudson et al. [71] successfully predicted bifurcations in experiments on fluid dynamics using nonlinear modeling techniques. Jones et al. [76] have analyzed a wide variety of time series using nonlinear techniques.

ACKNOWLEDGMENTS

We are grateful to Xiangdong He and David Wolpert for useful conversations and for help with the references. This work was performed under the auspices of the United States Department of Energy, with partial support from the National Institute of Mental Health Grant 1-R01-MH47184-01. Finally, we thank Jong Kim and John Stringer for organizing this informative workshop and for their encouragement and patience in awaiting our manuscript.

REFERENCES

1. H. D. I. Abarbanel, R. Brown, and M. B. Kennel, Lyapunov exponents in chaotic systems: Their importance and their evaluation using observed data. *Int. J. Mod. Phys. B* **5** 1347–1375 (1991).

2. H. D. I. Abarbanel and J. B. Kadtke. Information theoretic methods for choosing the minimum embedding dimension for strange attractors. UCSD Institute for Nonlinear Science, 1989.

3. N. B. Abraham, A. M. Albano, A. Passamante, and P. E. Rapp (eds.), *Measures of Complexity and Chaos. NATO Advanced Science Institute Series*, vol. 208. Plenum, New York, 1989.

4. H. Akaike, On an entropy maximization principle. In *Applications of Statistics* (P. Krishnaiah, ed.). North-Holland, Amsterdam, 1977.

5. P. Atten, J. G. Caputo, B. Malraison, and Y. Gagne, Détermination de dimension d'attracteurs pour différents écoulements. *J. de Mécanique Théor. Appl.* (Numéro spécial) 133 (1984).

6. R. Badii, G. Broggi, B. Derighetti, M. Ravani, S. Ciliberto, A. Politi, and M. A. Rubio, Dimension increase in filtered chaotic signals. *Phys. Rev. Lett.* **60** 979 (1988).

7. R. Badii and A. Politi, Intrinsic oscillations in measuring the fractal dimensions. *Phys. Lett. A* **104** 303 (1984).

8. R. Badii and A. Politi, Statistical description of chaotic attractors: The dimension function. *J. Statist. Phys.* **40** 725 (1985).

9. M. F. Barnsley and S. Demko, Iterated function systems and the global construction of fractals. *Proc. Roy. Soc. London Ser. A* **399** 243 (1985).

10. A. Ben-Mizrachi, I. Procaccia, and P. Grassberger, The characterization of experimental (noisy) strange attractors. *Phys. Rev. A* **29** 975 (1984).

11. J. H. Bentley, Multidimensional binary search trees used for associative searching. *Comm. ACM* **18**(9) 509–517 (1975).

12. J. H. Bentley, Multidimensional binary search trees in database applications. *IEEE Trans. Software Engineering* **SE-5** 333–340 (1979).

13. A. L. Berger, Obliquity and precision for the last 5,000,000 years. *Astron. Astrophys.* **51**(1) 127–135 (1976).

14. S. A. Billings, K. M. Tsang, and G. R. Tomlinson, Application of the narmax method to nonlinear frequency response estimation. Proceedings of the 6th International IMAC Conference, 1987.

15. G. E. P. Box and G. M. Jenkins, *Time Series Analysis Forecasting and Control.* Holden-Day, San Francisco, 1970.

16. J. L. Breeden and A. Hübler, Reconstructing equations of motion from experimental data with unobserved variables. *Phys. Rev. A* **42** 5817–5826 (1990).

17. L. Breiman and J. H. Friedman, Estimating optimal transformations for multiple regression and correlation. *J. Am. Statist. Assoc.* **80** 580 (1985).

18. W. A. Brock, Distinguishing random and deterministic systems. *J. Econ. Theory* **40** 168–195 (1986).

19. W. A. Brock, Causality, chaos, explanation and prediction in economics and finance. In *Beyond Relief: Randomness, Prediction and Explanation in Science*

(J. Casati and A. Karlquist, eds.). CRC Press, Boca Raton, 1991.

20. W. A. Brock and W. D. Dechert, A general class of specification tests: The scalar case. Proceedings of the Business and Economic Statistics Section of the American Statistical Association, 1988.

21. W. A. Brock and W. D. Dechert, Statistical inference theory for measures of complexity in chaos theory and nonlinear science. In *Measures of Complexity and Chaos* (N. B. Abraham, A. M. Albano, A. Passamante, and P. E. Rapp, eds.). Plenum, New York, 1989.

22. D. S. Broomhead, R. Indik, A. C. Newell, and D. A. Rand, Local adaptive Galerkin bases for large dimensional dynamical systems. *Nonlinearity* **4** 159–198 (1991).

23. D. S. Broomhead, R. Jones, and G. P. King, Topological dimension and local coordinates from time series data. *J. Phys. A* **20** L563–L569 (1987).

24. D. S. Broomhead and G. P. King, Extracting qualitative dynamics from experimental data. *Physica D* **20** 217 (1987).

25. D. S. Broomhead and D. Lowe, Multivariable functional interpolation and adaptive networks. *Complex Systems* **2** 321–355 (1988).

26. R. Brown, P. Bryant, and H. D. I. Abarbanel, Computing the Lyapunov spectrum of a dynamical system from an observed time series. *Phys. Rev. A* **43** 2787 (1991).

27. J. G. Caputo, Practical remarks on the estimation of dimension and entropy from experimental data. In *Measures of Complexity and Chaos* (N. B. Abraham, A. M. Albano, A. Passamante, and P. E. Rapp, eds.), pp. 99–110. Plenum, New York, 1989.

28. M. Casdagli, Nonlinear prediction of chaotic time series. *Physica D* **35** 335–356 (1989).

29. M. Casdagli, Chaos and deterministic versus stochastic nonlinear modelling. *J. Roy. Statist. Soc.* **54** 303–328 (1992).

30. M. Casdagli, A dynamical systems approach to modeling driven systems. In *Nonlinear Modeling and Forecasting* (M. Casdagli and S. Eubank, eds.), pp. 265–282. Addison-Wesley, Reading, MA, 1992.

31. M. Casdagli and S. Eubank (eds.), *Nonlinear Modeling and Forecasting*. Addison-Wesley, Reading, MA, 1992.

32. W. E. Caswell and J. A. Yorke, Invisible errors in dimension calculations: Geometric and systematic effects. In *Dimensions and Entropies in Chaotic Systems—Quantification of Complex Behavior. Springer Series in Synergetics*, vol. 32, p. 123. Springer-Verlag, Berlin, 1986.

33. A. Cenys and K. Pyragas, Estimation of the number of degrees of freedom from chaotic time series. *Phys. Lett. A* **129** 227 (1988).

34. J. D. Cowan and D. H. Sharp, Neural nets. *Quart. Rev. Biophys.* **21** 1988.

35. J. Cremers and A. Hübler, Construction of differential equations from experimental data. *Z. Naturforsch.* **42a** 797–802 (1987).

36. J. P. Crutchfield, J. D. Farmer, N. H. Packard, and R. S. Shaw, Chaos. *Scientific American* **254**(12) 46–57 (1986).

37. J. P. Crutchfield and B. S. McNamara, Equations of motion from a data series. *Complex Systems* **1** 417–452 (1987).

38. D. Currie, personal communication.

39. M. Denker and G. Keller, Rigorous statistical procedures for data from dynamical systems. *J. Statist. Phys.* **44** 67–93 (1986).

40. J. Deppisch, H.-U. Bauer, and T. Geisel, Hierarchical training of neural networks and prediction of chaotic time series. Unpublished, 1990.

41. J. P. Eckmann, S. Kamporst, and D. Ruelle, Recurrence plots of dynamical systems. *Europhys. Lett.* **4** 973–977 (1987).

42. J.-P. Eckmann and D. Ruelle, Ergodic theory of chaos and strange attractors. *Rev. Mod. Phys.* **57** 617 (1985).

43. B. Efron and R. Tibshirani, Bootstrap methods for standard errors, confidence intervals, and other measures of statistical accuracy. *Statist. Sci.* **1** 54–77 (1986).

44. S. Ellner, Detecting low-dimensional chaos in population dynamics data: A critical review. In *Does Chaos Exist in Ecological Systems* (J. Logan and F. H., eds.). University of Virginia Press, 1991.

45. S. Eubank and J. D. Farmer, An introduction to chaos and prediction. In *Proceedings of the Santa Fe Institute Summer School* (E. Jen, ed.), pp. 75–190. Addison-Wesley, Reading, MA, 1990.

46. S. J. Farlow, *Self-Organizing Methods in Modeling—GMDH Type Algorithms.* Marcel Dekker, New York, 1984.

47. J. D. Farmer, A rosetta stone for connectionism. *Physica D* **42** 153–187 (1990).

48. J. D. Farmer and J. J. Sidorowich, Predicting chaotic time series. *Phys. Rev. Lett.* **59** 845–848 (1987).

49. J. D. Farmer and J. J. Sidorowich, Exploiting chaos to predict the future and reduce noise. In *Evolution, Learning and Cognition* (Y. C. Lee, ed.), pp. 277–330. World Scientific, Singapore, 1988.

50. J. D. Farmer and J. J. Sidorowich, Optimal shadowing and noise reduction. *Physica D* **47** 373–392 (1991).

51. D. F. Findley, On the unbiased property of aic for exact or approximating linear stochastic time series models. *J. Time Series Anal.* **6** (1985).

52. A. M. Fraser, Information and entropy in strange attractors. *IEEE Trans. Information Theory* **IT-35** (1989).

53. A. M. Fraser, Reconstructing attractors from scalar time series: A comparison of singular system and redundancy criteria. *Physica D* **34** 391–404 (1989).

54. A. M. Fraser and H. L. Swinney, Independent coordinates for strange attractors from mutual information. *Phys. Rev. A* **33** 1134–1140 (1986).

55. J. H. Friedman, Multivariate adaptive regression splines. Technical Report 102, Department of Statistics, Stanford University, 1988.

56. H. Froehling, J. P. Crutchfield, J. D. Farmer, N. H. Packard, and R. S. Shaw, On determining the dimension of chaotic flows. *Physica D* **3** 605 (1981).

57. D. Gabor, Communication theory and cybernetics. *Trans. Inst. Radio Engineers* **CT-1**(4) 9 (1954).

58. N. Gershenfeld, An experimentalist's introduction to the observation of dynamical systems. In *Directions in Chaos*, vol. 2 (Hao Bai-Lin, ed.). World Scientific, Singapore, 1988.

59. J. F. Gibson, M. Casdagli, S. Eubank, and J. D. Farmer, An analytic approach to practical state space reconstruction. *Physica D* (1992). To appear.

60. J. P. Gollub and H. L. Swinney, Onset of turbulence in a rotating fluid. *Phys. Rev. Lett.* **35** 927 (1975).

61. P. Grassberger, Do climatic attractors exist? *Nature* **323** 609 (1986).

62. P. Grassberger, Information content and predictability of lumped and distributed dynamical systems. Technical Report WU-B-87-8, University of Wuppertal, 1987.

63. P. Grassberger and I. Procaccia, Characterization of strange attractors. *Phys. Rev. Lett.* **50** 346 (1983).

64. P. Grassberger and I. Procaccia, Measuring the strangeness of strange attractors. *Physica D* **9** 189–208 (1983).

65. H. S. Greenside, A. Wolf, J. Swift, and T. Pignataro, Impracticality of a box-counting algorithm for calculating the dimensionality of strange attractors. *Phys. Rev. A* **25** 3453 (1982).

66. F. R. Hampel, P. J. Rousseeuw, E. M. Ronchetti, and W. A. Stahel, *Robust Statistics*. Wiley, New York, 1986.

67. J. D. Hays and J. J. Morley, Towards a high-resolution, global, deep sea chronology for the last 750,000 years. *Earth and Planetary Sci. Lett.* **53** 279–295 (1981).

68. X. He and A. Lapedes, Nonlinear modeling and prediction using multilayer radial basis networks. Technical report LA-UR-91-1375, Los Alamos National Laboratory 1991.

69. X. He and Z. Zhu, Nonlinear time-series modeling by self-organizing methods. Beijing University, 1990.

70. A. Hübler and E. Lusher, Resonant stimulation and control of nonlinear oscillators. *Naturwissenschaften* **76** 67 (1989).

71. J. L. Hudson, M. Kube, R. A. Adomaitis, I. G. Kevrekidis, A. S. Lapedes, and R. M. Farber, Nonlinear signal processing and system identification: Application to time series from electrochemical reactions. *Chem. Eng. Sci.* **45** 2075–2081 (1990).

72. N. F. Hunter, Analysis of nonlinear systems using delay coordinate models. In 10th Annual Modal Analysis Conference Proceedings, 1990.

73. N. F. Hunter, Application of nonlinear time-series models to driven systems. In *Nonlinear Modeling and Forecasting*, (M. Casdagli and S. Eubank, eds.), pp. 467–492. Addison-Wesley, Reading, MA, 1992.

74. J. Imbrie and J. Z. Imbrie, Modeling the climatic response to orbital variations. *Science* **207**(29) 943–952 (February 1980).

75. J. Imbrie and N. J. Shackelton, Variations in the earth's orbit: Pacemaker of the ice ages. *Science* **194** 1121–1132 (1976).

76. R. Jones, Y. C. Lee, S. Qian, C. Barnes, K. Bisset, G. Bruce, G. Flake, K. Lee, L. Lee, W. Mead, M. O'Rourke, I. Poli, and L. Thode, Nonlinear adaptive networks: A little theory, a few applications. Technical Report LA-UR-91-0273, Los Alamos National Laboratory, 1991.

77. J. L. Kaplan and J. A. Yorke, Chaotic behavior of multidimensional difference equations. In *Functional Differential Equations and Approximations of Fixed Points* (H.-O. Peitgen and H.-O. Walther, eds.). *Lecture Notes in Math.*, vol. 730, p. 204. Springer-Verlag, Berlin, 1979.

78. E. J. Kostelich and J. A. Yorke, Noise reduction in dynamical systems. *Phys. Rev. A* **38**(3) 1649 (1988).

79. John R. Koza, Genetic programming: A paradigm for genetically breeding populations of computer programs to solve problems. Stanford University Computer Science Department, 1990.

80. A. Kumar and S. Mullick, Attractor dimension, entropy and modeling of phoneme time series. Indian Institute of Technology, 1990.

81. J. Kurths and A. A. Ruzmaikin. On forecasting the sunspot numbers. *Solar Physics* **126** 407–410 (1990).

82. H.-Y. Lai, Threshold modelling of nonlinear dynamic systems. In Proceedings of the 5th International Modal Analysis Conference, p. 1487, 1987.

83. L. D. Landau, On the problem of turbulence. *Akad. Nauk. Doklady* **44** 339 (1944).

84. L. D. Landau and E. M. Lifschitz, *Fluid Mechanics*. Addison-Wesley, Reading, MA, 1959.

85. A. S. Lapedes and R. Farber, Nonlinear signal processing using neural networks: Prediction and system modeling. Technical Report LA-UR-87-2662, Los Alamos National Laboratory, 1987.

86. A. S. Lapedes and R. M. Farber, How neural nets work. In *Neural Information Processing Systems* (D. Z. Anderson, ed.). AIP Press, 1988.

87. W. E. Larimore, System identification, reduced order filtering, and modelling via canonical variate analysis. In Proceedings of 1983 American Control Conference, 1983.

88. B. LeBaron, Nonlinear forecasts for the S & P stock index. In *Nonlinear Modeling and Forecasting* (M. Casdagli and S. Eubank, eds.), pp. 381–394. Addison-Wesley, Reading, MA, 1992.

89. W. Liebert, K. Pawelzik, and H. G. Schuster, Optimal embedding of chaotic attractors from topological considerations. Unpublished, 1989.

90. W. Liebert and H. G. Schuster, Proper choice of the time delay for the analysis of chaotic time series. *Phys. Lett. A* **142** 107–111 (1988).

91. E. N. Lorenz, Deterministic nonperiodic flow. *J. Atmospheric Sci.* **20** 130–141 (1963).

92. E. N. Lorenz, Atmospheric predictability as revealed by naturally occurring analogues. *J. Atmospheric Sci.* **26** 636–646 (1969).

93. G. Mayer-Kress (ed.), *Dimensions and Entropies in Chaotic Systems*. Springer-Verlag, Berlin, 1986.

94. G. Mayer-Kress, Application of dimension algorithms to experimental chaos. In *Directions in Chaos* (Hao Bai-Lin, ed.), vol. 1. World Scientific, Cleveland, OH, 1988.

95. W. Mead, R. Jones, Y. C. Lee, C. Barnes, G. Flake, L. Lee, and M. O'Rourke. Using CNLS-net to predict the Mackey–Glass chaotic time series. In *Nonlinear Modeling and Forecasting* (M. Casdagli and S. Eubank, eds.), pp. 39–72. Addison-Wesley, Reading, MA, 1992.

96. A. I. Mees, Modelling complex systems. Math Department, University of Western Australia, 1989.

97. A. I. Mees, P. E. Rapp, and L. S. Jennings, Singular-value decomposition and embedding dimension. *Phys. Rev. A* **36** 340 (1987).

98. F. Mitschke, Acausal filters for chaotic signals. *Phys. Rev. A* **41** 1169–1171 (1990).

99. M. Möller, W. Lange, F. Mitschke, N. B. Abraham, and U. Hübner, Errors from digitizing and noise in estimating attractor dimensions. *Phys. Lett. A* **138** 176 (1989).

100. J. Moody, Fast learning in multi-resolution hierarchies. In *Advances in Neural Information Processing* (D. Touretzky, ed.), p. 29. Morgan Kaufmann, 1989.

101. F. Moon, *Chaotic Vibrations*. Wiley, New York, 1987.

102. M. Mundt, W. B. Maguire, and R. P. Chase, Chaos in the sunspot cycle: Analysis and prediction. *J. Geophys. Res.* 1705–1716 (1991).

103. M. A. H. Nerenberg and C. Essex, Correlation dimension and systematic geometric effects. *Phys. Rev. A* **42** 7065–7074 (1990).

104. A. R. Osborne and A. Provenzale, Finite correlation dimension for stochastic systems with power-law spectra. *Physica D* **35** 357–381 (1989).

105. N. Packard, A genetic algorithm for the analysis of complex data. *Complex Systems* 4 (1990).

106. N. H. Packard, J. P. Crutchfield, J. D. Farmer, and R. S. Shaw, Geometry from a time series. *Phys. Rev. Lett.* **45** 712–716 (1980).

107. T. Poggio and F. Girosi, A theory of networks for approximation and learning. In *Foundations of Neural Networks*, (C. Lau, ed.), pp. 91–106. IEEE Press, Piscataway, NJ (1992)

108. H. Poincaré, *Science et Methode*. Bibliothèque Scientifique, 1908. (English translation by F. Maitland, Dover, 1952.)

109. A. Politi, G. D'Alessandro, and A. Torcini, Fractal dimensions in coupled map lattices. In *Measures of Complexity and Chaos* (N. B. Abraham, A. M. Albano, A. Passamante, and P. E. Rapp, eds.), pp. 409–424. Plenum, 1989.

110. D. Pollard, A simple ice sheet model yields 100 kyr glacial cycles. *Nature* **296** 334–338 (1982).

111. M. J. D. Powell, Radial basis function for multivariable interpolation: A review. Technical Report, University of Cambridge, 1985.

112. D. Prescott and T. Stengos, Do asset markets overlook exploitable nonlinearities? The case of gold. University of Guelph, Ontario, 1988.

113. M. B. Priestley, State dependent models: A general approach to nonlinear time series analysis. *J. Time Series Anal.* **1** 47–71, 1980.

114. J. Rissanen, Universal modeling and coding. *IEEE Trans. Information Theory* **IT-27** 12 (1981).

115. J. Rissanen, Stochastic complexity and modeling. *Ann. Statist.* **14** 1080 (1986).

116. D. Ruelle, personal communication.

117. D. Ruelle, Deterministic chaos: The science and the fiction. *Proc. Roy. Soc. London Ser. A* **427** 241–248 (1990).

118. D. Ruelle and F. Takens, On the nature of turbulence. *Comm. Math. Phys.* **20** 167–192 (1971).

119. D. Rummelhart and J. McClelland, *Parallel Distributed Processing*, vol. 1. MIT Press, Cambridge, MA, 1986.

120. D. A. Russell, J. D. Hanson, and E. Ott, Dimension of strange attractors. *Phys. Rev. Lett.* **45** 1175 (1980).

121. M. Sano and Y. Sawada, Measurement of the Lyapunov spectrum from chaotic time series. *Phys. Rev. Lett.* **55** 1082 (1985).

122. T. Sauer, J. Yorke, and M. Casdagli, Embedology. Technical Report, University of Maryland, 1990.

123. J. D. Scargle, Studies in astronomical time series analysis. IV. Modeling chaotic and random processes with linear filters. *Astrophys. J.* **359** 469–482 (1990).

124. W. M. Schaffer, S. Ellner, and M. Kot, The effects of noise on dynamical models of ecological systems. *J. Math. Biol.* **24** 479–523 (1986).

125. W. M. Schaffer and M. Kot, Nearly one-dimensional dynamics in an epidemic. *J. Theoret. Biol.* **112** 403–427 (1985).

126. W. M. Schaffer and M. Kot, Differential systems in ecology and epidemiology. In *Chaos*, (A. V. Holden, ed.), pp. 158–178. Princeton University Press, 1986.

127. J. Scheinkman and B. LeBaron, Nonlinear dynamics and stock returns. *J. Business* **62** 311–338 (1989).

128. G. Schwartz, Estimating the dimension of a model. *Ann. Statist.* **6** 461 (1978).

129. I. Schwartz, Multiple recurrent outbreaks and predictability in seasonally forced nonlinear epidemic models. *J. Math. Biol.* **21** 347–361 (1985).

130. T. Serre, J. R. Buchler, and M. J. Goupil, Predicting white dwarf light curves. Unpublished, 1990.

131. B. W. Silverman, *Kernel Density Estimation Techniques for Statistics and Data Analysis*. Chapman Hall, London, 1986.

132. L. A. Smith, Intrinsic limits on dimension calculations. *Phys. Lett. A* **133** 283 (1988).

133. L. A. Smith, J.-D. Fournier, and E. A. Spiegel, Lacunarity and intermittency in fluid turbulence. *Phys. Lett. A* **114** 465 (1986).

134. R. L. Smith, Estimating dimension in noisy chaotic time series. *J. Roy. Statist. Soc.* **54** 329–352 (1992).

135. R. L. Smith, Optimal estimation of fractal dimension. In *Nonlinear Modeling and Forecasting* (M. Casdagli and S. Eubank, eds.), pp. 115–136. Addison-Wesley, Reading, MA, 1992.

136. K. Stokbro and D. K. Umberger, Forecasting with weighted maps. In *Nonlinear Modeling and Forecasting* (M. Casdagli and S. Eubank, eds.), pp. 73–94. Addison-Wesley, Reading, MA, 1992.

137. M. Stone, Cross-validatory choice and assessment of statistical predictions. *J. Roy. Statist. Soc. Ser. B* **36** 111–120 (1974).

138. M. Stone, An asymptotic equivalence of choice of model by cross-validation and Akaike's criterion. *J. Roy. Statist. Soc. Ser. B* **39** 44–47 (1977).

139. M. Stone, Asymptotics for and against cross-validation. *Biometrika* **64** 29–35 (1977).

140. G. Strang, Wavelets and dilation equations: A brief introduction. *SIAM Rev.* **31** 614–627 (1989).

141. Suarez and Held, Note on modelling climate response to orbital parameter variations *Nature* **263** 46–47 (1976).

142. G. Sugihara, personal communication.

143. G. Sugihara and R. May, Nonlinear forecasting as a way of distinguishing chaos from measurement error in forecasting. *Nature* **344** 734–741 (1990).

144. F. Takens, Detecting strange attractors in fluid turbulence. In *Dynamical Systems and Turbulence* (D. Rand and L.-S. Young, eds.). Springer-Verlag, Berlin, 1981.

145. F. Takens, Invariants related to dimension and entropy. In *Atas do 13°*. Colóqkio Brasiliero de Matemática, Rio de Janeiro, 1983.

146. J. Theiler, Spurious dimension from correlation algorithms applied to limited time series data. *Phys. Rev. A* **34** 2427 (1986).

147. J. Theiler, Lacunarity in a best estimator of fractal dimension. *Phys. Lett. A* **135** 195 (1988).

148. J. Theiler, Estimating fractal dimension. *J. Opt. Soc. Am. A* **7** 1055–1073 (1990).

149. J. Theiler, Statistical precision of dimension estimators. *Phys. Rev. A* **41** 3038 (1990).

150. J. Theiler, Some comments on the correlation dimension of $1/f^\alpha$ noise. *Phys. Lett. A* **155** 480–493 (1991).

151. J. Theiler, B. Galdrikian, A. Longtin, S. Eubank, and J. D. Farmer, Testing for nonlinearity: The method of surrogate data. In *Nonlinear Modeling and Forecasting* (M. Casdagli and S. Eubank, eds.), pp. 163–188. Addison-Wesley, Reading, MA, 1992.

152. T. Tishby, A dynamical systems approach to speech processing. International conference on acoustics speech and signal processing, 1990.

153. H. Tong, *Threshold Models in Nonlinear Time Series Analysis. Lecture Notes in Statist.*, vol. 21. Springer-Verlag, New York, 1983.

154. H. Tong, *Non-Linear Time Series: A Dynamical System Approach*. Clarendon Press, Oxford, 1990.

155. H. Tong and K. S. Lim, Threshold autoregression, limit cycles and cyclical data. *J. Roy. Statist. Soc. Ser. B* **42**(3) 245–292 (1980).

156. B. Townshend, Nonlinear prediction of speech signals. *IEEE Trans. Acoustics, Speech, and Signal Processing* 1990.

157. B. Townshend, Nonlinear prediction of speech signals. In *Nonlinear Modeling and Forecasting* (M. Casdagli and S. Eubank, eds.), pp. 433–453. Addison-Wesley, Reading, MA, 1992.

158. A. Weigend, D. Rumelhart, and B. Huberman, Predicting the future: A connectionist approach. *Int. J. Neural Systems* **1** 193–209 (1990).

159. N. Wiener, *Nonlinear Problems in Random Theory*. Wiley, New York, 1958.

160. A. Wolf, J. B. Swift, H. L. Swinney, and J. A. Vastano, Determining Lyapunov exponents from a time series. *Physica D* **16** 285 (1985).

161. D. Wolpert, A new technique for improving the performance of any generalizer. Technical Report LA-UR-90-401, Los Alamos National Laboratory, 1990.

162. D. Wolpert, The relationship between Occam's razor and convergent guessing. *Complex Systems* **4** 319–368 (1990).

163. G. U. Yule, On a method of investigating periodicities in disturbed series with special reference to Wolfer's sunspot numbers. *Philos. Trans. Roy. Soc. London Ser. A* **226** 267–298 (1927).

CHAPTER 16

A STUDY OF FLUIDIZED-BED DYNAMICAL BEHAVIOR: A CHAOS PERSPECTIVE

S. W. TAM and M. K. DEVINE
Chemical Technology Division
Argonne National Laboratory
Argonne, IL 60439

An extensive search for chaotic behavior has been conducted on a large time-series data set for fluidized-bed pressure fluctuation. The nonlinear dynamical techniques employed include the correlation integral and the nearest-neighbor methods. A mutual information algorithm has been employed to obtain the "optimal" time delay necessary for the analysis of the experimental time series. A singular value decomposition technique has been used for noise reduction. The results indicate that despite the highly fluctuating appearance of the data, there is a nonrandom component present. For the data that have been examined, no low-dimensional (i.e., dimension 2–3) attractor has been found. Any strange attractor that may be present is likely to have at least moderate dimension (i.e., above 3–4). For the analysis of possibly chaotic time series from experimental data (in contrast to model-generated data), this study has revealed that the issues of data set size, noise, and moderate- to high-dimensional attractors are critical. Careful utilization of existing and developing nonlinear dynamical techniques is necessary to avoid drawing spurious conclusions.

16.1. INTRODUCTION

Fluidized beds have long been used as multiphase reactors in the chemical processing industries and as fossil-fuel combustion beds in the energy pro-

Applied Chaos, Edited by Jong Hyun Kim and John Stringer.
ISBN 0-471-54453-1 © 1992 John Wiley & Sons, Inc.

duction industries [1]. They operate on the principle of fluidization in which a fluid is allowed to flow through a solid phase (usually a system of particles of varying sizes) at a velocity above a minimum value U_{mf} at which the drag exceeds the net particle weights. At that stage the particles "fluidize," which is a state of motion in many ways similar in a reverse manner to sedimentation. The process provides efficient mixing among the phases in the system, resulting in good mass–heat transfer characteristics.

Depending on the extent to which the fluid-flow velocity V exceeds U_{mf}, the fluidized bed can be in different flow regimes. In the regimes of interest (including the so-called bubbling, slugging, and turbulent regimes), the fluidized-bed flow parameters tend to fluctuate in time. The flow parameters include particle velocities, pressures, and porosities. The presence of porosities (or bubbles) to different extents is a common characteristic of these flow regimes. Figure 16.1 is a plot of a typical time series data set for the pressure

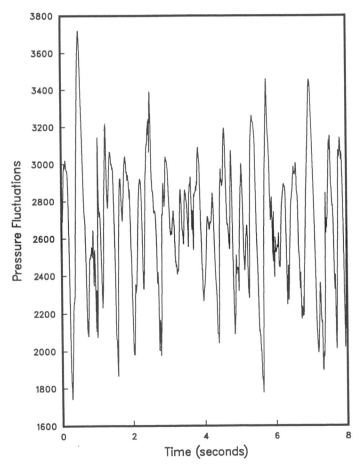

Figure 16.1. Pressure fluctuation data (in arbitrary units versus time; reference [2]).

fluctuation of a fluidized bed [2]. Both the qualitative features of the data and the fact that fluidization processes can be described by the nonlinear equations of multiphase flows prompt the following question. Do these fluidization data exhibit only stochastic features, or do they contain chaotic characteristics as well? The study aims to address, on a comprehensive basis, some of the issues raised in an earlier preliminary study [3] on this question. To address this question, we have employed a combination of nonlinear dynamical techniques in a detailed analysis of a relatively large data set for fluidized-bed pressure fluctuation.

The issues raised by this question have implications toward a basic understanding of a complex multiphase dynamical system. The problem of bubble formations and their stabilities and the question of erosion of the structural components of the bed are critical and yet unsettled issues. The resolutions of these issues would contribute toward further advances in industrial applications of fluidized-bed processes. The present study on fluidized-bed dynamical behavior from a chaos perspective represents a step in that direction.

16.2. EXPERIMENTAL DATA AND ANALYSIS METHODS

16.2.1. Experimental Data

The experimental pressure data were obtained by D. Dent et al. (at the Commonwealth Scientific and Industrial Organization, Australia) on a fluidized bed of 4×4 ft cross section filled to a depth of 4 ft with silica sand. The superficial gas velocity was 2 ft/s. The pressure data were obtained at a sampling rate of 500 digitizations per second with a 10 bit analog-to-digital converter [2]. The total data set represents an hour of continuous run [2]. In the present analysis, several subsets of the data of length between 15,000 and 50,000 points have been utilized.

16.2.2. Data Analysis Methods

The methods of analysis that we have employed include the Grassberger–Procaccia correlation integral (GP-CI) method [4] and the Badii–Politi nearest-neighbor (BP-NN) algorithm [5, 6]. The BP-NN approach, though much less widely used, has been shown to be very useful in analyzing attractors of moderate dimensions (i.e., $D > 2$–3). In addition, because we are analyzing actual experimental data, noise inevitably arises from a variety of sources. We have thus utilized a noise reduction technique called singular value decomposition (SVD), which is a well-known method in signal processing [7]. In the present analysis, SVD has been used only for processing the experimental data for subsequent analysis by the GP-CI algorithm. No attempt has been made to extract directly with SVD any dimensional information on the possible attractor underlying the experimental time-series data. A combined GP-SVD approach has been shown to be useful in analyzing model-generated data, including the Lorenz and the Rossler systems [8]. Here, we have integrated this composite approach with a mutual information analysis (see subsequent discussion) and applied them to an experimental time-series data set for fluidized-bed pressure fluctuation [2].

16.2.3. Embedding Technique

Given a scalar time series $x(t)$, one can construct d-dimensional vectors by the embedding technique with the following form [9–12]:

$$(x(t), x(t + T), \dots, x(t + (d - 1)T))$$

where d is the embedding dimension and T is the time delay. The geometrical structure generated with these vectors has the same dimensional characteristics as the correlation dimension (see discussion that follows) of the attractor generated from the dynamic underlying the original scalar time series. This possibility of reconstructing various important facets of the original (possibly multidimensional) attractor from just a scalar time series is important in that nonlinear dynamical technique is being applied to the analysis of "real world" experimental data. In fluidized-bed experiments, simultaneous measurements of several state variables are at best difficult and often practically impossible. The embedding technique makes possible the circumventing of such multivariable measurements.

16.2.4. Choice of Time Delay T

In the present work, we have chosen to obtain the "optimal" time delay T via the mutual information approach. This approach was originally suggested by Shaw [13] and has since been extensively tested [14]. Our experience has indicated that it is a very useful and practical approach. The curve for the mutual information $I(T)$ has a large peak around $T = 0$ due to self-correlation. For large T, this curve generally decays toward zero. However, very often this decay may be oscillatory and may occur over a long range, which indicates a nontrivial correlation. This is the case for our present pressure fluctuation data and is illustrated in Figure 16.2. As shown in the figure, the first minimum occurs at a T value of about 53, that is, a characteristic time of approximately $1/10$ s. This is the time delay used in subsequent analysis.

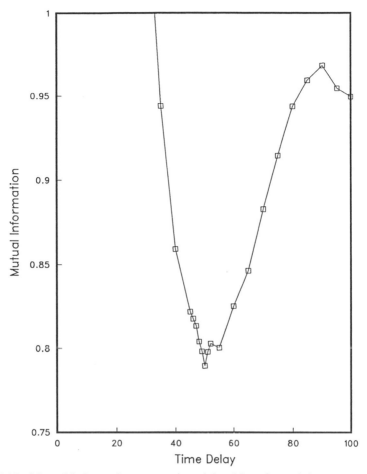

Figure 16.2. Mutual information versus time delay. Note first minimum near $T = 53$.

16.3. RESULTS AND DISCUSSION

16.3.1. Correlation Integral Approach

The correlation integral (CI) is defined as

$$C(l) = \lim_{N \to \infty} 1/N^2 \, \{\text{numbers of pairs of points on the attractor whose distance is less than 1}\}$$

Here N is the number of data points in the scalar time series. For some range of l, called the scaling region, $C(l)$ scales as l^μ, where μ is the correlation dimension [4]. A saturation of μ as the embedding dimension d

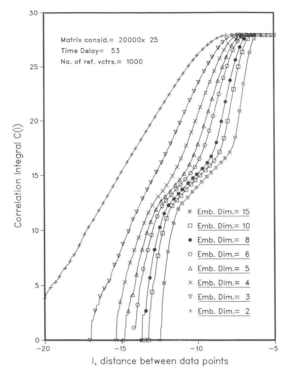

Figure 16.3. Correlation integral versus l distance between data points.

increases would be an indication that the experimental signal has a nonrandom component [4]. Before the pressure data were subjected to a CI analysis, they were first processed by SVD (singular value decomposition) [5] to reduce the effect of noise in the data. The SVD method also indicates the limit one is able to embed a given data set to higher dimension without degrading the data quality. The correlation integral $C(l)$ as a function of l is shown in Figure 16.3 in a log-log plot to base 2. The embedding dimension d varies between 2 and 15. The correlation dimension μ has been extracted from the region of $C(l)$ where it behaves as l^μ via a local slope analysis. This is the region prior to $C(l)$ bending over toward the flat horizontal structure of the saturation regime. The results of such an analysis clearly indicate that μ increases monotonically with d, and that within the range of d tested there is as yet no definite indication of a saturation. Furthermore, one can see from Figure 16.3 that the scaling region shrinks rapidly with increasing d. When d reaches 15, the scaling region is only marginally evident. At that stage, $C(l)$ exhibits almost continuous curvature, rendering an unambiguous determination of the scaling region difficult. This is consistent with the SVD analysis indicating the limitation of embedding to arbitrarily high dimensions arising

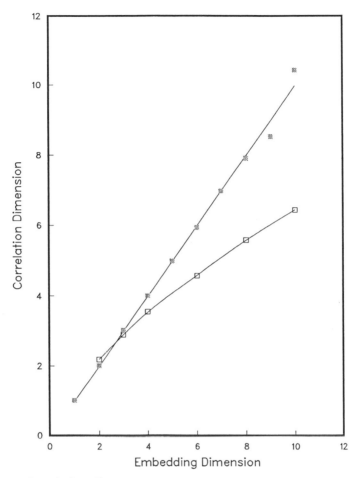

Figure 16.4. Correlation dimension versus embedding dimension. Empty square = pressure data; filled square = random number.

from decreasing signal-to-noise (S/N) ratios. The resulting μ as a function of d is displayed in Figure 16.4. Up to $d = 10$ there is no sign of saturation. However, the values of μ that have been obtained are significantly below those from a truly stochastic process, such as arising from a "good" random number generator. The present results seem to indicate the presence of a nonrandom component in the process underlying the time series data. Nevertheless, the S/N ratio and the rapidly shrinking scaling region with increasing d rule out an unambiguous determination of the saturation regime and, hence, an evaluation of the dimension of the corresponding attractor. If an attractor exists for the present data, it does not have a low-dimensional structure (i.e., μ in the range of 2–3).

16.3.2. The Badii–Politi Nearest-Neighbor Analysis

Several conclusions have been reached from the CI algorithm used to analyze the experimental time series. It is important to verify these by an independent method. There are several different kinds of generalized dimensions associated with a strange attractor. One of the most useful ones, aside from the correlation dimension, is the so-called information dimension D_I. One can analyze D_I by several approaches. A very convenient one that leads to a practical computational algorithm is that of Badii and Politi (BP) [5, 6]. The BP algorithm is as follows.

Consider N points lying on an attractor. These N points may be generated by the embedding technique, as for example from a sequence of measurements. Select any one point X_r among the N points as a reference point. Take any other n points from the N points that are not the same as X_r, where n is an integer that has N as an upper bound. Compute the minimum distance d_n between X_r and the n chosen points. Repeat the calculation for a sufficiently large number of different reference points X_r and obtain a quantity $\langle d_n \rangle$, which is the nearest-neighbor distance average over the set of X_r's. Note that the $\langle d_n \rangle$ so obtained is a function of n, the number of points that has been chosen. A scaling relationship exists between $\langle d_n \rangle$ and n in the form

$$\langle d_n \rangle \sim n^{-1/D_I} \tag{16.1}$$

As one can see, the scaling relationship provides a way to extract D_I. Although D_I and μ represent different measures of an attractor, they are related through an inequality [5, 6]

$$\mu \leq D_I \tag{16.2}$$

For actual experimental data, experience has indicated that the values of the two dimensions frequently are similar.

In principle, the nearest-neighbor distance is supposed to be used in the BP method. However, because of the presence of noise and the scarcity of data in the nearest-neighbor region for experimental data, it is best in practice to use a higher order near-neighbor separation rather than the nearest one. Although there is no firm theoretical guidance on the best order, our experience has indicated that higher embedding dimensions tend to require higher order. In the cases that we have examined, the necessary order seems to be around 100 to 300. Figure 16.5 gives the information dimension as a function of the embedding dimension. No saturation can be detected, but D_I is significantly below d for $d > 4$, indicating the presence of a nonrandom component in the pressure data. The difference between D_I and d becomes increasingly larger as d increases toward 10. The lack of

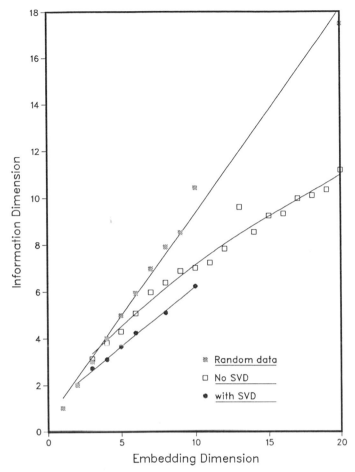

Figure 16.5. Information dimension versus embedding dimension for analysis with and without use of singular decomposition (SVD).

saturation precludes an unambiguous determination of the D_I for the attractor. The present analysis indicates that, for the data considered, the attractor (if present) does not have a low-dimensional structure (i.e., $D_I \sim 2$–3). This is consistent with the results from the CI analysis. For comparison, Figure 16.5 also displays the result of a BP analysis in which the data have not been preprocessed with SVD. The more noisy data give a D_I that is above that from the less noisy data, as expected.

A combined SVD-CI approach for the analysis of model data generated from the Lorenz and Rossler systems has been shown to be useful [8]. A similar approach was used by us on engineering time-series data on pressure fluctuation. However, as far as the present authors are aware, this is the first time that a combined SVD-BP analysis has been applied on a "real world" data set. Our results show an increasingly significant deviation from the Euclidean (embedding) dimension d, particularly for $d > 3$. In the regime where $d > 10$, the pure BP analysis (i.e., without SVD) yields a convex curve as d increases toward 25. Nevertheless, no definitive saturation is evident. Because of the increasing dominance of noise, as indicated by the SVD analysis, results from high embedding dimensions (e.g., d significantly above 10) need to be treated with caution.

16.4. CONCLUSION

The present analysis on the time series data for fluidized-bed pressure fluctuation indicated that several critical issues arise from the application of nonlinear dynamical techniques to the analysis of systems of "real world" interest (including technological, climatic, and medical-biological systems). These issues include the problem of noise, the presence of moderate to high-dimensional attractors, and the constraint of data size limitation. A small data set size and the presence of noise would invalidate embedding of data to arbitrarily high dimensions. This, in turn, would severely limit the possible detection of moderate- to high-dimensional attractors. These factors provide a serious challenge to the effectiveness of applying the available data analysis techniques to chaotic time series. Nevertheless, useful information such as the presence of nonrandom processes and the lower bound on the relevant dimensions of the possible attractor can be extracted, as the present analysis has shown. These results are consistent with the conclusions reached from a previous study [3].

The important question of the present study is, what is the implication of our CI and BP results for fluidized-bed processes? The present results have indicated the absence of a low-dimensional attractor for the data that we have examined. However, it is possible that, for such a complicated process as fluidization, other operating regimes are responsible for the time evolution of the state variables being characterized by low-dimensional chaos. Parameters such as flow velocities, particle size distributions, characteristic geometric extent of the bed, and their combination may well be responsible for moving the system from one regime to the other.

To proceed beyond what has been learned from the present study, one needs to utilize a wide set of data that would cover various regions of the phase space that comprise the operating regimes and fluidized beds. Furthermore, related flow processes such as fluidized-bed combustion may hold promise and need to be investigated. Techniques applicable to noisy attractors that are not necessarily low-dimensional need to be developed, validated, and applied. It is through these pathways that further understanding can be gained on the relationship between nonlinear dynamics and fluidized-bed flow processes.

ACKNOWLEDGMENTS

The authors would like to acknowledge D. Dent and R. La Nauze of CSIRO for making the pressure data available to them and J. Stringer of EPRI for stimulating discussions. This work is partially supported by EPRI.

REFERENCES

1. D. Geldart (ed.), *Gas Fluidization Technology*. Wiley, London 1986.
2. D. Dent, Private communication, 1989.
3. S. W. Tam and M. K. Devine, In *Measures of Complexity and Chaos* (N. B. Abraham, A. M. Albano, A. Passamante, and P. E. Rapp, eds.). Plenum, New York, 1989.
4. P. Grassberger and I. Procaccia, Measuring the strangeness of strange attractors. *Physica D* **9** 189 (1983).
5. E. J. Kostelich and H. L. Swinney, In *Chaos and Related Non-Linear Phenomena* (I. Procaccia and M. Shapiro, eds.). Plenum, New York, 1987.
6. R. Badii and A. Politi, Statistical description of chaotic attractors: The dimension function. *J. Stat. Phys.* **40** 725 (1985).
7. D. S. Broomhead and G. P. King, Extracting qualitative dynamics from experimental data. *Physica D* **20** 271 (1986).
8. A. M. Albano, J. Muench, C. Schwartz, A. I. Mees, and P. E. Rapp, Singular-value decomposition and the Grassberger–Procaccia algorithm. *Phys. Rev. A* **38** 3017 (1988).
9. N. Packard, J. P. Crutchfield, J. D. Farmer, and R. S. Shaw, Geometry from a time series. *Phys. Rev. Lett.* **45** 712 (1980).
10. F. Takens, In *Dynamical Systems of Turbulence* (D. A. Rand and L.-S. Young, eds.). Springer, Berlin, 1981.
11. H. Froehling, J. P. Crutchfield, D. Farmer, N. Packard, and R. S. Shaw, On determining the dimension of chaotic flow. *Physica D* **3** 605 (1981).
12. H. Whitney, Differentiable manifolds. *Ann. Math.* **37** 645 (1936).
13. R. S. Shaw, *The Dripping Faucet as a Model Chaotic System*. Aerial Press, Santa Cruz, CA, 1985.
14. A. M. Frazer and H. L. Swinney, Independent coordinates for strange attractors from mutual information. *Phys. Rev. A* **33** 1135 (1986).

CHAPTER 17

FORECASTING CATASTROPHE BY EXPLOITING CHAOTIC DYNAMICS*

H. B. STEWART and A. N. LANSBURY[†]
Mathematical Sciences Group
Department of Applied Science
Brookhaven National Laboratory
Upton, NY 11973

17.1. INTRODUCTION

Although chaotic behavior in dynamical systems has sometimes (even by Poincaré [1]) been viewed with dismay, in many instances chaotic dynamics can and should be regarded as presenting opportunities for understanding and for predicting over the short term. From the standpoint of the experimental dynamicist observing the behavior of a system producing a stream of data in the form of a time series, recent studies have used chaotic dynamical time series to examine information content [2], reconstruct multidimensional attractors [3, 4], reconstruct vector fields for the purpose of short term forecasting and noise reduction [5–8], and find unstable periodic motions [9, 10], which according to Poincaré are the only generally applicable means for understanding structure in phase space.

Our purpose here is to introduce a variation on the theme of short term forecasting from a chaotic time series. We show that for the lowest-dimensional chaotic attractors, it is possible to predict incipient catastrophes, or crises, by examining time series data taken near the catastrophic bifurcation threshold, but always remaining on the safe side of the threshold.

*This work was supported by the Applied Mathematical Sciences program of the U.S. Department of Energy under Contract No. DE-AC02-76CH00016.
†Permanent address: Department of Civil Engineering, University College, London.

Applied Chaos, Edited by Jong Hyun Kim and John Stringer.
ISBN 0-471-54453-1   © 1992 John Wiley & Sons, Inc.

More specifically, we assume that some dynamical system has been observed and a time series

$$x_1^1, x_2^1, x_3^1, \ldots, x_n^1$$

of the dynamical state variable x at times $j = 1, 2, 3, \ldots, n$ has been recorded, and that this dynamical system has a control parameter μ whose value was fixed at μ_1 throughout the observation of the preceding data. Reconstruction theorems [3] prove that if n is large enough, this single variable time series suffices to generate a complete geometric phase space portrait of the attractor.

Subsequently, with the control at a new fixed setting μ_2, a second time series $\{x_j^2, \; j = 1, 2, 3, \ldots, n\}$ is recorded. This process may be repeated for several values μ_i, $i = 1, 2, \ldots, I$ of the control. Suppose for simplicity that $\mu_1 < \mu_2 < \cdots < \mu_I$. We then ask the following questions: Under what circumstances is it possible to infer that a catastrophic change in dynamical behavior will occur at some critical threshold value μ^* slightly greater than μ_I? Can the threshold value μ^* be estimated from the data recorded for μ_i, $i = 1, 2, \ldots, I$? Our answer is that for dynamical systems with very simple chaotic attractors, this kind of prediction is indeed possible. In other words, chaotic attractors may contain in the record of their behavior sufficient information to predict dynamic catastrophes, without disturbing the dynamics and before any consequences of such a catastrophe can occur.

The question of forecasting bifurcations has already been studied for systems with a regular periodic attractor close to a local bifurcation, such as a fold catastrophe [11–15]. If the system is supposed to be an ideal deterministic one without noise or external perturbation, then the only observable information is the state space position of the periodic orbit at each of the control values μ_i, $i = 1, 2, \ldots, I$. It might be imagined that the generic parabolically folded shape of the path of a Poincaré section point as a function of μ could be used to extrapolate the value μ^* at which the location of the periodic orbit becomes infinitely sensitive to the value of μ. This approach has not been tried in practice, probably because the generic shape becomes evident only in extreme proximity to the bifurcation point, and perhaps also because the numerical problem of estimating μ^* from μ_i is poorly conditioned.

Fortunately, there are at least two modified approaches to the regular bifurcation prediction problem that have been tested successfully. The approach of Thompson and Virgin [11, 12] assumes that the system is accessible to small disturbances given to the steady state; the rate of decay of ensuing transients is shown to have a generic dependence on μ which can be exploited to predict the approaching bifurcation. For example, in reference [11] the jump to resonance of an electromagnetically driven steel beam was successfully predicted using as predictor the square of the frequency of transient beats against the steady oscillation.

A second approach assumes that the dynamical system operates in a noisy environment. Wiesenfeld [13] has shown how a system near an instability will appear to be very sensitive to excitation at certain frequencies, in ways which depend only on the type of bifurcation and its nearness in terms of a bifurcation parameter. This is explained by the progressively weaker damping of transients as the instability is approached, and produces characteristic bumps in the power spectrum called noisy precursors. Wiesenfeld deduced the forms of these precursors in reference [13], and in reference [14], Jeffries and Wiesenfeld confirmed the theory by experimental studies of an electrical circuit with a periodically driven *p-n* junction. It should be noted that Jeffries and Wiesenfeld related the noisy precursors to a natural bifurcation parameter, the inverse of experimentally measured transient decay time determined following external disturbances similar to the approach of Thompson and Virgin. This natural bifurcation parameter is presumed to be proportional to a generic control like the μ_i in the black box experimental scenario previously described. It would in principle be possible to predict incipient bifurcations using noisy precursors alone, without the need for external disturbances. Thus the noisy precursor approach might be used in situations where the dynamical system is not accessible to any form of experimental intervention, but is simply being observed while some control is drifting very slowly at a known rate.

In the present work, we consider the problem of forecasting catastrophic bifurcation of a chaotic attractor in both a mathematical prototype of chaotic dynamics, the Hénon map, and a more familiar nonlinear oscillator— Duffing's equation with twin-well potential. The behavior of this Duffing oscillator is important in many applications including structural vibrations [16, 17], breaking of chemical bonds [18], and capsize of ships [19]. The method we propose will be applicable to systems having chaotic attractors of lowest possible dimension, that is, slightly greater than 1 for discrete time dynamical systems. We hope that the ideas presented here may be generalized to higher-dimensional systems, but that will depend in part on better geometric phase space understanding of the possible types of catastrophic bifurcations of chaotic attractors, and in particular the types of unstable basic sets that can collide with attractors to trigger catastrophes [20–22] or crises [23].

17.2. CATASTROPHES IN THE HÉNON MAP

The Hénon map is defined by the iteration

$$x_{n+1} = -ax_n^2 + y_n$$
$$y_{n+1} = bx_n \tag{17.1}$$

of dynamical state variables x and y in a two-dimensional (x, y) phase space. The parameters a and b are the controls of this dynamical system, held at fixed values while equations (17.1) are iterated from initial conditions (x_0, y_0) to determine the long term behavior. The parameter b is related to the amount of damping or dissipation in the system, with b values close to zero corresponding to strong dissipation; the parameter a may be thought of as a stress parameter, similar to a forcing amplitude in a driven system, with greater a values corresponding to increased forcing or stress. For $b = 0$, equations (17.1) always give $y_{n+1} = 0$ after the first iteration, so the dynamics are described by the state variable x_n and the one-dimensional map

$$x_{n+1} = 1 - ax_n^2 \qquad (17.2)$$

Equation (17.2) is an interesting dynamical model in its own right; qualitatively similar modes have found applications in population biology [24] and economics [25] as well as the physical sciences (e.g., reference [26]).

The dynamical behavior of (17.2) is by now well known [23, 24] and depends in a complicated way on the value of the parameter a. For $a < -\frac{1}{4}$, there is no stable long term behavior, and $x_n \to -\infty$ from every initial condition. For $-\frac{1}{4} < a$ there is an unstable fixed point at $x^* = [-1 - \sqrt{(1 + 4a)}]/2a$; all initial conditions to the left of x^* diverge to $-\infty$, while for $-\frac{1}{4} < a < 2$ there is an interval of initial conditions to the right of x^* that lead to stable long term behavior, in the sense that iterates remain bounded for all time. For $-\frac{1}{4} < a < \frac{3}{4}$ the stable behavior is a fixed point; increasing the stress parameter a beyond $\frac{3}{4}$, the stable long term behavior bifurcates by successive period doubling and becomes chaotic. There are many small intervals of a values for which the unique stable behavior is periodic, repeating exactly after p iterates. As a approaches 2 these windows of periodic behavior become narrower and chaotic dynamics predominate: iterates wander over an interval of x values, and there is sensitive dependence on initial conditions, so that nearby initial conditions separate under iteration at an exponential rate.

At $a = 2$, the stress reaches a critical value and a catastrophe occurs: The interval of x values visited by long term iterates (in other words, the chaotic attractor) touches the unstable fixed point x^*. For any $a > 2$, chaotic motions cannot be sustained, and there is no stable long term behavior. This loss of stability of the chaotic attractor, which existed for $a < 2$, is associated with the attractor touching x^* at $a = 2$, and is an example of a blue sky catastrophe [21], or boundary crisis [23].

The behavior of equation (17.2) near $a = 2$ is illustrated in Figure 17.1, where successive iterates are plotted showing x_{n+1} versus x_n. The heavy dots are 1000 long term iterates computed from (2) after discarding the first 500 iterates. The dashed curve shows the locus of $f(x) = 1 - ax^2$; this locus intersects the 45° line $f(x) = x$ in two points, the leftmost being the unstable separator x^*, marked by a small circle.

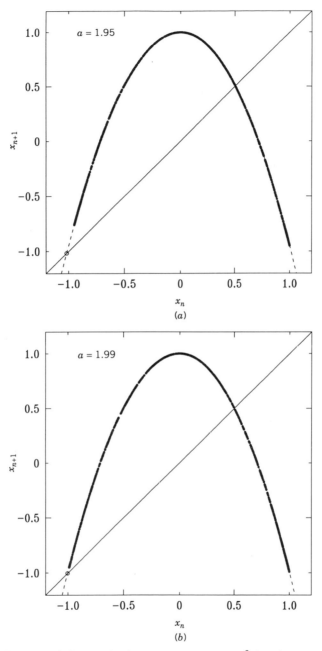

Figure 17.1. Iterates of the quadratic map $x_{n+1} = 1 - ax_n^2$ for the two values of the parameter a approaching catastrophe at $a = 2$.

Comparison of the two cases $a = 1.95$ and $a = 1.99$ in Figure 17.1 shows that the left edge of the attractor moves closer to x^* as the control a approaches 2. For the one-dimensional map (17.2) the left edge of a chaotic attractor is the second image of the critical point $x = 0$, that is $f(f(0)) = f(1) = 1 - a$. (We ignore the small intervals of a values for which the attractor is periodic.) It will be convenient to measure the distance from this point to x^* vertically in Figure 17.1; by this measure the distance

$$D(a) \equiv f(1 - a) - x^* = 1 - a(1 - a)^2 - \left[-1 - \sqrt{(1 + 4a)} \right] \Big/ 2a \quad (17.3)$$

which is a continuous function of a for $a > 0$ and goes to zero as $a \to 2$.

Because this distance varies smoothly with a, we might use $D(a)$ as a measure of how close the system is to the catastrophe at $a = 2$. Suppose, for example, that the data represented by heavy dots in Figure 17.1 were obtained from observation of a black box dynamical system at two different control settings a_1 and a_2. In this situation the broken curve representing $f(x)$ is not known. Nevertheless, we observe that the left edge of the attractor is close to the 45° line, or bisectrix, where $x_{n+1} = x_n$. Furthermore, we note that whatever $f(x)$ may be, it certainly has slope greater than 1 near the left edge of the attractor; therefore when the attractor edge reaches the bisectrix, the contact point must be an unstable separator. This implies that a crisis will occur, although we cannot, in general, say whether the attractor will explode (interior crisis) or lose stability altogether (boundary crisis). In any case, $D(a_1)$ and $D(a_2)$ can be estimated by extending the attractor locus to the bisectrix (by linear extrapolation, for example) to estimate the location of the unstable separator $x^*(a_1)$ and $x^*(a_2)$. Once $D(a_1)$ and $D(a_2)$ are estimated in this way, it is a simple matter to estimate by extrapolation the critical value a^* for which $D(a^*) = 0$. Thus by exploiting the simple chaotic structure of the dynamics and using the fact that any fixed point must lie on the bisectrix in an (x_n, x_{n+1}) plot, we can predict an incipient catastrophe.

In order for this forecasting method to be useful, two remaining obstacles must be overcome. The first is that for the Hénon map with $b \neq 0$, and for Poincaré maps of forced oscillators, such as the Duffing oscillator, the distance from attractor edge to the separator point is not a continuous function of a generic control. Because there are no closed form expressions for this distance generalizing (17.3), we shall turn to numerical evidence. In so doing we shall also obtain information about the second obstacle, which is that the unstable separator point may, in general, not be a fixed point, but can instead be a periodic point with period $p > 1$.

The dynamical behavior of the Hénon map for $b < 0$, and also the typical behavior of many simple dynamical systems like nonlinear oscillators, resembles the quadratic map in many ways. There is period doubling leading to chaos as the stress parameter a is increased with b fixed; small windows of a values in the chaotic regime can be found where only periodic attractors exist. Additionally, at some threshold $a^* = a^*(b)$, the main attractor loses stability, and for $a > a^*(b)$ all orbits diverge to infinity. (In oscillators, the system would not diverge but would typically jump to a different attractor representing qualitatively different behavior, as in the familiar jump to resonance [27].) The principal difference from the quadratic map is that more than one attractor may coexist for certain values of $a < a^*(b)$.

For $b < 0$ the Hénon map has two fixed points:

$$x^* = \left\{ (b - 1) - \sqrt{\left[(b - 1)^2 + 4a \right]} \right\} \bigg/ 2a$$

$$y^* = bx^* \tag{17.4}$$

and

$$\bar{x} = \left\{ (b - 1) + \sqrt{\left[(b - 1)^2 + 4a \right]} \right\} \bigg/ 2a$$

$$\bar{y} = b\bar{x} \tag{17.5}$$

By linearizing the Hénon map near (x^*, y^*), the eigenvectors and eigenvalues, or multipliers, can be computed. One multiplier is greater than 1, and both are positive, so (x^*, y^*) is an unstable saddle separator, analogous to x^* with $b = 0$. Likewise $(\bar{x}, \bar{y})$ is found to be (after the first period doubling) unstable with negative multipliers, analogous to the point in Figure 17.1 where $1 - ax^2$ crosses the bisectrix with negative slope.

In cases where more than one attractor exists, numerical simulations usually show one large chaotic attractor containing $(\bar{x}, \bar{y})$, which attracts most initial conditions, and periodic or small chaotic attractors with much smaller basins of attraction. We concentrate attention on this large attractor, which we call the main attractor, and determine $a^*(b)$ by the following numerical experiment. First we fix b at $b = -0.0001$, and $a = 1.5$. An orbit is started at $(x_0, y_0) = (x_0, bx_0)$, where $x_0 = \bar{x} + 0.0001$ and iterated 50,000 times. If the result stays to the right of (x^*, y^*), we consider that a^* has not been reached, increase a to 2.0, and check that the resulting orbit diverges within 50,000 iterates. We then conduct a binary search for two a values between 1.5 and 2.0, separated by less than 10^{-5}, such that the orbit started at (x_0, y_0) remains bounded for the smaller a and diverges for the larger a. This gives $a^* = a^*(b = -0.0001)$. We then increment b by -0.0001 and search again for $a^*(b)$.

The results of this experiment are shown in Figure 17.2. A point is plotted at $a^*(b)$ for each trial b value, and with a very few exceptions these dots fall on an arc emerging from $b = 0$, $a^*(0) = 2$. Further examination shows that as $a \to a^*(b)$, the main chaotic attractor just touches the saddle fixed point (x^*y^*), and loses stability in a boundary crisis, or blue sky catastrophe. This holds true for $0 > b \geq -0.08$, where $a^*(b)$ turns a corner in Figure 17.2.

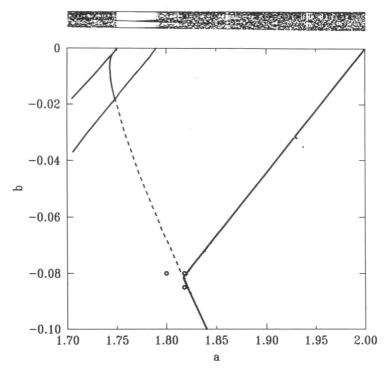

Figure 17.2. Numerical determination of catastrophe values $a^*(b)$ for the Hénon map; just above the (a, b) chart is a bifurcation diagram of the quadratic map for $1.7 < a < 2$.

A second experiment was then performed to measure the distance from attractor to (x^*, y^*) for $1.5 < a < a^*(b)$. The results are shown in Figure 17.3. Here the dashed arc represents the distance $D(a)$ for $b = 0$ computed from Equation 17.3, again ignoring the small regimes of periodic attractors; the heavy arcs consist of numerically determined distances from (x^*, y^*) to the main attractor. These distances were found by using equation (17.4) for (x^*, y^*) and taking the closest iterate (x_n, y_n), $1000 < n < 6000$. One technical detail should be noted: Only iterates (x_n, y_n) with $y_n > 0$ are tested for closeness to (x^*, y^*); iterates with $y_n < 0$ may be closer to (x^*, y^*) if $a \ll a^*$, but should not be counted because they correspond to points on the right half of the quadratic map and would be very far from (x^*, y^*) if the y coordinate were rescaled.

Figure 17.3 shows that the observed distance does go to zero in every case except $b = -0.09$. To exploit the distance as a predictor, we would like the heavy arcs in Figure 17.3 to be free of discontinuities. In fact they are not; two types of discontinuities are apparent. The first type is rare isolated dots lying above the general trend; this is due to small windows of a values where

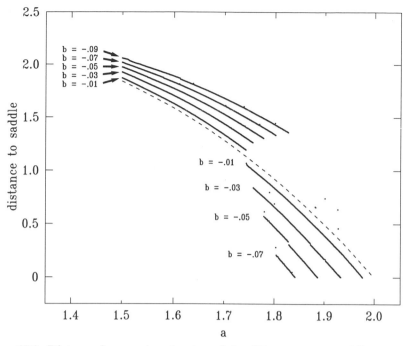

Figure 17.3. Distance from main attractor of the Hénon map to saddle separator (x^*, y^*) determined by numerical experiment.

the main chaotic attractor collapses to a stable periodic orbit. This type of discontinuity occurs even for $b = 0$, and would be easily recognized in experiments, because the general trends on either side of such windows are nearly identical; in real dynamical systems with parametric noise such discontinuities would be entirely suppressed.

A second kind of discontinuity is an apparent step jump in distance, caused by a sudden increase in size of the main attractor, that is, an attractor explosion or interior crisis. For each value $b = -0.01, -0.03, -0.05$, and -0.07 one large jump occurs, and any attempt to predict a^* using values of a to the left of this jump would be doomed to failure. This large jump is caused by an unstable period 3 orbit touching the chaotic attractor to trigger explosion; this explosion occurs along the dashed arc in Figure 17.2 emerging from the period 3 window of the quadratic map bifurcation diagram at $b = 0$, $a \simeq 1.75$.

Mathematical theory suggests that there are in fact infinitely many additional attractor explosions causing jumps in the distance to (x^*, y^*), but examination of Figure 17.3 shows only a very few jumps of much smaller magnitude. This suggests that in practice prediction of a^* is feasible if only the large jump can be taken into account. This is indeed possible, as we shall see shortly.

17.3. FORECASTING CATASTROPHE

Let us now return to the viewpoint discussed at the beginning, namely that of an experimental dynamicist presented with only time series data recorded from a black box dynamical system. We assume that more than one time series has been recorded, each at a different know value of μ_i of some control parameter. We emphasize that the dynamicist knows only the values of the μ_i and the corresponding data $\{x_j^i\}$, and desires to estimate the critical value $\mu^* < \mu_I$ at which a catastrophic change in the attractor will occur.

To apply distance to the unstable saddle as a practical predictor, it remains to find some means of locating this saddle using only such information as would be available in an experimental observation, namely a sequence of data $\{x_j^i\}$ on the attractor. Our strategy is to use μ values near μ^*, where many of the $\{x_j^i\}$ data correspond to points close to the unstable saddle. In general, this provides an opportunity to reconstruct dynamical rules (for example, following reference [5]) and use the reconstructed rules to estimate the location of the unstable saddle by extrapolation. Here we choose instead to use an equivalent graphic device based on the simple structure of the chaotic attractor and motivated by Figure 17.1.

To demonstrate this we turn for our example to the forced Duffing oscillator

$$\ddot{x} + k\dot{x} - x + x^3 = A \sin \omega t \qquad (17.6)$$

which describes damped, forced oscillations of system in a twin-well potential $v(x) = -x^2/2 + x^4/4$. This has been invoked as a model for mechanical vibrations of a vertical Euler support column loaded past the buckling point and shaken laterally [16]. We consider a region of the (ω, A, k) parameter space where nonlinear resonance produces two coexisting attractors in each potential well; this well-known phenomenon (see, e.g., reference [27]) has recently been shown to play an important role in the escape from confinement within a generic potential well with smooth potential barrier [28], having a wide range of applications from the breaking of molecular bonds [18] to the capsize of ships [19]. Typically in such situations the resonant motion becomes chaotic just before losing stability, and the chaotic attractor bears a very strong resemblance to the attractors of the Hénon map. It is this destabilization of the resonant chaotic motion that we aim to forecast.

Figure 17.4 shows 100 data points from each of two time series $\{x_j^1\}$ and $\{x_j^2\}$ computed from Equation 17.6 with $k = 0.5$, $\omega = 0.9$, and $A_1 = 0.348$ for the first series, and $A_2 = 0.350$ for the second series. The points are sampled stroboscopically at $t = 2\pi n$, $n = 1, 2 \dots$. Treating these as data from a black box dynamical system, we might infer from the slightly greater range of x values in the second case that it is under somewhat greater stress than the first case. However, it seems impossible to guess from this tradi-

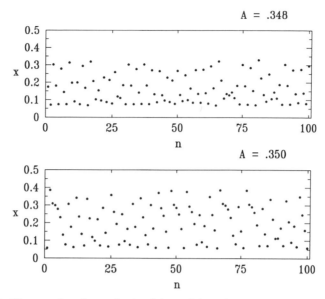

Figure 17.4. Time series data obtained by solving the twin-well Duffing equation [equation (7.6)] with $k = 0.5$ and sampling stroboscopically in phase with the sinusoidal forcing at $t = 2\pi n$, $n = 1, 2, \ldots, 100$ with (a) $A_1 = 0.348$ and (b) $A_2 = 0.350$.

tional presentation of the data whether the system is near a critical threshold of catastrophic change.

Figure 17.5 shows the same time series (extended to 500 points each) presented as plots of x_{n+1} versus x_n. Here we see clearly that the second case has a fingertip of the attractor very close to the first bisectrix or 45° line, and we infer that some sort of catastrophe is imminent. In fact a very small increase in A of less than 0.0002 causes the attractor to lose stability, with an ensuing rapid transient jump to another attractor, in this case the nonresonant periodic motion that oscillates with a much lesser response amplitude measured from the bottom of the potential well.

As an example of forecasting by extrapolation, we consider another set of parameter values, illustrated in Figure 17.6 showing 500 points on a resonant motion of equation (17.6) with $k = 0.35$, $\omega = 1$, and (a) $A = 0.260$ as well as (b) $A = 0.261$. The points are sampled stroboscopically at $t = 2\pi n$, $n = 1, 2, \ldots$, and plotted x_{n+1} versus x_n, with the bisectrix drawn. By comparing the two cases, we see that an attractor tip (with slope greater than 1) is moving toward the bisectrix near $x_{n+1} = x_n \simeq 0.5$. In each case we estimate by linear extrapolation from the tip to the point where the attractor will touch the bisectrix, and the distance from that point to the attractor tip. Comparing the distances obtained in the two cases, we estimate that the attractor will touch the bisectrix when $A = 0.2618$. In fact, numerical experi-

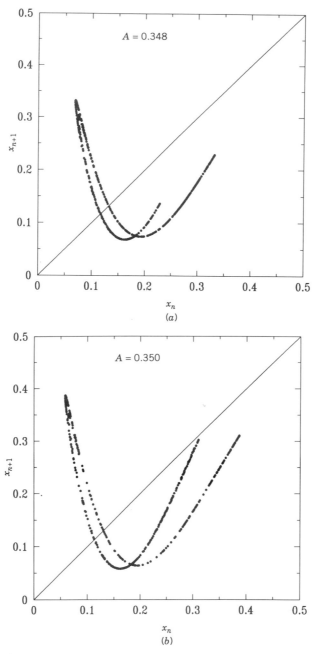

Figure 17.5. The same data as Figure 17.4, extended to 500 data points, and plotted x_{n+1} versus x_n, showing incipient catastrophe at $A_2 = 0.350$.

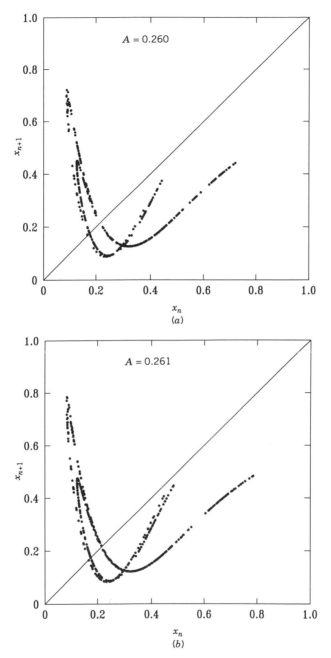

Figure 17.6. Chaotic attractors of the twin-well Duffing oscillator near catastrophe, plotted x_{n+1} versus x_n for (a) $A = 0.260$, $k = 0.35$, (b) $A = 0.261$, $k = 0.35$.

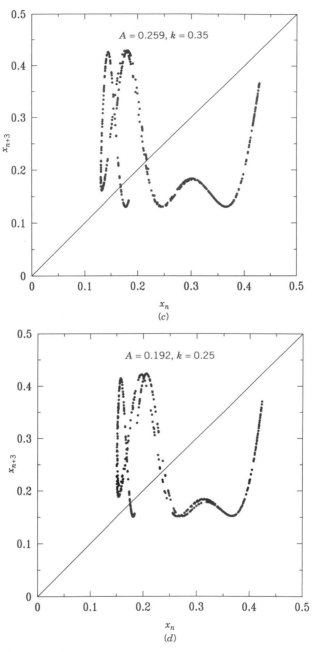

Figure 17.6. *(Continued).* Chaotic attractors of the twin-well Duffing oscillator near catastrophe. Plotted x_{n+3} versus x_n for (c) $A = 0.259$, $k = 0.35$ and (d) $A = 0.192$, $k = 0.25$.

ments show the attractor losing stability between $A = 0.2616$ and $A = 0.2617$, with an ensuing jump back to the nonresonant periodic motion.

It happens in this example that A values less than 0.260 are not useful because the attractor undergoes a substantial explosion in size at $A \simeq 0.2595$. As with the Hénon map, this explosion is triggered when the chaotic attractor touches an unstable periodic orbit of period 3. Thus to forecast the ultimate loss of stability correctly, we would need forewarning of this explosion. This can be achieved by plotting x_{n+3} versus x_n, as illustrated in Figure 17.6c with $A = 0.259$. Here we see the rightmost tip of the attractor near the bisectrix; by increasing A we could observe this tip moving closer and just touching the bisectrix when the attractor explodes.

Note that forecasting does not predict the outcome of the catastrophe, which as just shown might be either an attractor explosion or a complete loss of stability, that is, a blue sky catastrophe. To illustrate this point further, Figure 17.6d shows attractor points of equation (17.6) with $k = 0.25$, $\omega = 1$, and $A = 0.192$. As in Figure 17.6c, the attractor is approaching the bisectrix plotted x_{n+3} versus x_n, but in this case, upon touching the implied period 3 unstable orbit, the chaotic attractor will undergo a blue sky catastrophe, not an explosion as in Figure 17.6c. Indeed Figure 17.6d is a preview of a catastrophe studied earlier in reference [29].

Finally, we note that the simple graphical device of Figures 17.5 and 17.6 works because the attractor has a simple, thin shape, which is well approximated by a curve due to the extreme compression of fractal layers. In more general systems, this graphical device might not be applicable, but the underlying principle would still be valid: a chaotic attractor near incipient catastrophe is close in phase space to some singularity representing instability, and the distance from attractor to singularity can be estimated from the information contained in the data observable from dynamics on the attractor.

17.4. CONCLUSIONS

To summarize, we have shown that it is possible in simple dynamical systems to predict an incipient catastrophic change in dynamical behavior using only observed experimental data taken before the catastrophe. The prediction does not specify what form the post-catastrophe behavior will take. Also it is useful to know the period of the unstable motion suspected of triggering the catastrophe; in our examples, the period was either 1 or 3. Mathematical theory suggests that this forecasting procedure might be disrupted by numerous attractor explosions, but in practice these disruptions can, to a first approximation, be ignored, at least for the Hénon map attractor, which is the most commonly observed chaotic attractor in simple dynamical systems. Thus the record of information observable from a chaotic dynamical system may be sufficient to forecast catastrophe.

It should be noted that the specific graphical device of using attractor distance to the bisectrix of a first (or nth) return map can only be expected to work for a very limited class of dynamical systems in which the attractor has the appearance of a simple one-dimensional branched manifold. This graphical device should therefore not be understood as the essential point of this work, but rather as an easily understood proxy for a more sophisticated approach, which would consist of reconstructing a vector field from time series data, and then extrapolating the reconstructed vector field to a region surrounding the attractor. The motive for this extrapolation would be to find nearby loci of instability, such as unstable periodic motions, which could be associated with incipient bifurcation.

A number of potential difficulties can be anticipated with this approach to forecasting bifurcations. The reconstructed vector field must be extrapolated, and extrapolation is always risky. In lightly damped systems, incipient bifurcation may be associated with unstable periodic motions of long period, which may be very difficult to identify. In many cases, the locus of instability associated with catastrophic bifurcation may move in close to the attractor only at the last minute, when the bifurcation parameter is extremely close to its threshold: This difficulty is likely to be insurmountable in the case of canards [30] in slow–fast dynamical systems, and may be serious even in generic systems where the locus of instability has a very large multiplier in an expanding direction in phase space.

In spite of these potential limitations, the concept of forecasting bifurcations by exploiting chaotic dynamics has the advantage of requiring no external manipulation of the system, and may therefore warrant further investigation.

ACKNOWLEDGMENTS

The authors thank Michael Thompson, Kurt Wiesenfeld, and Ralph Abraham for stimulating discussions. We are grateful to Charles A. Norton for automating the computation of Figures 17.2 and 17.3, and to the Science and Engineering Research Semester Program of the U.S. Department of Energy for supporting him. H.B.S. acknowledges the support of the Department of Engery's Applied Mathematical Sciences program. A.N.L. acknowledges a travel grant from the Royal Society of London.

REFERENCES

1. H. Poincaré, *Les Méthodes Nouvelles de la Mécanique Céleste*. Gauthier-Villars, Paris, 1899.
2. R. Shaw, Strange attractors, chaotic behavior, and information flow. *Z. Naturforsch.* **36a** 80–112 (1981).

3. F. Takens, Detecting strange attractors in turbulence. In *Dynamical Systems and Turbulence*. Springer-Verlag, New York, 1980.

4. N. H. Packard, J. P. Crutchfield, J. D. Farmer, and R. Shaw, Geometry from a time series. *Phys. Rev. Lett.* **45** 712–716 (1980).

5. J. D. Farmer and J. J. Sidorowich, Exploiting chaos to predict the future and reduce noise. In *Evolution, Learning and Cognition* (Y. C. Lee, ed.). World Scientific, Singapore, 1988.

6. J. P. Crutchfield and B. S. McNamara, Equations of motion from a data series. *Complex Systems* **1** 417–452 (1987).

7. J. Cremers and A. Hübler, Construction of differential equations from experimental data. *Z. Naturforsch.* **42a** (1987), 797–802.

8. G. Sugihara and R. M. May, Nonlinear forecasting as a way of distinguishing chaos from measurement error in time series, *Nature* **344** 734–741 (1990).

9. P. Cvitanović, Periodic orbits as the skeleton of classical and quantum chaos. *Physica D* **5** 138–151 (1991).

10. G. B. Mindlin, X.-J. Hou, H. G. Solari, R. Gilmore, and N. B. Tufillaro, Classification of strange attractors by integers. *Phys. Rev. Lett.* **64** 2350–2353 (1990).

11. J. M. T. Thompson and L. N. Virgin, Predicting a jump to resonance using transient maps and beats. *Int. J. Nonlinear Mech.* **21** 205–216 (1986).

12. L. N. Virgin, Parametric studies of the dynamic evolution through a fold. *J. Sound Vibration* **110** 99–109 (1986).

13. K. Wiesenfeld, Noisy precursors of nonlinear instabilities. *J. Statist. Phys.* **38** 1071–1097 (1985).

14. C. Jeffries and K. Wiesenfeld, Observation of noisy precursors of dynamical instabilities. *Phys. Rev. A* **31** 1077–1084 (1985).

15. K. Wiesenfeld, Virtual Hopf phenomenon: A new precursor of period-doubling bifurcations. *Phys. Rev. A* **32** 1744–1751 (1985).

16. F. C. Moon and P. J. Holmes, A magnetoelastic strange attractor. *J. Sound Vibration* **65** 285–296 (1979).

17. P. J. Holmes, A nonlinear oscillator with a strange attractor. *Philos. Trans. Roy. Soc. A* **292** 419–448 (1979).

18. N. Bloembergen, Comments on the dissociation of polyatomic molecules by intense 10.6 μm radiation. *Optics Commun.* **15** 416–418 (1975).

19. J. M. T. Thompson, R. C. T. Rainey, and M. S. Soliman, Ship stability criteria based on chaotic transients from incursive fractals. *Phil. Trans. Roy. Soc. A* **332** 149–167 (1990).

20. E. C. Zeeman, Bifurcation and catastrophe theory. In *Papers in Algebra, Analysis, and Statistics*. Amer. Math. Soc., Providence, RI, 1982.

21. R. H. Abraham, Chaostrophes, intermittency, and noise. In *Chaos, Fractals, and Dynamics*. Dekker, New York, 1985.

22. H. B. Stewart and J. M. T. Thompson, Towards a classification of generic bifurcations in dissipative dynamical systems. *Dyn. Stab. Systems* **1** 87–96 (1986).

23. C. Grebogi, E. Ott, and J. A. Yorke, Crises, sudden changes in chaotic attractors, and transient chaos. *Physica D* **7** 181–200 (1983).

24. R. M. May, Simple mathematical models with very complicated dynamics. *Nature* **261** 459–467 (1976).

25. R. H. Day, Irregular growth cycles. *Amer. Econ. Rev.* **72** 406–414 (1982).

26. R. Shaw, *The Dripping Faucet as a Model Chaotic System.* Aerial Press, Santa Cruz, CA, 1984.

27. A. H. Nayfeh and D. T. Mook, *Nonlinear Oscillations.* Wiley, New York, 1979.

28. J. M. T. Thompson, Chaotic phenomena triggering escape from a potential well. *Proc. Roy. Soc. A* **421** 195–225 (1989).

29. H. B. Stewart, A chaotic saddle catastrophe in forced oscillators. In *Dynamical Systems and Circuits.* (F. Salam and M. Levi, eds.), pp. 138–149. SIAM, Philadelphia, 1988.

30. M. Diener, The canard unchained, or how fast/slow dynamical systems bifurcate. *Math. Intelligencer* **6** 38–49 (1984).

PART V

GENERAL TOPICS

CHAPTER 18

THE POWER OF CHAOS

PREDRAG CVITANOVIĆ
Niels Bohr Institute
Blegdamsvej 17
DK-2100 Copenhagen Ø
Denmark

18.1. INTRODUCTION

Nonlinear physics presents us with a perplexing variety of complicated fractal objects and strange sets. Notable examples include strange attractors for chaotic dynamical systems, regions of high vorticity in fully developed turbulence, and fractal growth processes. The word "chaos" has in this context taken on a narrow technical meaning: To say that a certain system exhibits chaos means that the system obeys deterministic laws of evolution but that the outcome is highly sensitive to small uncertainties in the specification of the initial state. In a chaotic system any open ball of initial conditions, no matter how small, will in finite time spread over the extent of the entire asymptotically accessible phase space. Once this is grasped, the focus of theory shifts from attempting precise prediction (which is impossible) to description of the geometry of the space of possible outcomes, and evaluation of averages over this space. By now most of us appreciate the fact that the phase space of a generic nonlinear dynamical system is an infinitely interwoven mixture of islands of stability and regions of chaos. Possible trajectories are qualitatively of three distinct types: They are either asymptotically unstable (positive Lyapunov exponent), asymptotically marginal (vanishing Lyapunov exponent), or asymptotically stable (negative Lyapunov exponent). Here we concentrate on the unstable orbits that build up chaos. Furthermore, we concentrate on the quantum mechanics of systems that are classically chaotic.

Today this physics is perhaps not as "applicable" as the chaos-theory–inspired potential improvements in weather prediction, but I suspect that

Applied Chaos, Edited by Jong Hyun Kim and John Stringer.
ISBN 0-471-54453-1 © 1992 John Wiley & Sons, Inc.

Bohr's treatment of electrons as planetary orbits around atomic nucleus did not appear particularly "applicable" in its day either. What we describe here is very much in the spirit of early quantum mechanics, and were physicists of that period as familiar with classical chaos as we are today, this physics would have been developed in the 1920s. The main idea is this: In the Bohr–de Broglie visualization of quantization, one places a wave instead of a particle on a Keplerian orbit around the hydrogen nucleus. The quantization condition is that only those orbits for which such a wave is a standing wave are allowed; from this follow the Balmer spectrum, colors in the world around us, and the more sophisticated quantum theory of Schrödinger and others. Today we are very aware of the fact that elliptic orbits are a peculiarity of the Kepler problem and that chaos is the rule; so we can generalize the Bohr quantization to chaotic systems? The answer was provided by Gutzwiller [1, 2]: A chaotic system can be quantized by placing a wave on each of the infinity of unstable periodic orbits. We shall develop here the theory of such averages over infinities of unstable orbits.* However, before launching into the requisite mathematics, we want to emphasize that the periodic orbits in quantum mechanics are not just a theorist's toy; the theory itself arose from studies of eminently applicable semiconductors, and unstable periodic orbits have been measured in experiments [6] on the very paradigm of Bohr's atom, the hydrogen atom, this time in an external field. Not that far in the future, classical mechanics textbooks will have a chapter on the Keplerian motions followed by a chapter on chaos, and quantum mechanics textbooks will have a chapter on the Bohr atom followed by a chapter on a classically chaotic atom.

In retrospect many triumphs of both classical and quantum physics seem a stroke of luck: A few integrable problems, such as the harmonic oscillator and the Kepler problem, though "nongeneric," have gotten us very far. Success has lulled us into a habit of expecting simple solutions to simple equations—an expectation shattered for many by the recently acquired ability to scan numerically the phase space of nonintegrable dynamical systems. The initial impression might be that all our analytic tools have failed us and that chaotic systems are amenable only to numerical statistical investigations. However, as we show here, we already possess a perturbation theory of the deterministic chaos of predictive quality comparable to that of the traditional perturbation expansions for nearly integrable systems. This theory is based on the observation that the motion in dynamical systems of few degrees of freedom is often organized around a few *fundamental* cycles. These short cycles capture the skeletal topology of the motion in the sense that any long orbit can be approximately pieced together from the fundamen-

---

*The important role played by periodic orbits was already noted by H. Poincaré [3] and has been at the core of much of the mathematical work on the theory of the dynamical systems ever since. We refer the reader to the classic text of Ruelle [4] and the reprint selection [5] for a summary and references to the mathematical literature.

tal cycles. Computations with such systems require techniques reminiscent of statistical mechanics; however, the actual calculations are crisply deterministic throughout. The strategy will be to expand averages over chaotic phase space regions in terms of short unstable periodic orbits, with the small expansion parameter being the nonuniformity of the flow (here referred to as "curvature") across neighborhoods of periodic points. One of the surprises is that the quantum mechanics of classically chaotic systems is very much like like the classical mechanics of chaotic systems; one needs nearly the same zeta functions and cycle expansions, with the same group theory factorizations and dependence on the topology of the classical flow. To get some feeling for how and why unstable cycles come about, we start by playing a game of pinball.

18.2. PINBALL CHAOS

A physicist's pinball [7–11] game is a pinball reduced to its bare essentials: three disks in a plane (see Figure 18.1).

Pointlike pinballs are shot at the disks from random starting positions and angles; they spend some time bouncing between the disks and then escape. For a physicist a good game consists of predicting the asymptotic lifetime (or the escape rate) of a pinball to many significant digits. The unstable cycles as a skeleton of chaos are almost visible here: a good strategy for keeping the ball bouncing as long as possible is to aim it as close as possible to a periodic trajectory. Short periodic trajectories are easily drawn and enumerated—some examples are given in Figure 18.1—but it is rather hard to extract the systematics of the orbits from their physical space trajectories.

A clearer picture of the dynamics is obtained by constructing a phase space Poincaré section. We start [12] by exploiting the sixfold symmetry of the disks and restrict the pinball to bouncing in the fundamental domain (Figure 18.1b). The whole pinball plane can be retiled by six copies of the fundamental domain, but the details are inessential for present considerations. We define our Poincaré section by marking x_i, the position of the ith bounce off the bottom wall (Figure 18.1b), and $\sin \phi_i$, the momentum component parallel to the bottom wall; $(x, \sin \phi)$ coordinates are a convenient choice, because they are phase-space volume-preserving [13].

We next mark the initial conditions $\Omega_{.\epsilon_1}$ that do not escape in one bounce. There are two strips of survivors, because the trajectories originating from one disk can hit either of the other two disks or can escape. We label the two strips with $\epsilon = 0, 1$. There are four strips $\Omega_{.\epsilon_1\epsilon_2}$ that survive four bounces, and so forth. Another way to look at the survivors after two bounces is to plot $\Omega_{\epsilon_1 \cdot \epsilon_2}$, the intersection $\Omega_{.\epsilon_2}$ with the strips Ω_{ϵ_1}. obtained by time reversal ($\sin \phi \rightarrow -\sin \phi$). Provided that the disks are sufficiently well separated, what emerges is a complete binary Cantor set with the usual Smale horseshoe [14] foliation and symbolic dynamics.

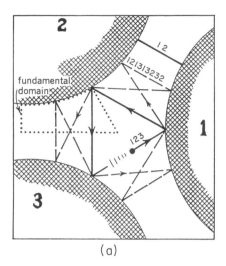

(a)

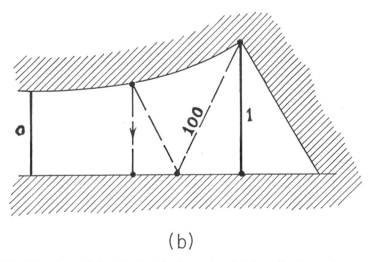

(b)

Figure 18.1. The three-disk pinball with the ratio of disk radius to center separation a: $R = 1:2.5$: (a) the three disks, with $\overline{12}$, $\overline{123}$, and $\overline{121232313}$ cycles indicated; (b) the fundamental domain, that is, a wedge consisting of a section of a disk, two segments of symmetry axes acting as straight mirror walls, and an escape gap. The preceding cycles restricted to the fundamental domain are now the two fixed points $\overline{0}$ and $\overline{1}$ and the $\overline{100}$ cycle.

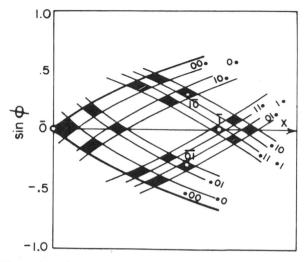

Figure 18.2. The Poincaré section of the phase space for the fundamental domain pinball, Figure 18.1b. Indicated are the fixed points $\bar{0}, \bar{1}$ and the 2-cycle $\overline{01}, \overline{10}$, together with strips that survive $1, 2, \ldots$ bounces. Iteration corresponds to bit shift; for example, region $\ldots 01.01 \ldots$ maps into $\ldots 010.1 \ldots$.

After n iterations the survivors are divided into 2^n distinct neighborhoods: The ith neighborhood consists of all points with itinerary $i = \epsilon_1 \epsilon_2 \epsilon_3 \ldots \epsilon_n$ $\epsilon_i = \{0, 1\}$. Each such patch contains a periodic point $\overline{\epsilon_1 \epsilon_2 \epsilon_3 \ldots \epsilon_n}$ with the basic block infinitely repeated. Periodic points are skeletal in the sense that as we look further and further, they stay put forever, while the finite covers shrink onto them. The periodic points are dense on the asymptotic repeller and their number increases exponentially with cycle length. As we shall see, this exponential proliferation of cycles is not as dangerous as it might seem.

Before continuing with the pinball as an illustration of the origin and structure of unstable cycles, we turn briefly to the role of cycles in more general settings.

18.3. CYCLES AS THE SKELETON OF CHAOS

Consider a general three-dimensional flow, sketched in Figure 18.3. To be interesting, the flow should be recurrent; otherwise it is a transient state that we cannot observe for long times. If the flow is recurrent, we can cut it by a Poincaré section; if it is a map of a compact disk domain onto itself, it must have at least one fixed point. Now consider the ways in which the flow can deform the neighborhood of a fixed point. There are essentially two possibilities: the neighborhood can return wrapped around the fixed point (the fixed

Recurrent flows: stable, winding

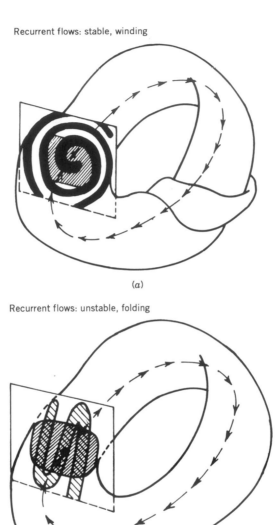

Recurrent flows: unstable, folding

(b)

Figure 18.3. A recurrent flow around (a) an elliptic fixed point and (b) a hyperbolic fixed point.

point is stable or elliptical—see Figure 18.3a) or squeezed, stretched, and folded (the fixed point is unstable or hyperbolic—see Figure 18.3b).

In the traditional approach, we use integrable motions of Figure 18.3a as zeroth-order approximations to physical systems and account for weak non-linearities perturbatively. We tend to think of a dynamical system as a smooth system whose evolution can be followed by integrating a set of differential equations. When this is actually followed through to asymptotic times, we discover that the strongly nonlinear systems show an amazingly rich structure that is not at all apparent in their formulation in terms of differential equations. However, hidden in this apparent chaos is a rigid skeleton, a tree of *cycles* (periodic orbits) of increasing lengths and self-similar structure. The new insight is that the zeroth-order approximations to harshly chaotic dynamics should be very different from those for the nearly integrable systems: A good starting approximation here is the stretching and kneading of a baker's map of Figure 18.3b, rather than the winding of a harmonic oscillator of Figure 18.3a.

For low-dimensional deterministic dynamical systems the cycles (periodic orbits) provide a possibly optimal invariant description of a dynamical system, with the following virtues:

1. Cycle symbol sequences are *topological* invariants: They give the spatial layout of a strange set.

2. Cycle eigenvalues are *metric* invariants: They give the scale of each piece of a strange set.

3. Cycles are *dense* on the asymptotic nonwandering set.

4. Cycles are ordered *hierarchically*: Short cycles give good approximations to a strange set, longer cycles only refinements. Errors due to neglecting long cycles can be bounded and typically fall off exponentially with the cutoff cycle length.

5. Cycles are *structurally robust*: eigenvalues of short cycles vary slowly with smooth parameter changes.

6. Asymptotic averages (such as generalized dimensions, escape rates, quantum mechanical eigenstates, and other "thermodynamic" averages) can be computed from short cycles by means of *cycle expansions*.

Points 1 and 2: That the cycle topology and eigenvalues are invariant properties of dynamical systems follows from elementary considerations. If the same dynamics is given by a map f in one set of coordinates and by a map g in the next, then f and g (or any other good representation) are related by a reparametrization and a coordinate transformation $f = h^{-1} \circ g \circ h$. Because both f and g are arbitrary representations of the dynamical system, the explicit form of the conjugacy h is of no interest, only the properties invariant under any transformation h are of general import. The most obvious invariant properties are topological; a fixed point must be a fixed point in any representation, a trajectory that exactly returns to the initial point (a cycle) must do so in any representation (Figure 18.4). Furthermore, a good representation should not mutilate the data; h must be a smooth transformation that maps nearby cycle points of f into nearby cycle points of g. This smoothness guarantees that the cycles are not only topological invariants, but that their linearized neighborhoods are also metrically invariant. In particular, the cycle eigenvalues [eigenvalues of the Jacobians $df^{(n)}(x_k)/dx$ of periodic orbits $f^{(n)}(x_k) = x_k$] are invariant.

Point 3: The cycles are intuitively expected to be *dense* because on a connected chaotic set a typical trajectory is expected to behave ergodically and to pass infinitely many times arbitrarily close to any point on the set, including the initial point of the trajectory itself. Generically, one expects to be able to move the initial point gently in such a way that the trajectory returns precisely to the initial point (Figure 18.5). This is by no means guaranteed to work, and it must be checked for the particular system at hand. A variety of ergodic (but insufficiently mixing) counterexamples can be constructed—the most familiar being a quasiperiodic motion on a torus.

Point 5: An important virtue of cycles is their *structural robustness.* Many quantities customarily associated with dynamical systems depend on the

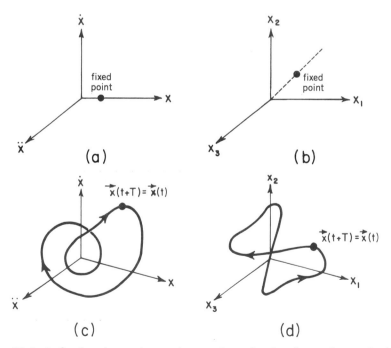

Figure 18.4. A fixed point and a cycle remain a fixed point and a cycle in any representation of a dynamical system. Here (*a*) and (*c*) phase space is built from x and its derivatives; (*b*) and (*d*) might be the same trajectories in the time-delay coordinates.

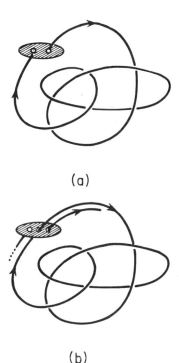

(a)

(b)

Figure 18.5. (*a*) A close recurrence of an unstable trajectory (*b*) can be exploited to locate a nearby cycle.

notion of "structural stability" [14], that is, robustness of strange sets to small parameter variations. This is certainly not a property of generic dynamical systems, such as the Hénon map [15] $(x, y) \rightarrow (1 - ax^2 + y, bx)$. For example, although numerical studies indicate that for $a = 1.4$ and $b = 0.3$ the attractor is "strange," parameter variation as minute as changing a to $a = 1.39945219$ destroys this attractor and replaces it with a stable cycle of length 13. Still, the short unstable cycles of length less than 13 are structurally robust in the sense that they are only slightly distorted by such parameter changes, and averages computed using them as a skeleton are insensitive to small deformations of the strange set. In contrast, lack of structural stability wreaks havoc with long time averages such as Lyapunov exponents, for which there is no guarantee that they converge in any finite numerical computation.

The theoretical advance that we will concentrate on here is *point 4:* We now know how to control the errors due to neglecting longer cycles. As we explain in Section 18.5, even though the number of invariants is infinite (unlike, for example, the number of Casimir invariants for a compact Lie group) the dynamics can be well approximated to any finite accuracy by a small finite set of invariants. The origin of this convergence is geometrical, as we now show by returning to our game of pinball.

18.4. PINBALL ESCAPE RATE

Consider Figure 18.2 again. In each bounce the initial conditions get thinned out, yielding twice as many thin strips as at the previous bounce. The phase-space volume is preserved by the flow, so they are contracted along the stable eigendirections and are ejected along the unstable eigendirections; the total fraction of survivors after n bounces is proportional to

$$\Gamma_n = \sum_i^{(n)} l_i \tag{18.1}$$

where i is a binary label of the ith strip, and l_i is the width of the ith strip. One expects the sum (18.1) to fall off exponentially with n and to tend to a limit

$$\Gamma_n = e^{-n/T_n} \to e^{-n/T} = e^{-n\gamma} \tag{18.2}$$

T is the *asymptotic lifetime* of a random initial pinball; $\gamma = 1/T$ is the pinball *escape rate*.* We shall now show that this asymptotic escape rate can be extracted from a highly convergent *exact* expansion by reformulating the sum (18.1) in terms of unstable periodic orbits.

Each neighborhood i in Figure 18.2 contains a periodic point $\bar{i}$. The finer the intervals, the smaller is the variation in flow across them, and the strip width l_i is well approximated by the contraction around the periodic point, $l_i = a_i/|\Lambda_i|$. Here Λ_i is the expanding eigenvalue of the linearized map evaluated on the periodic point i, and a_i is a prefactor defined by

$$a_i = l_i |\Lambda_i| \tag{18.3}$$

Now we make the only approximation in our derivation of the ζ function: For large n the prefactors $a_i \approx O(1)$ are overwhelmed by the exponential growth of Λ_i, so we neglect them. This is called the *hyperbolicity* assumption. The a_i reflect the particular distribution of incoming pinballs. The asymptotic trajectories are strongly mixed by bouncing chaotically between the disks, and we expect them to be insensitive to smooth variations in the initial distribution. If the hyperbolicity assumption is justified, we can replace l_i in (18.1) by $1/\Lambda_i$ and form a formal sum over all periodic orbits of all lengths:

$$\Omega(z) = \sum_{n=1}^{\infty} z^n \sum_i^{(n)} |\Lambda_i|^{-1}$$

$$= \frac{z}{|\Lambda_{\bar{0}}|} + \frac{z}{|\Lambda_{\bar{1}}|} + \frac{z^2}{|\Lambda_{\overline{00}}|} + \frac{z^2}{|\Lambda_{\overline{01}}|} + \frac{z^2}{|\Lambda_{\overline{10}}|} + \frac{z^2}{|\Lambda_{\overline{11}}|}$$

$$+ \frac{z^3}{|\Lambda_{\overline{000}}|} + \frac{z^3}{|\Lambda_{\overline{001}}|} + \cdots \tag{18.4}$$

*A lucid introduction to escape from repellers is given in reference [16]. For a review of transient chaos, see references [17] and [18]. The ζ-function formulation is given in reference [19].

As for large n, the nth-level sum (18.2) tends to the limit $e^{-n\gamma}$, the escape rate γ is determined by the smallest $z = e^{\gamma}$ for which (18.4) diverges:

$$\Omega(z) \approx \sum_{n=0}^{\infty} (ze^{-\gamma})^n = \frac{1}{1 - ze^{-\gamma}} \qquad (18.5)$$

This observation is the motivation for introducing the sum (18.4). Rather than attempting to estimate the escape rate from the $n \to \infty$ limit of preasymptotic sums (18.1), we shall determine γ from the singularities of (18.4).

If a trajectory retraces itself r times, its expanding eigenvalue is Λ_p^r, where p is a *prime* cycle. A prime cycle is a single traversal of the orbit; its label is a nonrepeating symbol string. There is only one prime cycle for each cyclic permutation class. For example, $p = \overline{0011} = \overline{1001} = \overline{1100} = \overline{0110}$ is prime, but $\overline{0101} = \overline{01}$ is not. The stability of a cycle is (by the chain rule) the same everywhere along the orbit, so each prime cycle of length n_p contributes n_p terms to the sum (18.4). Hence (18.4) can be rewritten as

$$\Omega(z) = \sum_p n_p \sum_{r=1}^{\infty} \left(z^{n_p} |\Lambda_p^{-1}| \right)^r = \sum_p \frac{n_p z^{n_p} |\Lambda_p^{-1}|}{1 - z^{n_p} |\Lambda_p^{-1}|}$$

where the index p runs through all distinct *prime* cycles. The $n_p z^{n_p}$ factors in the sum suggest rewriting it as a logarithmic derivative

$$\Omega(z) = -z \frac{d}{dz} \ln \zeta(z)$$

The resulting infinite product

$$\frac{1}{\zeta(z)} = \prod_p (1 - t_p), \qquad t_p = z^{n_p} |\Lambda_p^{-1}| \qquad (18.6)$$

is an example of a dynamical ζ function.* The name is motivated by the (purely formal) similarity of the infinite product to the Euler product representation of the Riemann ζ function.

The preceding derivation of the ζ-function formula for the escape rate has one shortcoming—it estimates the fraction of survivors as a function of the number of pinball bounces. However, the physically interesting quantity is the mean lifetime; giving the same weight to all paths with the same number of bounces overestimates the contributions of the long trajectories and underestimates the short trajectories (remember, the flight paths between disks are of different lengths). The correct weight [19, 20] is obtained by

---

*See reference [4], Section 7.23.

replacing the discrete "topological" time n_p by the actual cycle period T_p in (18.6):

$$t_p = e^{T_p \gamma} |\Lambda_p^{-1}| \tag{18.7}$$

Perhaps more surprisingly, (18.6) also yields *quantum* resonances, with the quantum amplitude associated with a given cycle

$$t_p = \frac{1}{\sqrt{\Lambda_p}} \exp\left(\frac{i}{\hbar} S_p(E) + i\pi m_p\right) \tag{18.8}$$

essentially (in a somewhat vague sense) the square root of the classical weight (we refer the reader to references [10–12] for detailed discussions).

Expression (18.6) is the main result of this section; the problem of estimating the asymptotic escape rates from infinite n sums such as (18.1) is now reduced to a study of the singularities of the ζ function (18.6). The escape rate is related by (18.5) to a divergence of $\Omega(z)$, and $\Omega(z)$ diverges whenever $1/\zeta(z)$ or $\zeta(z)$ has a zero. Glancing back, we see that the derivation is very general and should work for any average over any strange set that satisfies two conditions: (1) The weight associated with a part of the set is multiplicative along the trajectory. (2) The set is organized in such a way that the nearby points in the symbolic dynamics have nearby weights.

We conclude this section by noting that if one is interested in correlation spectra [19] rather than just the leading resonances, the correct function to study is not (18.6), but a Selberg-type product discussed elsewhere [21, 22]. For present purposes, $1/\zeta$ is good enough, and a much more transparent illustration of the shadowing of long cycles by short cycles is discussed later.

18.5. CYCLE EXPANSIONS AND CURVATURES

How are the formulas such as (18.6) used? We start by computing the lengths and eigenvalues of shortest cycles. In our pinball example this can be done by elementary geometrical optics; in general potentials or maps, this requires some numerical integrations and Newton's method searches for periodic solutions. The result is a table like Table 18.1.

The next step is the key step in our approach [23, 24]: We observe that the expansion of the Euler product (18.6)

$$1/\zeta = 1 - \sum_f t_f - \sum_n c_n \tag{18.9}$$

allows a regrouping of terms into dominant *fundamental* contributions t_f and decreasing *curvature* corrections c_n. For example, if the strange set is labeled by binary symbol sequences, as is the pinball of Figure 18.3, then the

TABLE 18.1. Lengths and Eigenvalues of Shortest Cycles
(for Ratio of Disk Radius to Center Separation $a:R = 1:6$)

p	T_p	Λ_p
0	4.0	9.89897948557
1	4.26794919243	-11.7714551964
10	8.31652948517	-124.094801992
010	12.3217466162	-1240.54255704
101	12.580807741	1449.54507485
0100	16.3222764744	-12295.706862
1011	16.8490718592	-17079.0190089
1001	16.5852429061	14459.97595
00100	20.3223300257	-121733.838705
$\vdots$	$\vdots$	$\vdots$

Euler product (18.6) is given by

$$1/\zeta = (1 - t_0)(1 - t_1)(1 - t_{10})(1 - t_{100})(1 - t_{101})(1 - t_{1000})$$
$$\times (1 - t_{1001})(1 - t_{1011})(1 - t_{10000})(1 - t_{10001})$$
$$\times (1 - t_{10010})(1 - t_{10011})(1 - t_{10101})(1 - t_{10111}) \cdots \quad (18.10)$$

The curvature expansion is obtained by multiplying out the Euler product and grouping together the terms of the same total symbol string length:

$$1/\zeta = 1 - t_0 - t_1 - [t_{10} - t_1 t_0] - [(t_{100} - t_{10} t_0) - (t_{101} - t_{10} t_1)]$$
$$- [(t_{1000} - t_0 t_{100}) + (t_{1110} - t_1 t_{110})$$
$$+ (t_{1001} - t_1 t_{001} - t_{101} t_0 + t_{10} t_0 t_1)] - \cdots \quad (18.11)$$

The fundamental cycles t_0 and t_1 have no shorter approximates; they are the "building blocks" of the dynamics in the sense that all longer orbits can be approximately pieced together from them. We call the sum of all terms of same total length n [grouped in brackets in (18.11)] the nth curvature correction c_n, for geometrical reasons we shall soon try to explain. It is also often possible to pair off individual longer cycles and their shorter approximants [grouped in parentheses in (18.11)]. If all orbits are weighted equally $(t_p = z^{n_p})$, such combinations cancel exactly; if orbits of similar symbolic dynamics have similar weights, the weights in such combinations will almost cancel.

Given the curvature expansion (18.11), the calculation is straightforward. We substitute the eigenvalues and lengths of prime cycles (for the example at hand, up to 5 pinball bounces—a total of 14 cycles) into the curvature expansion (18.11) and obtain a polynomial approximation to $1/\zeta$. We then vary the exponent γ in (18.7) and determine the escape rate γ by finding the

TABLE 18.2. Escape Rates [Equation (18.11)]

n	Full Three Disks	Fundamental Domain
1		0.407693
2	0.43578	0.410280
3	0.40491	0.410336
4	0.40945	0.410338
5	0.41037	0.410338
6	0.41034	

smallest γ for which (18.11) vanishes. The zeros are easily determined by standard numerical methods (such as the Newton algorithm), with accuracy as good as seven significant digits for the pinball example considered here.

The convergence can be illustrated by listing γ (Table 18.2) computed from truncations of (18.11) to different maximal cycle lengths (the ratio of disk radius to disk–disk center separation is $a : R = 1 : 6$). The first column gives the maximal cycle length used, the second the estimate of the classical escape rate from the full three-disk cycle expansion, the third from the fundamental domain expansion [12].

For comparison, a numerical simulation of reference [9] yields $\gamma = 0.410\ldots$, and the $n = 2, 3$ approximations of reference [9] yield 0.3102 and 0.4508, respectively.

If one has some experience with numerical estimates of dimensions, one realizes that the convergence here is very impressive; only three input numbers (the two fixed points $\overline{0}$ and $\overline{1}$ and the 2-cycle $\overline{10}$) already yield the escape rate to four significant digits! We have omitted an infinity of unstable cycles; so why does approximating the dynamics by a finite number of cycle eigenvalues work so well?

A typical curvature expansion term in (18.11) is a *difference* of a long cycle $\{ab\}$ minus its shadowing approximation by shorter cycles $\{a\}$ and $\{b\}$:

$$
t_{ab} - t_a t_b = t_{ab}\left(1 - \frac{t_a t_b}{t_{ab}}\right) = t_{ab}\left(1 - \left|\frac{\Lambda_{ab}}{\Lambda_a \Lambda_b}\right| \exp[(T_a + T_b - T_{ab})\gamma]\right)
$$

$$(18.12)$$

To understand why this should be small compared to t_{ab}, try to visualize the description of a chaotic dynamical system in terms of cycles as a tessellation of the dynamical system, with smooth flow approximated by its periodic orbit skeleton, each "face" centered on a periodic point, and the scale of the face determined by the linearization of the flow around the period point (see Figure 18.6).

The orbits that follow the same symbolic dynamics, such as $\{ab\}$ and a "pseudo-orbit" $\{a\}\{b\}$, lie physically close; longer and longer orbits resolve

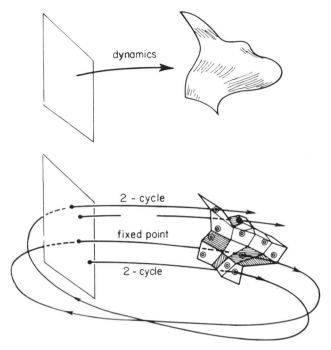

Figure 18.6. Approximation to (*a*) a smooth dynamics by (*b*) the skeleton of periodic points, together with their linearized neighborhoods.

the dynamics with finer and finer resolution in the phase space. If the weights associated with the orbits are multiplicative along the flow (for example, products of derivatives) and the flow is smooth, the term in parentheses in (18.12) falls off *exponentially* with the cycle length, and therefore the curvature expansions are expected to be highly convergent.

We have used here the curvature expansions to evaluate pinball escape rates, but the technique is much more general. It is applicable [24] to a broad class of "thermodynamical" averages such as those used in the extraction of generalized dimensions, as well as to the quantum periodic orbits sums [10, 11] to which we now turn.

18.6. QUANTUM CYCLE EXPANSIONS

In this section we illustrate the structure of quantum resonances by considering fcw simple cycle expansions [11].

The simplest conceivable system with exponentially growing errors has only one periodic orbit (for example, there is only one unstable orbit bouncing between two disks in a plane), and the cycle expansion (18.11) is simply

$$\frac{1}{\zeta(k)} = 1 - \frac{\exp(iLk + i\pi m)}{\sqrt{\Lambda}} \tag{18.13}$$

where Λ is the expanding eigenvalue of the cycle Jacobian, L is the cycle length, m is the Maslov index (here $m = 2$ for the two-disk reflections), and the free motion action is $S(k) = \int \mathbf{p} \cdot d\mathbf{q} = \hbar k L$. The resonances $1/\zeta(k_n) = 0$ are given by

$$k_n = \frac{\pi}{L}(m + 2n) - i\frac{\lambda}{2}, \qquad n \in Z \tag{18.14}$$

where $\lambda = \ln|\Lambda|/T$ is the cycle Lyapunov exponent (for constant-velocity, $v = 1$, motion the cycle period T equals the cycle length L). The real part of the wave number k goes through a resonance for each n, the number of nodes in a standing wave, and the imaginary part indicates that the resonance is unstable, with the decay rate proportional to the cycle Lyapunov exponent.

The $1/\zeta(k)$ function is only the leading part of the Gutzwiller quantum mechanical periodic orbit expansion; the full expression is given by the Selberg-type zeta function $Z(k) = \prod 1/\zeta_r(k)$, which for a single fixed point takes a particularly simple form:

$$Z(k) = \prod_{r=0}^{\infty} \left(1 - \frac{\exp(iLk + i\pi m)}{\sqrt{\Lambda}\,\Lambda^r}\right) \tag{18.15}$$

with spectrum

$$k_{r,n} = \frac{\pi}{L}(m + 2n + rs) - i\lambda\left(\frac{1}{2} + r\right), \qquad n \in Z, r = 0, 1, 2, \ldots \tag{18.16}$$

where $s = 0$ for positive Λ, and $s = 1$ for negative Λ. Hence the full Gutzwiller spectrum forms a lattice in the complex k plane, spaced horizontally by π/L and vertically by λ. In practice, only the resonances $\text{Im}(k) = -\lambda/2$ closest to the real axis are important. The more-distant resonances, which give exponentially small contributions, are as we shall see, of considerable theoretical interest. The exact Green's functions for disks in a plane are known [25–27] and this could be a good playground for studying effects of the nonleading terms in semiclassical expansions. In the present context it suffices to make a mental note of the fact that the simplest conceivable problem already has an infinity of resonances in the negative $\text{Im}(k)$ direction. It is also instructive to plot $|Z(k)|$ or the phase of $Z(K)$ in the complex k plane. For $\text{Im}(k) > \lambda/2$, the $e^{iLk}/\Lambda^{1/2+r}$ term in (18.15) is always smaller than 1, $Z(k)$ is of order 1 throughout the upper complex k half-plane, and there are no resonances there. However, for large negative $\text{Im}(k)$, the second terms in (18.15) factors grow exponentially in magnitude, and the nice regular lattice of zeros (18.16) hides a wildly varying $Z(k)$. Also note that although the cycle expansions of $1/\zeta(k)$ can yield numerically very accurate leading resonance spectra [10, 11], in contrast to $Z(k)$, they cannot be extended to the nonleading resonances [28, 29]; the breakdown of analyticity is accompanied with generation of spurious zeros along the border of analyticity.

This example illustrates qualitatively the quantum mechanical Gutzwiller spectrum; one expects infinite families of zeros, with the convergence of the theory controlled by their imaginary parts. The imaginary parts converge for reasons we have already discussed in context of cycle expansions on classical strange sets [the imaginary k axis corresponds to the real γ axis in (18.7)]; complex phases only add oscillations that do not increase the magnitude of terms in cycle expansions and thus do not destroy their convergence.

Although unstable, the two-disk system is not chaotic: It has only one orbit, rather than an exponentially proliferating number (positive entropy) of unstable orbits. The effect of entropy can be illustrated by another simple model, a dynamical system with complete binary symbolic dynamics and uniform Lyapunov exponent for all orbits. As we have seen, if the symbolic dynamics is a subshift of finite type, a cycle expansion (18.6) for a chaotic system (an infinity of unstable cycles) consists of fundamental cycles and curvature corrections. If the dynamics is binary and expansion is uniform, the curvature corrections vanish and the factors in (18.15) become simply

$$\frac{1}{\zeta_r(k)} = 1 - 2t$$

$$t = \frac{e^{iLk}}{\sqrt{\Lambda}\,\Lambda^r} \tag{18.17}$$

Although there is most likely no quantum system with such simple cycle structure (in classical dynamics a tent map is an example [24] of such system), we need this model only to illustrate the qualitative effect of positive entropy on quantum spectra.

The factor 2 in (18.17) reflects the fact that in this model the number of topologically distinct unstable trajectories grows as 2^t; the spectrum has the same lattice structure as for the single unstable cycle, but all the resonances are now shifted *upward* in the $\text{Im}(k)$ direction by the entropy $h = \ln 2/T$:

$$k_{r,n} = \frac{\pi}{L}(m + 2n + rs) - i\left(\lambda\left(\frac{1}{2} + r\right) - h\right). \tag{18.18}$$

Positive entropy *lowers* the decay rate; the physical reason is that for a chaotic system the local instability of the system (Lyapunov exponent) is being compensated by the backscatter into the system (the number of orbits that have not escaped by time t, i.e., the entropy). In classical bound systems probability conservation amounts to the relation $\lambda = h$ [30], but the quantum mechanical situation is subtler: The factor $\frac{1}{2}$ in $\lambda/2 - h$ in (18.18) reflects the fact that the quantum mechanical probability equals (amplitude)^2. The probability (or unitarity) conservation constraints in quantum chaotic context are still poorly understood. Although the main effect of the positive entropy is to bring the leading resonance closer to the Re(k) axis, the whole lattice of nonleading resonances in the lower complex k half-plane *remains*, and the spacing between the nonleading resonances is still given by the Lyapunov exponent.

We emphasize this, because even if the bound-state leading resonances moved exactly onto the real axis, it is hard to imagine a mechanism by which contributions of individual cycles of wildly differing phases and magnitudes would conspire to eliminate all the nonleading resonances in the complex plane. On the other hand, existence of such resonances is a direct violation of quantum mechanical unitarity that guarantees that the bound-state energy eigenvalues (for example, the hydrogen spectrum) are real. So even if cycle expansions yield the energy eigenvalues with good accuracy, we expect that the semiclassical saddle-point approximation that was the basis for the Gutzwiller cycle formulas will always lead to spurious subexponential eigenvalues in the complex energy plane.

It therefore comes as something of a shock to learn that for the genuine Selberg zeta [31],* interpretable as the quantum mechanical spectrum of a Laplacian on a space of negative constant curvature, this unlikely conspiracy takes place, and the cycle expansion in this context is *exact*.

The preceding examples are simplistic, but they do capture the essential features of quantum resonance spectra for problems with sufficiently simple symbolic dynamics; one gest qualitatively right estimates of resonance locations and separations, beat oscillations in spectra [11], and so on. We illustrate this by a brief discussion of the anisotropic Kepler problem.

18.7. ANISOTROPIC KEPLER PROBLEM

In this section we apply the cycle expansions technique to the anisotropic Kepler problem, given by the Hamiltonian

$$E = -\frac{1}{2} = \frac{u^2}{2\mu} + \frac{v^2}{2\nu} - \frac{1}{\left(x^2 + y^2\right)^{1/2}} \tag{18.19}$$

---

*See, for example, reference [32].

This model has been studied extensively by Gutzwiller; we refer the reader to his excellent monograph [2] for an overview of theoretical and experimental work related to the problem.

Although in the isotropic Kepler problem an orbit returns precisely to the initial position and velocity after one Kepler period, in the anisotropic Kepler problem a generic orbit misses the initial point and continues wandering chaotically through the phase space—numerical evidence [2] even suggests that chaotic trajectories take the full measure of the available phase space.

The anisotropic Kepler problem arises in solid-state physics, where electrons associated with donor impurities in silicon and germanium semiconductor crystals have anisotropic (and experimentally measurable) effective masses. For example [33], for silicon the anisotropy mass ratio is $\mu/\nu = 4.810$, and for germanium $\mu/\nu = 19.48$. The quantum mechanical spectra are known to high accuracy, both theoretically [34] and experimentally [35]. Furthermore, Gutzwiller [36] has obtained by Ising model techniques rather good estimates for dozens of the lowest eigenvalues, using 50 cycles of topological length 10. Hence it is possible to test the spectra obtained by the cycle expansion methods against both the exact spectra and those obtained by trace formulas.

The anisotropic Kepler problem is also uniquely suited to testing the validity of quantum cycle expansions for a deeper theoretical reason: Gutzwiller has provided strong evidence that, for anisotropies as large as the physically interesting ones, all cycles (computed so far) are unstable and in 1–1 correspondence with a simple symbolic dynamics. For a Hamiltonian system such a claim is very surprising; we know of no other example of a bound Hamiltonian system with simple symbolic dynamics. However, as explained previously, for simple symbolic dynamics the cycles can be combined into "shadowed" curvature combinations, and cycle expansions are expected to converge well.

We have carried out a preliminary check of cycle expansions for the anisotropic Kepler problem, and the results are promising. The cycle expansion evaluation of spectra for the anisotropic Kepler problem differs from Gutzwiller's periodic orbit calculation [36] in several important aspects:

1. The zeta functions that we expand are already asymptotic, arranged in such a way that the spectrum is dominated by short cycles, with long cycles contributing exponentially small curvature corrections. This is expected to lead to exponentially better convergence than that obtainable from the Gutzwiller trace sums, which require extrapolation to asymptotics from finite cycle lengths.

2. The backbone of our calculation are collision orbits, which serve as fundamental cycles; those were omitted from previous calculations.

3. The fact that series of cycles accumulate to limits with finite stability and action might cause cycle expansions to diverge. However, our preliminary results indicate that the Maslov phases induce cancellations

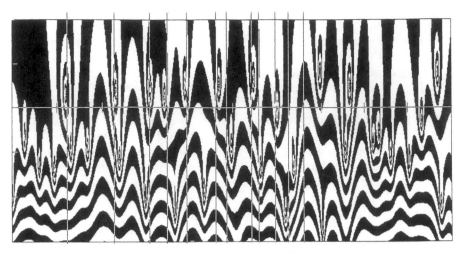

Figure 18.7. The symmetric A_1 subspace $\log|1/\zeta_{A_1}(k)|$ as a function of the complex wave number k for the anisotropic Kepler problem with anisotropy $\mu/\nu = 5$, estimated from the cycle expansion (18.11) truncated to the fundamental domain cycles up to topological length 6. If $n < \log|1/\zeta(k)| \leq n + 1$, the complex k point is marked black for n odd, white for n even. The zeros of $|1/\zeta(k)|$ are the centers of the rapidly varying elliptic regions. The quantum mechanical eigenvalues from reference [34] lie on the real k axis at values indicated by vertical lines. Note that the $\mathrm{Re}(k)$ and $\mathrm{Im}(k)$ scales differ by a factor of about 50 and that the leading resonances lie on the real axis within the numerical convergence of the cycle expansions (estimated to about 5%).

    between successive terms in the cycle expansion and that the remainders fall off exponentially.

4. Symmetries are used to restrict the dynamics to a fundamental domain, to factorize the spectra, and to simplify the symbolic dynamics.

Inserting the x axis collision orbit together with a total of 18 cycles gleaned from the tables of reference [37] into the cycle expansions (18.11), we obtain the spectrum of Figure 18.7. Cycle expansion of the Selberg-type zeta function [35, 22] yields (within the accuracy of the present calculation) the same leading eigenvalues.

These results are preliminary in the sense that we have not rechecked the Maslov and symmetry factors used and that a new calculation of cycles [39] dictated by the fundamental domain symbolic dynamics is needed. If all works well, the accuracy of semiclassical expansions based periodic orbit theory should be improvable by orders of magnitude beyond the original results of reference [36] and place us into position to attack more seriously the problem of unitarity, the WKB corrections, and so on.

Briefly, cycle expansions for the anisotropic Kepler problem are set up as follows [39].

18.7.1. Scale Conventions

The homogeneity of the potential (18.19) can be used to scale out the energy dependence and bring the action to a form analogous to the quantum phase in the pinball problem (18.13)

$$\frac{1}{\hbar} S_p(E) = kT_p(E) \tag{18.20}$$

where the wave number k is the square root of the energy expressed in Rydberg units,

$$k = \sqrt{\frac{E_R}{E}}, \qquad E_R = -\frac{m_0 e^4}{2\kappa^2 \hbar^2} \tag{18.21}$$

and e = electric charge. For silicon $m_0 = (m_1 m_2)^{1/2} = 0.4177$ electron mass, and the dielectric constant $\kappa = 11.4$. The quantum cycle weight (18.8) is given by

$$t_p = \frac{1}{\sqrt{\Lambda_p}} \exp(ikT_p + i\pi m_p) \tag{18.22}$$

with m_p the Maslov index of the p cycle.

18.7.2. Symbolic Dynamics

Effective use of cycle expansions requires identification of the shortest, fundamental cycles and control of the symbolic dynamics for long cycles.

Anisotropy $\mu > \nu$ causes the orbits that cross the y axis to fall faster toward the x axis, forcing them to cross the x axis before the next y-axis crossing. Because the motion along the x axis is sluggish compared with the motion across it, the orbit can cross (for sufficiently large [40, 41] anisotropy $\mu > \frac{9}{8}$) the x axis any number of times before the next y axis crossing. Let x_i be the initial x-axis crossing, and x_{i+1} the next crossing in the same direction. Then any trajectory can be pieced together from the following four types of topologically distinct segments [37], distinguished by whether the segment does or does not cross the y axis:

$$
\begin{aligned}
11 &\leftarrow x_i > 0, & x_{i+1} > 0 \\
01 &\leftarrow x_i > 0, & x_{i+1} < 0 \\
10 &\leftarrow x_i < 0, & x_{i+1} > 0 \\
00 &\leftarrow x_i < 0, & x_{i+1} < 0
\end{aligned} \tag{18.23}
$$

In terms of the labels $\{11, 10, 01, 00\}$, the dynamics is a complete four-symbol dynamics. This symbolic dynamics, introduced by Gutzwiller and Devaney, has not been proven to be of covering type: It is known that for every symbol sequence there exists at least one orbit, but it is only conjectured [2] that this orbit is unique. The conjecture is based on investigations of the foliation of the phase space, together with much numerical evidence [42], but it runs very much against what we know about generic Hamiltonian systems. Indeed, elliptic islands have been found [43] for small anisotropies (not much above $\mu/\nu > \frac{9}{8}$), and in another model that has for a long time looked chaotic, careful symbolic dynamics work has ferreted out stability islands [44]. Such islands, if they exist for $\mu/\nu \approx 5\text{–}20$, are certainly very small, and we shall proceed as though the anisotropic Kepler problem is fully chaotic.

18.7.3. Symmetrization

Our cycle expansions start with a rather technical, but extremely useful step for setting up actual calculations [12]: We restrict the dynamics to the fundamental domain. The symmetry of the two-dimensional anisotropic Kepler problem is C_{2v}, the group of x- and y-axis reflections and rotations by 180°. We take a quadrant in the x–y plane as the fundamental domain. A global trajectory is reduced to the fundamental domain by reflecting the trajectory back into the fundamental domain at every symmetry line crossing; we split the trajectory into segments between successive x-axis crossings and mark the segment by symbol "i" if it includes a y-axis reflection, and "o" otherwise. The global symbolic dynamics $\ldots s_n s_{n+1} s_{n+2} \ldots$ (18.23) is reduced to the fundamental domain symbolic dynamics [13] by a simple rule; $a_n = o$ if $s_n = s_{n+1}$, and $a_n = 0$ otherwise. For example $0111 \to iioo$ and $0000011111 \to ioooo$.

This reduces the global unrestricted four-symbol dynamics to the simplest imaginable symbol dynamics, the unrestricted binary dynamics (just as in the three-disk problem discussed previously).

Reduction to the fundamental domain also alters the cycle weights and eigenvalues by group-theoretic factors, and there are further subtleties concerning the proper weight for the o-collision orbit, which is also a boundary orbit [45]. In our preliminary calculation, the total effect of the Maslov phases and symmetry factors in the A_1 symmetric subspace is an overall $(-1)^{n_p}$ sign factor in (18.22), with n_p = topological length of the fundamental domain cycle.

18.7.4. Fundamental Cycles

The next step in construction of cycle expansions is identification of the fundamental cycles. In our formulation of the anisotropic Kepler problem, there are six shortest global orbits that play the role of fundamental cycles.

The first type of fundamental cycle is rather obvious. The Kepler problem has a pair of counterrotating circular orbits; in the anisotropic Kepler problem there is a corresponding pair of counterrotating "pseudocircular" orbits. In the fundamental domain, the two global pseudocircular orbits reduce to the i fixed point, an orbit starts on the x axis and is bounced back into itself by a reflection off the y axis. In the crudest attempt [46] are periodic orbit quantization, one simply uses the action of these orbits to find "standing wave" configurations; for reasons explained in the previous section, this is a fair attempt at getting a rough estimate of the spectrum, although omission of the infinity of other orbits necessarily yields energy eigenvalues with negative imaginary parts.

The next class of fundamental cycles comprises the four collision orbits along the x and y axes. Although they have been ignored in the previous trace calculations [37], they play the central role in our cycle expansions for the anisotropic Kepler problem, helping us to control the asymptotics of periods and stabilities of long cycles. The anisotropy destroys all Keplerian ellipses with the exception of these four collision orbits, whose periods are given by the third Kepler law [47]:

$$T = \frac{2\pi}{a^{3/2}m^{1/2}} \tag{18.24}$$

where m is the mass and a is the major semiaxis. With the $E = -\frac{1}{2}$ scale convention, the collision orbits reach the boundary of the x–y space at $r = 2$, so the major semiaxis is $a = 1$. For the silicon anisotropy ratio $\mu/\nu = 4.810$ (with mass scale fixed by setting $\mu\nu = 1$) the collision orbit periods are

$$T_x = 2\pi\mu^{1/4} = 9.305 \ldots$$

$$T_y = \frac{2\pi}{\mu^{1/4}} = 4.2427 \ldots \tag{18.25}$$

The stabilities of the collision orbits are also known in analytic form; Yoshida [48] gives a pretty formula for the expanding eigenvalue,

$$\Lambda_y = 1 + 2c^2 + 2|c|\sqrt{1 + c^2}, \qquad c = \cos\left(\frac{\pi}{2}\sqrt{9 - \frac{8\mu}{\nu}}\right) \tag{18.26}$$

We have not checked this formula numerically, but if it is correct, it implies that the stability of the collision orbit along the y axis is finite, but so large (of order 10^7 for silicon) that we can safely omit it for the time being from cycle expansions.

18.7.5. Cycle Accumulation Sequences

The x-axis collision orbit is densely enveloped by infinite sequences of orbits with arbitrarily large numbers of successive x-axis crossings (long

...$ooooo$... subsequences). We identify this collision orbit with the $\bar{o}$ fixed point, of well-defined Kepler period $T_o = T_x$. Numerical results indicate that the periods of the sequence of orbits $io, ioo, io \ldots oo$ that twine around the $\bar{o}$ collision orbit converge to

$$T_{io^n} = T_x - \frac{\text{constant}}{(3.10\ldots)^n} \tag{18.27}$$

Similarly, sequences of cycles of type $iioo \ldots oo$ hug closer and closer to an x-collision orbit followed by a y-collision orbit, and their periods converge to $T_y + T_x = 13.59737$ with the same geometric factor. However, the stability exponents in this sequence are very large, as for the y-axis collision orbit, and longer cycles can be safely omitted.

Numerical work indicates that also the stabilities of the io^n cycles converge geometrically to a finite limit $\mu_{io^\infty} = 3.?$, although not nearly as smoothly as for the cycle periods.

As is clear from references [29] and [49], if there are many orbits of comparable stability, they have to be summed before exponential convergence of the cycle expansion can work. An infinity of orbits of type $io, ioo, io \ldots oo$, all of the same stability and action, presages disaster; the cycle expansion might not converge. However, we are saved by the fact that for each bounce there is a Maslov prefactor -1. The geometric convergence (18.27) of the actions and the stabilities leads to geometric convergence of the alternating infinite series $\Sigma t_{\ldots o^n}$. Accumulations of such sequences suggest that just as in other problems with cycle accumulations [20, 49], the natural labels are not binary, but the numbers of consecutive occurrences [41] of symbol o.

Putting all these pieces together with the numerical values of actions and stabilities for all cycles up to given length (here up to length 6) we obtain a cycle expansion for $1/\zeta(k)$ and extract the resonance spectrum plotted in Figure 18.7. By estimates of Section 18.6, the nonleading resonances should have $\text{Im}(k)$ of order $-\lambda$. For the silicon anisotropy $\lambda \approx 0.2$, in qualitative agreement with the typical nonleading resonance distance to the real axis in Figure 18.7. For $1/\zeta(k)$ the nonleading resonances are outside the region of convergence, and the nonleading zeros in this figure are spurious effects of the finite convergence radius on the truncated cycle expansions. This is partially fixed by using $Z(k)$ cycle expansions, which for hyperbolic systems are entire functions in the k complex plane. Still, in all fairness, the nonleading resonances are uncomfortably close to the real axis, and further refinements of the theory are needed. In order to improve the spectrum we need either to go beyond the semiclassical leading terms or find a smart way of imposing unitarity. In this connection we note that the expansions advocated in reference [50] are actually the same as those we have already been using [22, 51, 52]. We find that the conjecture that Riemann–Siegel–type formulas exist is unlikely to hold in this context; the Riemann–Siegel formulas use the self-duality of Riemann and other zeta functions, but there is no evidence of such symmetry for generic Hamiltonian flows zeta functions.

18.8. SUMMARY

We have illustrated here the utility of cycles and their curvatures by the simple pinball example of Figure 18.1 and by a preliminary check of the spectrum of the anisotropic Kepler problem. The cycle expansions such as (18.9) outperform the pedestrian methods such as the nth-level trace sum (18.1) because the higher terms in the cycle expansion (18.9) are deviations of longer primitive cycles from their approximations by shorter cycles $c_{10} = -t_{10} + t_1 t_0$, $c_{1001} = -t_{1001} + t_1 t_{001} + t_{101} t_0 - t_{10} t_0 t_1, \ldots$, which vanish exactly in polygonal approximations and are expected to fall off exponentially for smooth hyperbolic flows.

In our approach the lessons of classical periodic orbit theory play a crucial role insofar as they dictate expansions that would probably seem less than obvious if one started with semiclassical expansions of quantum Green's functions: in particular, omission of the collision orbits would destroy shadowing and would make cycle expansions not any more convergent than the trace formulas used previously. Much of the spectrum can be qualitatively understood already in terms of the fundamental cycles. Comparison with the exact spectrum of [34] and Gutzwiller's best result [36] shows that, with considerably less labor, cycle expansions reproduce the spectrum to comparable accuracy; one basically constructs a polynomial of low order (fifth, for example) in the topological cycle length and determines its zeros.

To summarize, a motion on a strange attractor can be approximated by shadowing the orbit by a sequence of nearby periodic orbits of finite length. This notion is made precise here by decomposing orbits into primitive cycles and evaluating associated curvatures. A curvature measures the deviation of a longer cycle from its approximation by shorter cycles; the smoothness of the dynamical system implies exponential falloff for (almost) all curvatures. The technical prerequisite for implementing this shadowing is a good understanding of the symbolic dynamics of the classical dynamical system. The resulting curvature expansions offer an efficient method for evaluating classical and quantum periodic orbit sums; accurate estimates can be obtained by using as input the actions and eigenvalues of as few as 2–14 prime cycles.

ACKNOWLEDGMENTS

This chapter is based on common work with R. Artuso, E. Aurell, F. Christiansen, B. Eckhardt, and H. H. Rugh. We acknowledge stimulating exchanges with P. Dahlqvist, G. Russberg, and D. Wintgen. The author is grateful to the Carlsberg Foundation for support, to Edificio Celi for hospitality in Milano, and to M. C. Gutzwiller, for whom the anisotropic Kepler cycle expansions described were handcrafted, for many illuminating discussions and support at the IBM T. J. Watson Research Center.

REFERENCES

1. M. C. Gutzwiller, *J. Math. Phys.* **8** 1979 (1967); **10** 1004 (1969); **11** 1791 (1970).

2. M. C. Gutzwiller, *Chaos in Classical and Quantum Mechanics*. Springer, New York, 1990.

3. H. Poincaré, *Les Méthodes Nouvelles de la Méchanique Céleste*. Gauthier-Villars, Paris, 1892–1899.

4. D. Ruelle, *Statistical Mechanics, Thermodynamic Formalism*. Addison-Wesley, Reading, MA, 1978.

5. R. S. MacKay and J. D. Miess, *Hamiltonian Dynamical Systems*. Adam Hilger, Bristol, 1987.

6. A. Holle, J. Main, G. Wiebusch, H. Rottke, and K. H. Welge, *Phys. Rev. Lett.* **61** 971 (1988).

7. D. L. Rod, *J. Diff. Equations* **14** 129 (1973). R. C. Churchill, G. Pecelli, and D. L. Rod, *J. Diff. Equations* **17** 329 (1975). R. C. Churchill, G. Pecelli, and D. L. Rod, in *Como Conference Proceedings on Stochastic Behavior in Classical and Quantum Hamiltonian Systems* (G. Casti and J. Ford, eds.), pp. 112–123. Springer, Berlin, 1976.

8. B. Eckhardt, *J. Phys. A* **20** 5971 (1987).

9. P. Gaspard and S. A. Rice, *J. Chem. Phys.* **90**, 2225, 2242, 2255 (1989).

10. P. Cvitanović and B. Eckhardt, *Phys. Rev. Lett.* **63** 823 (1989).

11. P. Cvitanović, B. Eckhardt, and P. Scherer, unpublished.

12. P. Cvitanović and B. Eckhardt, submitted to *Nonlinearity*.

13. G. D. Birkhoff, *Acta Math.* **50** 359 (1927). (Reprinted in reference [6].)

14. S. Smale, *Bull. Am. Math. Soc.* **73** 747 (1967).

15. M. Hénon, *Comm. Math. Phys.* **50** 69 (1976).

16. L. P. Kadanoff and C. Tang, *Proc. Natl. Acad. Sci.* **81** 1276 (1984).

17. T. Tél, in *Directions in Chaos* (B.-L. Hao, ed.), vol. 3, p. 149. World Scientific, Singapore, 1988.

18. S. Bleher, C. Grebogi, and J. A. Yorke, *Physica* **46** 87 (1990).

19. D. Ruelle, *J. Statist. Phys.* **44** 281 (1986). W. Parry and M. Pollicott, *Ann. Math.* **118** 573 (1983).

20. P. Cvitanović and B. Eckhardt, *J. Phys. A* **24** L237 (1991).

21. F. Christiansen, G. Paladin, and H. H. Rugh, *Phys. Rev. Lett.* **65** 2087 (1990).

22. P. Cvitanović, *Physica D* **51** 138 (1991).

23. P. Cvitanović, *Phys. Rev. Lett.* **61** 2729 (1988).

24. R. Artuso, E. Aurell, and P. Cvitanović, *Nonlinearity* **3** 325 (1990).

25. A. Wirzba, *Chaos 2*, 89 (1992).

26. W. Franz, *Theorie der Beugung Elektromagnetischer Wellen*. Springer-Verlag, Berlin, 1957.

27. D. S. Jones, *The Theory of Electromagnetism*, Chapter 8. Pergamon, Oxford, 1964.

28. D. Ruelle, *Inventiones Math*. **34** 231 (1976).

29. R. Artuso, E. Aurell, and P. Cvitanović, *Nonlinearity* **3** 361 (1990).

30. Ya. B. Pesin, *Uspekhi Mat. Nauk* **32** 55 (1977) [*Russian Math. Surveys* **32** 55 (1977)].

31. A. Selberg, *J. Ind. Math. Soc*. **20** 47 (1956).

32. A. Terras, *Harminic Analysis on Symmetric Spaces and Applications I*. Springer-Verlag, Berlin, 1985. H. P. McKean, *Comm. Pure Appl. Math*. **25** 225 (1972); **27** 134 (1974).

33. B. W. Levinger and D. R. Frankel, *J. Phys. Chem. Solids* **20** 281 (1961). J. C. Hensel, H. Hasegawa, and M. Nakayama, *Phys. Rev*. **138** 225 (1965).

34. D. Wintgen, H. Marxer, and J. S. Briggs, *J. Phys. A* **20** L965 (1987).

35. H. Navarro, E. E. Haller, and F. Keilmann, *Phys. Rev. B* **37** 1082 (1988).

36. M. C. Gutzwiller, *Physica D* **5** 183 (1982).

37. M. C. Gutzwiller, in *Classical Mechanics and Dynamical Systems* (R. L. Devaney and Z. H. Nitecki, eds.), pp. 69–90. Marcel Dekker, New York, 1981.

38. A. Voros, *J. Phys. A* **21** 685 (1988).

39. F. Christiansen and P. Cvitanović, *Chaos* **2** 61 (1992).

40. M. C. Gutzwiller, *J. Math. Phys*. **18** 806 (1977).

41. R. Devaney, Transverse heteroclinic orbits. In *The Structure of Attractors in Dynamical Systems. Lecture Notes in Math*. **668** 271. Springer-Verlag, Berlin 1978.

42. M. C. Gutzwiller, *Physica D* **38** 160 (1989).

43. R. Broucke, in *Dynamical Astronomy* (V. Szebehely and B. Balasz, eds.) pp. 9–20. University of Texas Press, Austin, 1985.

44. P. Dahlqvist and G. Russberg, *Phys. Rev. Lett*. **65** 2837 (1990).

45. B. Lauritzen, *Phys. Rev. A* **43** 603 (1990).

46. M. C. Gutzwiller, *J. Math. Phys*. **12** 343 (1971).

47. J. Kepler, *Harmonice Mundi*. Linz, 1619.

48. H. Yoshida, *Celestial Mech*. **40** 51 (1987).

49. P. Cvitanović, G. H. Gunaratne, and M. J. Vinson, *Nonlinearity* **3** 873 (1990).

50. M. V. Berry and J. P. Keating, *J. Phys. A* **23** 4839 (1990).

51. F. Christiansen, P. Cvitanović, and H. H. Rugh, *J. Phys. A* **23** 713L (1990).

52. P. Dahlqvist and G. Russberg, *J. Phys. A* **24** 4763 (1991).

CHAPTER 19

SUDDEN CHANGE IN THE SIZE OF CHAOTIC ATTRACTORS: HOW DOES IT OCCUR?

YING-CHENG LAI,* CELSO GREBOGI,† and JAMES A. YORKE‡
University of Maryland
College Park, MD 20742

Boundary and interior crises are typical phenomena in nonlinear dynamical systems. In the case of an interior crisis, there is a sudden increase in the size of a chaotic attractor as the parameter passes through a critical value. Because unstable periodic orbits are believed to be dense on the attractor, at the interior crisis there is also a sudden increase in the number of unstable periodic orbits on the attractor. In this chapter, we investigate the origin of these unstable periodic orbits that become part of the attractor at the crisis. We demonstrate that at the interior crisis the chaotic attractor collides with a chaotic saddle already existent when the crisis occurs. This chaotic saddle is an invariant and nonattracting set having horseshoe-type dynamics. We also investigate the origin and evolution of the chaotic saddle.

19.1. INTRODUCTION

Almost all the sudden changes of chaotic attractors [1], as a parameter of the system varies, are due to crises [2–4]. One usually distinguishes between boundary crises and interior crises. In the former case, a chaotic attractor,

---

*Department of Physics and Astronomy, and Laboratory for Plasma Research.
†Laboratory for Plasma Research, Department of Mathematics, and Institute for Physical Science and Technology.
‡Department of Mathematics, and Institute for Physical Science and Technology.

Applied Chaos, Edited by Jong Hyun Kim and John Stringer.
ISBN 0-471-54453-1 © 1992 John Wiley & Sons, Inc.

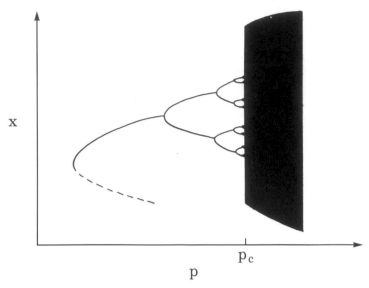

Figure 19.1. A typical bifurcation diagram of a chaotic dynamical system with an interior crisis at $p = p_c$, where p is the parameter to be varied and x is one of the phase space variables.

together with its basin of attraction, is suddenly destroyed after a collision with an unstable periodic orbit on the basin boundary as the parameter passes through a critical value. While in the latter case, which we study in this chapter, a chaotic attractor suddenly increases its size in phase space as a parameter passes through the crisis value. Because unstable periodic orbits are believed to be dense on chaotic attractors, at the interior crisis there is also a sudden increase in the number of unstable periodic orbits. Figure 19.1 shows schematically a bifurcation diagram of some typical dynamical system exhibiting an interior crisis, where the horizontal axis p denotes one of the parameters to be varied and the vertical axis x is one of the dynamical variables. At the crisis value of the parameter $p = p_c$, there is a sudden increase in the extent of the variable x.

As an example of interior crisis, consider the Ikeda map [5], which is an idealized model of a laser cavity, as shown in Figure 19.2. The map is given by

$$z_{n+1} = A + Bz_n \exp\left[ik - \frac{ip}{1 + |z_n|^2} \right] \tag{19.1}$$

where $z = x + iy$ is a complex number, $|z|$ is the laser magnetic field amplitude, and angle(z) is the phase of the magnetic field of the laser pulse. The parameter A is related to laser input amplitude, B is the coefficient of

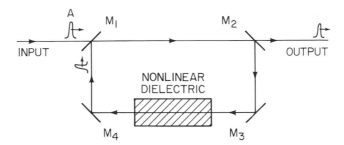

Figure 19.2. The Ikeda map can be viewed as arising from a string of light pulses of amplitude A entering at the partially transmitting mirror M_1, the time interval between the pulses is adjusted to be the round-trip travel time in the system. Let $|z_n|$ be the amplitude and angle (z_n) be the phase of the nth pulse just to right of mirror M_1. Then the terms in (19.1) have the following meaning: $(1 - B)$ is the fraction of energy in a pulse transmitted or absorbed in the four reflections from M_1, M_2, M_3, and M_4; k is the round-trip phase shift that would be experienced by the pulse in the absence of the nonlinear medium; $-p/(1 + |z_n|^2)$ is the phase shift due to the presence of the nonlinear medium.

the partially reflecting mirror, k is the laser empty cavity phase detuning, and p measures the detuning due to the nonlinear medium. We choose the following values of parameters: $A = 0.85$, $B = 0.9$, $k = 0.4$, and p is the parameter to be varied. Figure 19.3 shows both the chaotic attractor below and above the interior crisis in phase space. The crisis occurs at the parameter value $p_c = 5.169789\ldots$. Figure 19.3a shows a four-piece chaotic attractor at $p = 5.1697$ just below the crisis value. Figure 19.3b shows a single larger chaotic attractor just above the crisis at $p = 5.1700$.

In this chapter we address the following question: Where does the incremental portion of the chaotic attractor come from as p increases through p_c? Consequently, where do the new unstable periodic orbits come from? We argue that the incremental portion comes from a chaotic saddle that already exists for parameter values below the crisis. This chaotic saddle is an invariant and nonattracting set having horseshoe-type dynamics and resembles the new portion of the larger attractor just above the crisis in phase space, as shown in Figure 19.3c and 19.3d. At the interior crisis this chaotic saddle (Figure 19.3c) collides with the smaller attractor (Figure 19.3a), and together they make up the larger attractor (Figure 19.3b) above the crisis. We argue then that the unstable periodic orbits added to the attractor at the crisis come from this invariant chaotic saddle formed before the crisis. In Section 19.2, we demonstrate that the chaotic saddle results from an infinite number of chaotic attractors that are created by saddle-node bifurcations followed by period-doubling bifurcations and are destroyed by boundary crisis [2, 3] leaving behind chaotic saddles in phase space. These saddles link together through a complex chain of crossings of stable and unstable mani-

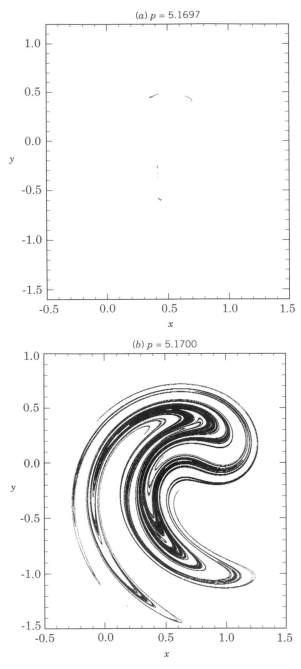

Figure 19.3. The chaotic attractors: (a) before (p = 5.1697) and (b) after (p = 5.1700) the interior crisis in phase space for the Ikeda map at A = 0.85, B = 0.9, k = 0.4; the crisis occurs at $p = p_c$ = 5.169789 . . . .

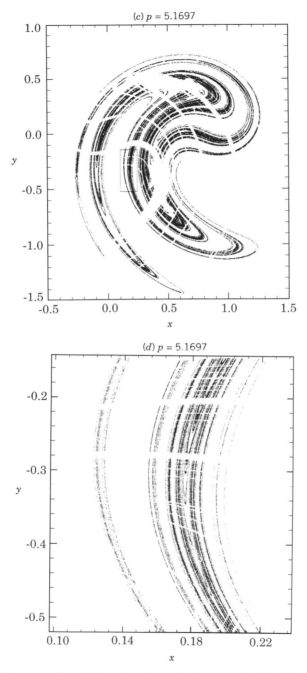

Figure 19.3. *(Continued).* (*c*) The chaotic saddle for *p* = 5.1697; (*d*) blowup of the boxed region in (*c*).

folds to form the *chaotic saddle* that collides with the attractor at the interior crisis. In Section 19.3, we use the PIM-triple [6, 7] method (described in the appendix) to find the chaotic saddle for a parameter value slightly less than the crisis value and we compare it with the incremental portion of the attractor just above the crisis. A second example of interior crisis in the Ikeda map is also given as a further illustration of our results. In Section 19.4, we give the conclusion. The appendix describes the PIM-triple algorithm.

19.2. ORIGIN OF THE CHAOTIC SADDLE

In this section, we described the formation of the chaotic saddle that collides with the attractor at the interior crisis. We illustrate our arguments with the Ikeda map because we believe this map has properties that are typical of two-dimensional diffeomorphisms and three-dimensional flows.

19.2.1. The Evolution of the Main Attractor

Figure 19.4 shows a bifurcation diagram at $A = 0.85$, $B = 0.9$, $k = 0.4$, and p varying in the range $[2.5, 5.5]$, where the horizontal axis denotes the parameter p and the vertical axis denotes the dynamical variable x. The

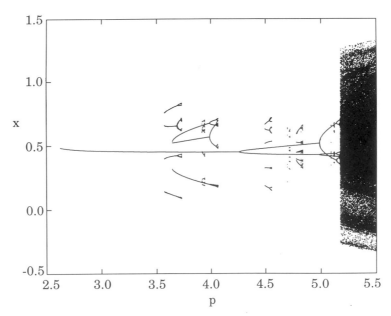

Figure 19.4. The bifurcation diagram of the Ikeda map at $A = 0.85$, $B = 0.9$, $k = 0.4$, and p in the range $[2.5, 5.5]$; the vertical axis is the dynamical variable x. The interior crisis of the main attractor occurs at $p_c = 5.169789\ldots$.

bifurcation diagram is obtained by varying p gradually in the range $[2.5, 5.5]$; at each fixed p value we choose a number of initial conditions, say 100, distributed along a line segment from $(x_a, y_a) = (-0.5, -1.5)$ to $(x_b, y_b) = (1.5, 1.0)$. For each initial condition, we first iterate the map for a sufficiently large number of times to get rid of the transient, then plot the subsequent few hundred iterates. The main attractor is created by a period-1 saddle-node bifurcation at $p_1 \simeq 2.58$, followed by a period-doubling cascade that accumulates at $p \simeq 5.16$. The chaotic attractor evolves into a four-piece attractor, as shown in Figure 19.3a. At the interior crisis, the four-piece attractor merges into a single piece, as shown in Figure 19.3b.

19.2.2. Evolution of Other Coexisting Attractors

Besides the main attractor, there are an infinite number of other attractors [8]. The creation and destruction of some of these chaotic attractors can be seen in Figure 19.4. Most of them exist only for a very narrow interval of the parameter p and cannot be seen in the scale of the figure. However, the bifurcations of some low-period attractors still can be seen. The most notable ones are the saddle-node bifurcations of period 3 at $p_3 \simeq 3.6397$, of period 4 at $p_4 \simeq 3.5854$, and of period 5 at $p_5 \simeq 4.4865$. The subsequent destructions of the corresponding attractors result from boundary crises. At the boundary crises, the small chaotic attractors are converted into small chaotic saddles. In general, the larger the period of the saddle-node bifurcation, the smaller the parameter range in which the corresponding attractor can exist. Actually, when the period tends to infinity, the measure of the set of parameter values in parameter space for the existence of the corresponding attractor tends to zero [8].

To illustrate the evolution of one of these attractors, we choose the period-3 attractor as an example. Figure 19.5 is a blowup of part of the bifurcation diagram of Figure 19.4 for p in the range of $[3.60, 4.10]$. In this parameter range, a six-piece chaotic attractor evolves from a period-3 saddle-node bifurcation. This six-piece attractor is destroyed at $p_{3c} = 4.0672\ldots$ as the result of a boundary crisis. In the parameter range of $p \in [p_3, p_{3c}]$, the main attractor is a period-1 stable fixed point. When the period-3 saddle-node bifurcation occurs, it creates a basin of attraction for this attractor by cutting a chunk out of the the basin of the main attractor [9]. At the boundary crisis, the attractor, which is a six-piece attractor at $p = p_{3c}$, collides with the basin boundary. For $p > p_{3c}$, the six-piece chaotic attractor is converted into a chaotic saddle having horseshoe-type dynamics containing an infinite number of unstable saddle orbits [3] (periodic and aperiodic). This period-3 chaotic saddle, together with other saddles formed by similar process, will eventually make up the *chaotic saddle*, like the one in Figure 19.3c, that will collide with the main attractor at the interior crisis $p = p_c$.

We still lack a complete understanding of the dynamics of most of those infinite number of saddle orbits. However, there is better understanding of a

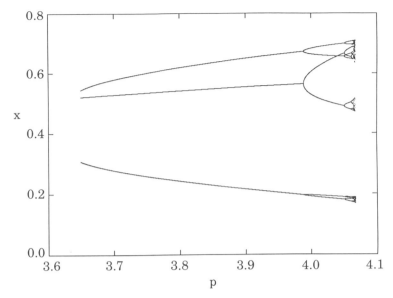

Figure 19.5. A blowup of part of the bifurcation diagram in Figure 19.4*a* for the parameter *p* in the range [3.6, 4.1] in which a six-piece chaotic attractor evolves from a period-3 saddle-node bifurcation (at $p_3 \simeq 3.6397$) and is then destroyed by boundary crisis at $p_{3c} \simeq 4.0672$.

particular class, the so-called *simple Newhouse orbits* [9, 10]. These orbits are a subset of the periodic saddles created at the original saddle-node bifurcations. They have the property that the associated eigenvalues of the linerized map are both positive. Simple Newhouse orbits proceed once around a circuit along the unstable manifold, as shown in Figure 19.6. Although there are an infinite number of simple Newhouse orbits, they are still a comparatively small subset of all the saddles created in the evolution of the system. Nevertheless, it is sufficient for us to understand how they link together and how they contribute to the formation of the large chaotic saddle that collides with the main attractor at $p = p_c$. In the following, we restrict ourselves to the evolution of simple Newhouse orbits to demonstrate the origin of this chaotic saddle.

19.2.3. The Origin of the Chaotic Saddle

As the parameter is increased from the period-1 saddle-node bifurcation value, the main attractor undergoes a period-doubling bifurcation. At the first period doubling, the node becomes unstable and is transformed into a period-1 flip saddle (with eigenvalue -1). This flip saddle is responsible for connecting different simple Newhouse orbits and for forming the chaotic saddle. Numerically, we observed that the stable and unstable manifolds of

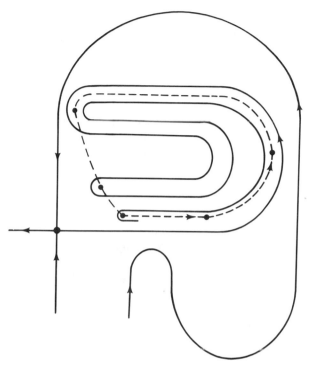

Figure 19.6. Schematic illustration of a period-5 simple Newhouse orbit; dashed arrows indicate the sequence in which points on the orbit occur.

simple Newhouse orbits cross each other and form a chain of heteroclinic crossings. In particular, the unstable manifolds of all those simple Newhouse orbits cut across the stable manifold of the period-1 flip saddle. All these heteroclinic crossings form a complicated chaotic saddle having horseshoe-type dynamics. As the parameter p increases, more and more simple Newhouse orbits are created and they add into this chaotic saddle which becomes more and more complicated. As the main attractor and the chaotic saddle grow in size in the phase space as p increases, eventually they collide with each other at the interior crisis. This results in a sudden increase in the size of the chaotic attractor and a sudden increase in the number of the unstable periodic orbits on the attractor. Thus, the chaotic saddle provides the new unstable periodic orbits added to the main attractor at the crisis.

To justify this scenario further, we proceed as follows.

19.2.3.1. Crossing of the Unstable Manifolds of Simple Newhouse Orbits with the Stable Manifold of the Period-1 Flip Saddle. We observe

numerically that the unstable manifolds of period-3, period-4, period-5, and other higher-period simple Newhouse orbits cut across the stable manifold of the period-1 flip saddle. In fact, this is true for all values of p for which both the flip saddle and the simple Newhouse orbits exist. We denote the crossing of the stable manifold of the period-1 flip saddle ($1fs$) with the unstable manifold of the period-N simple Newhouse periodic orbit (Nu) by

$$1fs \times Nu, \qquad N = 3, 4, 5, 6, \ldots \qquad (19.2)$$

19.2.3.2. Crossing of the Stable Period-N Manifold with Unstable Period-(N + 1) Manifold.

We have also examined numerically the stable and unstable manifolds of the simple Newhouse orbits, and we have observed the following: The stable manifold of the period-N orbit always crosses the unstable manifold of the period-$(N + 1)$ orbit. In general, we have

$$Ns \times (N + 1)u, \qquad N = 3, 4, 5, \ldots \qquad (19.3)$$

19.2.3.3. Heteroclinic Manifold Crossings.

We now review the significance of the crossings of a stable manifold and an unstable manifold when each is associated with different periodic saddle. Whenever we have

$$Ms \times Nu \qquad (19.4)$$

for the period-M and period-N saddles, the following then holds [9]:

$$\overline{Ms} \supseteq Ns$$
$$\overline{Nu} \supseteq Mu \qquad (19.5)$$

where the overbar denotes closure, as schematically shown in Figure 19.7. This says that the closure of the period-M stable manifold contains the period-N stable manifold and the period-N unstable manifold contains the

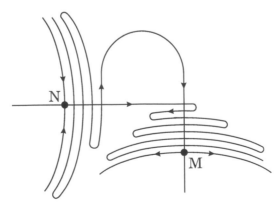

Figure 19.7. Heteroclinic crossings for a period-M saddle orbit and a period-N saddle orbit; only one point of each periodic orbit is shown.

period-M unstable manifold. Thus from (19.2), (19.4), and (19.5) we have

$$\overline{1fs} \supseteq Ns \tag{19.6}$$

and

$$\overline{Nu} \supseteq 1fu \tag{19.7}$$

for any period-N simple Newhouse orbit. Similarly, (19.3)–(19.5) imply the following:

$$\overline{Ns} \supseteq (N+1)s \tag{19.8}$$

and

$$\overline{(N+1)u} \supseteq Nu \tag{19.9}$$

for the period-N and period-$(N+1)$ simple Newhouse orbits. For example, for $N = 3$ and 4, we have

$$\overline{3s} \supseteq 4s, \qquad \overline{4u} \supseteq 3u \tag{19.10}$$

Combining (19.6)–(19.9), we have

$$\overline{1fs} \supseteq \overline{3s} \supseteq \overline{4s} \supseteq \overline{5s} \cdots \tag{19.11}$$

and

$$\overline{1fu} \subseteq \overline{3u} \subseteq \overline{4u} \subseteq \overline{5u} \cdots \tag{19.12}$$

Therefore, in particular, (19.11) implies that the stable manifold of the period-1 flip saddle is extremely complex and convoluted because its closure includes the stable manifolds of an infinite number of simple Newhouse orbits.

19.2.3.4. Crossing of the Unstable Manifold of the Period-1 Flip Saddle with the Stable Manifolds of Simple Newhouse Orbits. Numerically, we also find that the unstable manifold of the period-1 flip saddle cross the stable manifold of the period-N simple Newhouse orbit, that is,

$$Ns \times 1fu, \qquad N = 3, 4, 5, \ldots \tag{19.13}$$

The following relation is implied from (19.4) and (19.5):

$$\overline{Ns} \supseteq 1fs \tag{19.14}$$

and

$$\overline{1fu} \supseteq Nu \tag{19.15}$$

Combining (6), (7), (14) and (15), we have

$$\overline{1fs} = \overline{Ns}$$

$$\overline{1fu} = \overline{Nu} \tag{19.16}$$

As the parameter p increases toward the interior crisis value $p = p_c$, the period N for which (19.11) and (19.12) holds also increases. At $p = p_c$ we expect (19.16) to hold for all values of N. Hence, at the interior crisis the following equality holds:

$$\overline{1fs} = \overline{3s} = \overline{4s} = \overline{5s} = \cdots \tag{19.17}$$

and

$$\overline{1fu} = \overline{3u} = \overline{4u} = \overline{5u} = \cdots \tag{19.18}$$

19.2.3.5. The Structure of the Chaotic Saddle That Collides with the Main Attractor at the Interior Crisis. Equations (19.17) and (19.18) tell us that as p approaches the critical value for the interior crisis the stable and unstable manifolds of the period-1 flip saddle are equivalent to those of all the simple Newhouse orbits. The crossing of the stable and unstable manifolds of these saddle orbits (the period-1 flip saddle plus all the simple Newhouse orbits) implies the existence of horseshoe-type dynamics, which in turn implies the existence of an infinite number of unstable periodic and aperiodic orbits. The chaotic saddle is thus the closure of these unstable periodic and aperiodic orbits. For the parameter p below the crisis value, both the stable and unstable manifolds of all these saddle orbits have a Cantor structure [11].

19.3. NUMERICAL RESULTS

In this section, we calculate numerically the chaotic saddle by using the PIM-triple procedure proposed by Nusse and Yorke [6]. The details of the PIM method can be found in the Appendix. In the following we give two examples of interior crises.

19.3.1. Ikeda Map for $A = 0.85$, $B = 0.9$, $k = 0.4$, and p Varying near $p_c = 5.169789\ldots$

This case was discussed in Section 19.2. Figures 19.3c and 19.3d show the chaotic saddle at $p = 5.1697$, which is a parameter value slightly less than the crisis value. Figure 19.3d is a blowup of that part of the chaotic saddle indicated by a box in Figure 19.3c. The expected Cantor structure is obvious. Comparing Figure 19.3b with Figure 19.3c, we see that the incremental

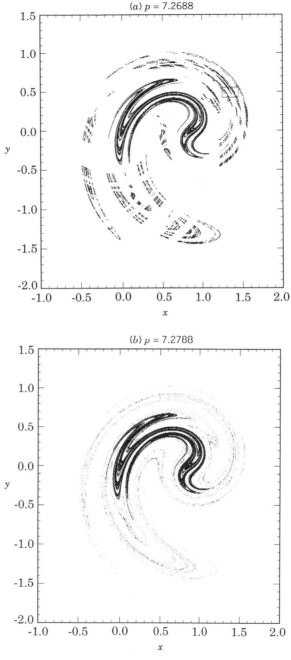

Figure 19.8. The second example of the Ikeda map at $A = 0.85$, $B = 0.9$, $k = 0.4$, and p varying around the interior crisis value $p_c = 7.26884894\ldots$: (a) the strange saddle together with the chaotic attractor at $p = 7.2688$ before the crisis; (b) the larger chaotic attractor at $p = 7.2788$ after the crisis.

portion of the chaotic attractor, after the crisis, comes from the chaotic saddle. The main attractor, corresponding to the parameters of Figure 19.3c, is shown in Figure 19.3a. Observe that the main attractor falls in the gaps in Figure 19.3c.

19.3.2. Ikeda Map for $A = 0.85$, $B = 0.9$, $k = 0.4$, and p Varying near $p_c = 7.268848\ldots$

This case has been studied [4] as an example of crisis-induced intermittency in which a single (small) chaotic attractor suddenly busts into a much larger one due to collision with a chaotic saddle. Figure 19.8a shows the small attractor (in the center) together with the chaotic saddle at $p = 7.2688$ before the crisis and Figure 19.8b shows the single (larger) chaotic attractor at $p = 7.2788$ after the crisis.

19.4. CONCLUSION

In this chapter, we have studied the mechanism for the sudden enlargement of the chaotic attractor at the interior crisis. We argue that the interior crisis is due to the collision of a small chaotic attractor with a chaotic saddle as a system parameter changes. The chaotic saddle is formed before the crisis as results of creations and destructions of an infinite number of attractors. When these attractors are destroyed they leave behind small chaotic saddles, which are linked together to form the large chaotic saddle that collides with the main attractor at the crisis. This saddle has horseshoe-type dynamics and is responsible for the sudden increase in the number of unstable periodic orbits in the main attractor. Hence, the incremental portion of the chaotic attractor, after the crisis, comes from the chaotic saddle formed before the crisis; the new unstable periodic orbits on the chaotic attractor also come from the chaotic saddle.

APPENDIX: PIM-TRIPLE PROCEDURE

In this appendix, we briefly describe the PIM-triple procedure [6, 7] to find a trajectory on chaotic saddles existing in a given region in the phase space. A PIM- (proper interior maximum) triple is three points (a, b, c) in a straight line segment such that the interior point c (i.e., c is between a and b) has the maximum escape time (out of the region) as compared with the escape of both a and b.

The steps in the PIM-triple procedure are as follows:

1. Specify a region of interest. If there is (are) attractor(s) coexisting in the region, isolate the attractor(s) with circles of appropriate radii. If a

trajectory asymptotes to any attractor, this trajectory is considered to have escaped from the region.

2. Choose a line segment L in the above region that straddles the stable manifold of the chaotic saddle. Distribute uniformly a number (say, 30) of points on the line segment and calculate the escaping time for each point. Choose the point with maximum escaping time and two points on both side of this point. The three points so obtained constitute a PIM-triple.

3. Use the PIM-triple as the new line segment and repeat step 2 until the length of the PIM-triple so obtained is less than, say, 10^{-9} (denoted by I_0).

4. Iterate I_0 (three points together) forward under the map. Denote the new interval by I_1. If the length of I_1 exceeds 10^{-9}, we apply step 3 until its length is less than 10^{-9}.

5. Repeat step 4 to find a sequence of PIM-triple intervals $[I_n]_{n \geq 0}$.

6. Choose any one of the three points in the sequence $[I_n]_{n \geq n_0}$ as an approximation of a trajectory on the chaotic saddle, where n_0 is the number of preiterates.

ACKNOWLEDGMENTS

One of the authors (Y. C. Lai) wishes to thank M. Ding for useful discussions. This work was supported by the U.S. Department of Energy (Scientific Computing Staff, Office of Energy Research).

REFERENCES

1. Y. Ueda, *Ann. N.Y. Acad. Sci.* **357** 422 (1980).
2. C. Grebogi, E. Ott, and J. A. Yorke, *Phys. Rev. Lett.* **48**, 1507 (1982).
3. C. Grebogi, E. Ott, and J. A. Yorke, *Physica D* **7** 181 (1983).
4. C. Grebogi, E. Ott, F. Romeiras, and J. A. Yorke, *Phys. Rev. A* **36** 5365 (1987).
5. S. M. Hammel, C. K. R. T. Jones, and J. V. Moloney, *J. Opt. Soc. Am. B* **2** 552 (1985).
6. H. E. Nusse and J. A. Yorke, *Physica D* **36** 137 (1989).
7. J. A. Yorke, Dynamics—A Program for IBM PC Clones for Interactive Studies of Dynamics. Unpublished, 1989.
8. L. Tedeschini-Lalli and J. A. Yorke, *Comm. Math. Phys.* **106** 635 (1986).
9. C. Grebogi, E. Ott, and J. A. Yorke, *Physica D* **24** 243 (1987).
10. S. Newhouse, *Publ. Math. IHES* **50** 101 (1980).
11. G. H. Hsu, E. Ott, and C. Grebogi, *Phys. Lett. A* **127** 199 (1988).

CHAPTER 20

THE FUTURE OF CHAOS

O. E. RÖSSLER
Division of Theoretical Chemistry
University of Tübingen
7400 Tübingen
Germany

The "new science" of chaos is assessed with an eye to the future. The first root of chaos is Poincaré's invention of the "qualitative" as an exact category. Chaos is the most conspicuous entity found so far. It lives in three dimensions. Further prototypic dynamical "gestalts" are bound to be discovered in low-dimensional systems of more than three dimensions. They will play an equally important role in the applied sciences. A generalized Shil'nikov theorem will be presented as a conjecture. Chaos' second major root is ancient mythology. Anaxagoras' Mind and Laplace's Demon illustrate the peculiar rationality that accompanies chaos from the beginning. The fundamental limitation to both measurement and predictability, implicit in chaos, is fully understandable from the inside. Both aspects—the endo and the exo —are complementary. This "both reductionist and holistic" character of chaos is unique. It may give rise to a new synthesis of unprecedented scope.

Two theses concerning the future of chaos are presented. First, chaos will be fusioned with statistics into a universal auxiliary discipline of growing importance in the 21st century [1]. Second, chaos will spawn a new fundamental science that nevertheless is closely related to the applicational spirit of the first thesis.

The rationale for the first thesis, growing applicational importance of chaos, lies in a single fact which apparently has never been formulated as such: Nature prefers gestalts. Chaotic attractors are not just distinctively beautiful to human perception. They represent an objective gestalt, that is, a quality that differs discretely and demonstrably from other qualities. Pirsig [2]

Applied Chaos, Edited by Jong Hyun Kim and John Stringer.
ISBN 0-471-54453-1 © 1992 John Wiley & Sons, Inc.

credits Poincaré with having made the *qualitative* exact. Indeed Poincaré's greatest contribution to dynamics appears to be that he identified and named the limit cycle in two dimensions. The limit cycle not only is the first nontrivial attractor in history, but also the first living proof that quality—topological equivalence—is capable of uniting an infinity of quantitatively disjoint systems and phenomena.

Poincaré [3], as is well known, achieved the first complete classification of two-dimensional flows, in terms of a small number of qualitative limiting structures and their combination (which he called the phase portrait). However, into the third dimension, Poincaré barely ventured a peek, despite the fact that he did discover the homoclinic point (essential for chaos) in a two-dimensional cross section through a locally three-dimensional flow (cf. reference [4]).

Therefore it was a surprise that the third dimension *continues* to show the pattern discovered by Poincaré in two dimensions: There are only very few discretely different gestalts there, too. Chaos was the example that drove this point home. Its naming as a reality of its own [5] soon turned out to be felicitous because a new Poincaréan category happened to have been hit upon [6].

In this way it turned out that Poincaré had discovered a principle even more general and important than he himself could have anticipated: The existence of a discrete hierarchy of dynamical entities (loosely speaking, attractors and boundaries) as a function of dimension number [7].

The pertinent entities, moreover, are even *more* robust than one might have guessed from what holds true in two dimensions. For example, topological equivalence, used exclusively by Poincaré, proves too fine-grained a property when it comes to classifying chaotic attractors. The latter in general are "nonminimal" [7]. Furthermore, their robustness goes so far that even the addition of further dimensions (in the form of small "stray capacitances" or fast intermediate reactants, etc.) does not affect recognizable qualities of these structures (including both shape and sound [8]). Even after many dimensions have been added and many further changes been made in the equations (with many further couplings and many new high-order terms) so that the equations are quantitatively unrelated according to the usual standards, the qualitative prototypes have a tendency to pop up again in full garb, with virtually all the characteristics first encountered under the most idealized conditions surviving.

Just like a portrait of a person (a *portrait de face*) remains discretely distinguishable, in terms of both individuality and expression, across the most severe alterations, for reasons that have to do with human gestalt perception, so are these dynamical creatures (in a *portrait de phase*) incredibly robust toward the most radical transformations—for reasons that lie entirely within nature itself.

This fact explains the unprecedented success of chaos in the applicational domain (predicted in reference [8]). Chaos speaks to us in an unmistakable fashion under the most varied disguises, often long before methods allowing us to confirm this intuition quantitatively have been set up in response. Hearts and vocal cords and societies, dripping faucets, car engines, reaction mixtures, geysers, X-ray bursters, and coupled electric power generators, all obey the invisible director's baton. The dynamical prototypes of chaos and its higher-order analogs [9] happen to dominate the phenomenological world in an exact qualitative fashion. It is as if nature were confined to using only a small alphabet of shapes. This is the surprise and the power of the chaos paradigm.

In this vein, it for example recently turned out that *hyperchaos* (with more than one direction of stretching and folding and, hence, exponential divergence on the attractor [10]) has a new fingerprint, which is not a number, or set of numbers, but a picture [11]. Indeed, any self-affine (or self-similar) fractal of the Sierpinski type [11a] is hidden in and can be retrieved from a hyperchaotic system, such as can be expected to live in a system of coupled power generators.

The principle is worth to be illustrated here with a global-analytic example. Recall that a simple chaotic flow (like that of Figure 20.1a [12]) in general contains in its parameter space a point at which the central saddle

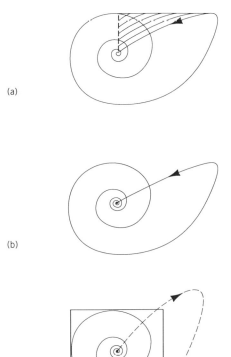

(a)

(b)

(c)

Figure 20.1. Relation between spiral chaos and Shil'niko's repellor: (*a*) spiral chaos in the equation of reference [12]; (*b*) an infinite-period trajectory obtainable in the same equation at one particular set of parameters; (*c*) equivalent Shil'nikov flow. (See text.)

focus is hit exactly by a recurrent trajectory of infinite period (Figure 20.1*b*). This situation is, under time reversal, equivalent to a situation described by Shil'nikov [13] (Figure 20.1*c*). Here a two-dimensional stable focus is unstable in a third dimension, such that through this third dimension an infinite-period homoclinic connection becomes possible. The latter mathematically implies a Smale horseshoe and hence chaos if the focal plane is not too weakly repelling [13]. It also implies a complicated (Cantor-stripes-type) intermingling of two neighboring basins if the Shil'nikov repellor is made part of a basin boundary, cf. reference [14].

This well-known mathematical situation possesses a next-higher-dimensional analog (Figure 20.2). In Figure 20.2*a*, we have a simple hyperchaotic motion that is based, not on an expanding spiral in two dimensions as in Figure 20.1*a*, but on an expanding screw in three dimensions [10]. Reinjec-

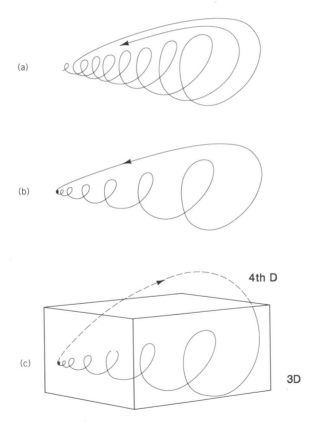

Figure 20.2. Relation between screw-based hyperchaos and a hyper-Shil'nikov limit set: (*a*) hyperchaos in the equation of reference [10]; (*b*) an infinite-period trajectory obtainable in the same equation at one particular set of parameters; (*c*) equivalent hyper-Shil'nikov flow.

tion using one more dimension for the roundabout way leads to expansion in two directions rather than merely in one. Again, only a finite neighborhood of the expanding linear fixed point is hit by the recurrent flow. However, as in Figure 20.1*b*, it is again possible to find a point in the parameter space of the underlying equation for which the fixed point is revisited exactly (Figure 20.2*b*). Time reversal then again reveals a prototypic situation, analogous to Shil'nikov's: Figure 20.2*c*. Here a stable node-focus in three dimensions is, via an added unstable manifold in a further dimension, reconnected to itself through the fourth dimension. The result is a "hyper-Shil'nikov repellor" if certain conditions on the local eigenvalues are met (cf. [14a]). It again is mathematically well characterized and once more implies in a structurally stable fashion the existence of two things: (a) a hyperhorseshoe (folded towel [10]) and hence a product of two subshifts of finite type; (b) a basin boundary running through the repellor that is self-similar (Cantor-like) in *two* directions and hence genuinely self-similar or self-affine. Both subresults still need to be elaborated in analogy to Shil'nikov's own results [15].

So far, the new basin-boundary structure has only been demonstrated numerically in a four-dimensional flow, which is more complicated than that of Figure 20.2*a* [16]. It would be desirable to confirm the result for the simpler equation of reference [10]. Analytical determination of the hyper-Shil'nikov point in that equation (or a piecewise-linear analog [16a]) may facilitate the numerical work.

Let us now turn to some broader issues.

Nature prefers continuous rules. Much like a bundle of everywhere parallel spaghetti, the trajectories of the most varied dynamical systems arising in nature tend to behave locally in a trivial (linear-differentiable) fashion. It is therefore possible to enter, with the mind, the underlying spaces in three-dimensions and in color, so as if they were virtual realities. The shapes found then become familiar friends down to their last cranny. The mixing of dough (in real space) and chaotic dynamics (in phase space) are not just mutual analogs, they are identical. Chaos *is* the simplest "3-D blender" [18].

The essence of chaos has already been seen with the mind's eye by Anaxagoras in ancient Athens (the same person who is the inventor of mathematical physics and the theory of the infinitely small and the self-similar [19]). His mental model of the cosmos—to view it as a giant mixing (or rather, unmixing) process in three-space—is identical to chaos-theoretical pictures developed much later.

Boscovich's universe of the 18th century is in the same category. It was already couched in terms of Newton's equations [20]. A single potential function, which was wiggly at small distances between the point-shaped elementary particles and which became identical to Newton's gravitational potential at larger distances, allowed Boscovich to write down the first complete, classical universe. Even though the intervening two millennia had added to its sophistication, Boscovich's universe still preserves some of the most striking features of Anaxagoras' model. Anaxagoras used a description in "dualist" terms. There was an outside mind ("Nous") and an inside tumble ("Chaos")—there is only circumstantial evidence for his using this latter term [7, 19, 21]. The Nous controlled the Chaos. Today, one usually speaks of Laplace's demon in the same context. (Laplace innocuously picked his famous phrase from Boscovich's earlier book [22].)

We have now entered the "demonic" side of chaos. The latter is no less important than the "synthetic" Poincaréan side emphasized previously. It is necessary to realize that the demonic side is only a very short distance away from an even older level in the history of chaos, the mythological one. In mythology Chaos is female and the controlling Mind is male. Comparative mythology, including the Hindu and Chinese cosmogonies (all of the mixing type), points in the same direction [19].

The Yin–Yang polarity of the myths is preserved in Anaxagoras' scientific picture, as well as in Boscovich's picture and in reality. The mixing chaos is a challenge to the mind. The principle of exponential divergence in mixing is responsible for the fact that both measurement and prediction become almost powerless, as is illustrated so well by Lorenz's butterfly [23]. To overcome this problem, Anaxagoras introduced the Mind as a deus ex machina. The Nous unmixes the whole perfect mixture even though the latter has already reached an invariant state. To picture this, Anaxagoras had to take recourse to the invention of what much later was called transfinite mathematics (and the "long line" of topology [21]). Because every substance had been participating in the mixing process, the Mind had to be declared "too fine" to be miscible itself.

The limitation theorem implicit in chaos, first seen by Popper [24], is of the same character as Gödel's, in one respect. The fact that a limitation result has to be accepted is made up for by its having been accessible in the first place. The gain on the metalevel exceeds the loss on the direct level. Both levels are intimately linked together, like Yin and Yang are.

This parallel to Gödel's theorem does not yet exhaust the present situation. An even more detached look can be taken. In our own time, a picture very much like Anaxagoras' has been invented independently by Bohm [25], under the name of the implicate order. To an internal observer, this order is irrevocably diffracted [26]. Nevertheless the whole "holomovement" itself is postulated to be uniquely defined (although in a nonlocal manner) when looked at in its own right. Only this privileged access is denied to us as creatures living on the inside.

Bohm is not aware of Hamiltonian chaos but only quotes a quasiperiodic "(un)mixing machine" [25]. Nevertheless there is no doubt that what he has in mind is a picture of chaos. Bohm's proposal amounts to a revival of the "two physics" concept which goes back to Boscovich. The *one* physics here is the familiar one—that in which the weather and also quantum mechanics are so notoriously unpredictasble. This is the world of the observer inside the understanding-defying chaos. The *other* physics is the *same* physics once more, but looked at from the outside. Here, a privileged outside Archimedean point is assumed. It allows one to understand *why* it is that the poor creatures inside are so limited in their capabilities. The "endo" and "exo" distinction of postmodern physics [28] thus really is not new. It unites the historical phases of chaos theory: the mythological, the Anaxagorean, and the Boscovichian.

The fact that chaos is a gestalt distinguishes it from other quantitative theories. It puts chaos on the edge, giving it both a reductionist and a holistic face. It is possible both to tune in audiovisually on a chaotic attractor [8] and simultaneously to understand with the mind its fundamental quantitative irreproducibility. This "swtiching" of perspectives is both entirely natural and entirely nontrivial.

Maybe one should not take pictures and gestalts too seriously. On the other hand, everything that is qualitative about chaos can be made quantitative according to Poincaré's principle. The existence of an even more outside point of view—the "ojo de diós" (eye of God) perspective—is, nevertheless, only intimated by chaos theory. It is certainly possible to pretend that a formulation of nature as a whole is accessible in principle, despite the fact that we cannot leave this universe in reality. If *testable* predictions could be reached, this would amount to a new fundamental level of understanding.

To conclude, there are three promises in chaos. First but not least, a great many tenured positions may be waiting in the next century, across all academic departments having anything to do with real-life data. Second, steady progress in the discovery of new "animals"—of the attractor and the boundary type—can be expected, first in four dimensions (cf. reference [29]) but then also in five, perhaps. The beauty of these gestalts will only be surpassed by their usefulness. Third, a new synthesis of the "reductionist" and the "holistic" face of science can be expected.

The computer kids of the future may put it differently. Chaos makes it legal to trust one's own spatial intuition and to get lost in and come back from virtual realities of all kind (starting with flight simulators on the PC and ending up with an appreciation of Levinas' multilevel proof that death is unacceptable).

To summarize, chaos has two kinds of root and two kinds of future. The one strand is medium-term and applied; it is almost entirely due to Poincaré. The other is long-term and fundamental; it goes back to prehistoric people whose originality survives only in the language and the images they have created.

ACKNOWLEDGMENTS

I thank Jong Hynn Kim, Bruce Stewart and Peter Weibel for discussions and stimulation, and J. O. R. for inspiration.

REFERENCES

1. R. Abraham, personal communication, 1984.
2. R. M. Pirsig, *Zen Buddhism and the Art of Motorcycle Maintenance*. Bantam Books, Toronto, 1974.
3. H. Poincaré, Sur les courbes définies par une équation différentielle. *J. de Math. Pures et Appl.* **1** 167–244 (1885).
4. O. E. Rössler, Chaos. In *Structural Stability in Physics* (W. Guttinger and H. Eikemeier, eds.), pp. 290–309. Springer-Verlag, New York, 1979.
5. T. Y. Li and J. A. Yorke, Period three implies chaos. *Amer. Math. Monthly* **82** 985–992 (1975).
6. O. E. Rössler, Continuous chaos. In *Synergetics—A Workshop* (H. Haken, ed.), pp. 184–197. Springer-Verlag, New York, 1977.
7. O. E. Rössler, The chaotic hierarchy. *Z. Naturforsch.* **36**a 788–802 1983.
8. O. E. Rössler, Different types of chaos in two simple differential equations. *Z. Naturforsch.* **31a** 1664–1670 (1976).
9. O. E. Rössler, Chaos and bijections across dimensions. In *New Approaches to Nonlinear Problems in Dynamics* (P. J. Holmes, ed.), pp. 477–486. SIAM, Philadelphia, 1980.
10. O. E. Rössler, An equation for hyperchaos. *Phys. Lett. A* **71** 155–158 (1979).
11. O. E. Rössler and J. L. Hudson, Self-similarity in hyperchaotic data. In *Brain Dynamics, Progress and Perspectives* (E. Basar and T. H. Bullock, eds.), pp. 113–121. Springer-Verlag, New York, 1989.
11a. B. Mandelbrot, *The Fractal Geometry of Nature*. Freeman, San Francisco, 1982.
12. O. E. Rössler, An equation for continuous chaos. *Phys. Lett. A* **57** 397–398 (1976).
13. L. P. Shil'nikov, The existence of a denumerable set of periodic motions. *Sov. Math. Dokl.* **6** 163–166 (1965).
14. P. Gaspard and G. Nicolis, What can we learn from homoclinic orbits in chaotic dynamics? *J. Stat. Phys.* **31** 499–518 (1983).
14a. O. E. Rössler, Singular-perturbation homoclinic hyperchaos. In *Engineering Applications of Dynamics of Chaos* (W. Szemplinska-Stupnicka and H. Troger, eds.), pp. 139–147. Springer-Verlag, Vienna, 1991.
15. L. P. Shil'nikov, A contribution to the problem of the structure of an extended neighborhood of a rough equilibrium state of saddle-focus type. *Math. USSR Sb.* **10** 91–102 (1970).
16. O. E. Rössler, J. L. Hudson, M. Klein, and C. Mira, Self-similar basin boundary in a continuous system. In *Nonlinear Dynamics in Engineering Systems* (W. Schiehlen, ed.), pp. 265–273. Springer-Verlag, New York, 1990.
16a. O. E. Rössler, J. L. Hudson, and R. Rössler, Homoclinic hyperchaos: An explicit example. *Physica D* (in press).
17. P. Hanusse, Computer-detected self-similarity. Paper presented at Zeinisjoch Meeting, P. J. Plath, organizer, February 1990.
18. O. E. Rössler, Chaotic behavior in simple reaction systems. *Z. Naturforsch.* **31a** 259–264 (1976).

19. O. E. Rössler, Anaxagoras' idea of the infinitely exact chaos. In *Chaos in Education, Teaching Nonlinear Phenomena*, vol. 2 (G. Marx, ed.), pp. 99–113. National Center for Educational Technology, Veszprem, 1987.

20. O. E. Rössler, Boscovich covariance. In *Beyond Belief: Randomness, Prediction and Explanation in Science* (J. L. Casti and A. Karlqvist, eds.), pp. 65–87. CRC Press, Boca Raton, FL, 1991.

21. O. E. Rössler, Long-line attractors. In *Iteration Theory and Its Functional Equations* (R. Liedl, L. Reich, and G. Targonski, eds.), pp. 149–161. *Lecture Notes in Math.* vol. 1163. Springer-Verlag, Berlin, 1985.

22. K. Stiegler, in *Proceedings of the 13th International Congress on the History of Science, Moscow 1971*, Section 6, p. 307, 1974. (Quoted later in S. G. Brush, *The Kind of Motion We Call Heat*, Book 2, p. 595. North-Holland, Amsterdam, 1976.)

23. E. N. Lorenz, The problem of deducing the climate from the governing equations. *Tellus* **16** 1–11 (1964).

24. K. R. Popper, Indeterminism in classical physics and quantum physics. *Brit. J. Philos. Sci.* **1** 117–133 (1951).

25. D. Bohm, *Wholeness and the Implicate Order*. Routledge & Kegan Paul, London, 1980.

26. D. Bohm, The implicit order (in German). In *Andere Wirklichkeiten* (R. Kakuska, ed.), pp. 65–107. Dianus-trikont, Munich, 1984.

26a. O. E. Rössler, Macroscopic behavior in a simple chaotic Hamiltonian system. In *Dynamical Systems and Chaos* (L. Garrido, ed.), pp. 67–77. *Lecture Notes in Phys.*, vol. 179. Springer-Verlag, Berlin, 1983.

27. O. E. Rössler, Boscovich's observer-centered explanation of the nonclassical nature of reality. In *Symposium on the Foundations of Modern Physics, Abstracts*, pp. 153–156. Department of Physical Sciences, University of Turku, Finland, Report Series, June 1990.

28. O. E. Rössler, Endophysics. In *Real Brains, Artificial Minds* (J. L. Casti and A. Karlqvist, eds.), pp. 25–46. North-Holland, New York, 1987.

29. O. E. Rössler, R. Wais, and R. Rössler, Singular-continuous Weierstrass-function attractors. In *2nd Int. Conf. Fuzzy Logic* (T. Yamakawa, ed.), in press.

APPENDIX

DISCUSSIONS

The following pages contain the discussions to 17 of the 20 papers included in this proceedings. The paper by A. L. Goldberger was presented on videotape, and the author was not present; O. E. Rössler was also not present. The paper by Lai, Grebogi, and Yorke excited considerable discussion, but this proved impossible to transcribe from the recording.

The last session of the workshop was a general discussion; its principal objective was to identify potential applications of chaos theory to problems of concern to the electric utility industry and to point out directions for future research.

The discussions were recorded and were transcribed after the meeting. The transcriptions were edited to eliminate some of the duplications and misspellings and then were sent to the speakers for review. The speakers were asked to correct errors of fact and to reword to clarify the meaning where that was necessary, but to change as little as possible, to retain the "live" atmosphere of the discussions. In the main, this was done, although in a few cases the speakers felt that the transcription of their remarks was so unclear that a more detailed rewriting was necessary. Where that has been done, it has been indicated in the text, because in one or two cases the rewritten comment makes some of the subsequent discussion seem a little redundant. To minimize delays in publishing, a further review of the transcripts by the discussants was not undertaken, and the editors apologize for any errors that have persisted.

Discussion Participants

Hilda A. Cerdeira, International Centre for Theoretical Physics, Trieste, Italy

Hsueh-Chia Chang, Department of Chemical Engineering, University of Notre Dame, Notre Dame, Indiana

Gary T. Chapman, Department of Mechanical Engineering, University of California at Berkeley, Berkeley, California

Predrag Cvitanović, Niels Bohr Institute, Copenhagen, Denmark

Robert de Paola, Division of Cardiothoracic Surgery, Children's Hospital of Philadelphia, Philadelphia, Pennsylvania

John J. Dorning, Department of Nuclear Engineering, University of Virginia, Charlottesville, Virginia

Walt Esselman, Electric Power Research Institute, Palo Alto, California

J. Doyne Farmer, Prediction Company, 234 Griffin Street, Santa Fe, New Mexico

Leon Glass, Department of Physiology, McGill University, Montreal, Canada

Michael Gorman, Department of Physics, University of Houston, Houston, Texas

Celso Grebogi, Laboratory for Plasma Research, University of Maryland, College Park, Maryland

Alex Harrison, Winders, Barlow & Morrison Inc., Englewood, Colorado

Laurence Keefe, Nielsen Engineering and Research, Mountain View, California

Jong H. Kim, Electric Power Research Institute, Palo Alto, California

William F. Lawkins, Mathematical Sciences Section, Oak Ridge National Laboratory, Oak Ridge, Tennessee

Paul Linsay, Massachusetts Institute of Technology, Cambridge, Massachusetts

John Maulbetsch, Electric Power Research Institute, Palo Alto, California

J. Kevin McCoy, Battelle Columbus Division, Columbus, Ohio

F. Allan McRobie, Centre for Nonlinear Dynamics and Its Applications, Department of Civil Engineering, University College, London, England

Julio M. Ottino, Department of Chemical Engineering, Northwestern University, Evanston, Illinois

Mihir Sen, Aerospace and Mechanical Engineering Department, University of Notre Dame, Notre Dame, Indiana

H. Bruce Stewart, Department of Applied Science, Brookhaven National Laboratory, Associated Universities, Inc., Upton, Long Island, New York

John Stringer, Electric Power Research Institute, Palo Alto, California

Shiu-Wing Tam, Argonne National Laboratory, Argonne, Illinois

J. Michael T. Thompson, Centre for Nonlinear Dynamics and Its Applications, Department of Civil Engineering, University College, London, England

Solomon C. S. Yim, Department of Civil Engineering, Oregon State University, Corvallis, Oregon

Discussion: Bridging the Gap between the Science of Chaos and Its Technological Applications. By J. J. Dorning and J. H. Kim

Esselman. This is a philosophical question. There are things such as heat transfer surfaces in a boiler, where we really want to understand the criteria to avoid bifurcation. There are other cases such as the fluidized bed, where we have to operate in a chaotic regime. I'd like you to comment as to how we can develop an understanding of actual physical systems such as these which will allow us to operate in the proper regime.

Dorning. Let me broaden the question a little bit. Rather than just chaos, we're really concerned about nonlinearity in general in engineering systems; thus, separate from chaos there may be cases where we don't even want periodicity. Since you posed this as a philosophical question, let me try to give you a philosophical answer—and that is that in the context of boiling channels, where in fact the old-time engineers knew what they were doing, they put nozzles in these channels to keep them stable at the expense of an increase in the pressure drop and some associated additional pump work. So, that problem was in a sense empirically solved a long time ago. The solution may not be optimum; in fact, maybe we should not even be operating some of these systems in equilibrium. For example, two-phase heat exchangers have a better net heat transfer coefficient if they're operated in an oscillatory mode; there are some examples in which they actually are forced sinusoidally just to stir them up. So, when one wants to operate most efficiently and therefore perhaps in a nonlinear regime, understanding nonlinear phenomena is very important so that one can then make an intelligent decision based on economics or other engineering criteria. For example, one might think of some sort of an engineering device that currently operates at equilibrium or maybe in some oscillatory mode that would get better performance by behaving aperiodically or chaotically.

Gorman. You showed an oscillating reactor . . .

Dorning. Yes, that was experimental data from a nuclear reactor in Sweden.

Gorman. Right, and under what conditions was that a desirable condition?

Dorning. According to today's conventional wisdom in the industry, none! That oscillation actually was recorded during a series of experiments that were run when the reactor was shut down. I think it was shut down for refueling.

Gorman. What is oscillating?

Dorning. All the phase variables were oscillating. Of particular interest is the coolant velocity in the heated channels. In fact, these so-called density-wave instabilities occur in heated channels with two-phase flow, even when such a channel is not in a reactor. Loosely speaking, boiling water nuclear reactors inherit these stability problems from the nonlinear dynamics of the two-phase flow in their boiling channels. However, the oscillating quantity that I showed was the neutron power, averaged over the whole reactor. Of course, there is concern about the monitoring in these reactors in which averaging over the whole reactor might yield a lovely average while locally the power could be very high and components could be at risk of melting. Is oscillatory operation desirable? Well, historically these are not intentionally operated in an oscillatory mode. When you buy one, you get certification from the manufacturer saying if you operate where we tell you, it will not oscillate. But in fact this happened in a reactor in the United States while it was operating, and its various safety features triggered and it shut down. However, there were some subsequent articles and newspaper items suggesting that the electric utility that owned the reactor had burned the fuel in such a way that it took it out of the parameter range that the vendor who built it told them to operate it in. I don't know all the details about it. There was no liability suit or anything, and it wasn't a serious incident. Now maybe it would be desirable to operate these reactors at higher powers; even if when you drive them up to a higher power, they're going to oscillate. There is a Nuclear Regulatory Commission diagram that has flow rate versus power, and a boiling water reactor is not allowed to be operated in certain regions of that diagram. That diagram is a rather restrictive two-dimensional projection of a very high-dimensional phase space. But, maybe, if we understood the physics better, it might be possible to demonstrate that in fact one could operate in this forbidden oscillatory region without anything bad happening. Now, as an engineer, I'm not advocating that, believe me! But in fact there might be some advantages.

Glass. The comment that oscillation in some sense may be a more efficient way to run things I think is very provocative. There is a certain prejudice among biologists that nature has discovered the best way of doing everything and many biological functions operate on a periodic basis. For example, the heart, respiration, and so forth, are more or less periodic.

Tam. A lot of the current so-called advanced reactor concepts use passive control, in the sense that in the event of a loss of coolant or pumping power heat could be removed basically by (natural) convection—and the question is, how is this concept affected by the sort of instabilities you have been discussing? How do they impact on the questions of safety? Should the reactor community be concerned?

Dorning. Well, I guess the people who reviewed my proposals over the past couple of years are not too concerned. Joking aside, these proposals were concerned with these convective flows in the so-called advanced reactors, which are analyzed using big thermohydraulics codes. I don't think that these big thermohydraulics codes can do the simple problems, which makes me worry about their ability to do the complex problems. I think this is something that should be studied. Now on the other hand let me also indicate that a lot of the complexity or uniqueness associated with Rayleigh–Bénard convection, at least in my opinion, is related to the fact that there is this nice uniform temperature along the bottom. One has to work hard to get this uniform temperature to reap some of this complexity. On the other hand, if you go to higher heat fluxes and higher temperatures, even with nonuniform temperatures a lot of complex dynamics occurs. Now the engineers that design these reactors are of course not fools; but one tends to develop an intuitive picture of the flow around a hot body within a pool, and one all too often is satisfied that the big computer codes are working when they compute a flow that is consistent with this mental image. I don't mean to say that all of this is wrong, but on the other hand I certainly can easily imagine all sorts of complexity of flows that could be very serious. Thermostratification is something that we certainly should be worried about: maybe it's not a really persistent Rayleigh–Bénard problem, but for some periods of time it is, and there might be very high temperature gradients.

Kim. Let me make a comment on your question. The most important thing is whether we'll be able to remove the heat, and in all that's published and all the experiments we have done so far, we have never seen anything where the heat removal was of any concern—and so, that is a practical answer to your question. We don't see any problems in terms of removing the heat. That's our view.

Discussion: Global Integrity in Engineering Dynamics. By F. A. McRobie and J. M. T. Thompson

Dorning. You mentioned the one-dimensional model, which I presume is the roll and not the pitch. I presume naval architects are also concerned about the pitch, and in particular the combination of the pitch and the roll. The real problem is thus two-dimensional, and I can easily see the pitch combining with the roll in some sort of resonance: As the bow goes down and you're rolling to the starboard side, it's going to enhance the roll to the starboard, I should think. Do you consider those at all? Or, have you thought about the additional complications that that would introduce?

Thompson. We've thought about them, and we're planning to look at them but we haven't looked at them yet. Why we got interested in this problem is that the navy has frigates on towed array duties, and when you're on a towed array duty you have to keep going on a dead straight line so you may get beam seas. A towed array is a sonar device, it is antisubmarine I guess. So when you're on this sort of duty you have to keep going, even if the seas are coming from the side, even if you've got very strong beam seas. Now, I gather a merchant ship would normally change course and simply wouldn't keep going under these conditions, but as I understand it these frigates have to keep on this towed array duty, so they keep going—and apparently they might roll through enormous angles, trying to keep on course. Now under those conditions with a beam sea, there probably isn't much coupling, it probably is primarily a rolling action. But I do agree, there are all sorts of other circumstances where there would be interaction between different modes.

Chapman. To follow that up, I have two comments. Actually, with the beam sea probably the more important thing is the plunging motion, which actually changes the stability rolling torque, because as you go up and down you change the buoyancy center relative to where the center of gravity actually is to some extent. That probably influences the rolling motion, probably more than the pitch motion does.

Dorning. Well certainly in the beam seas, since there is no pitch motion.

Chapman. And in fact, that leads to the question. I like your criteria very much and you show two models that seem to be robust. What's the real world say?

Thompson. I don't know. We haven't carried this investigation very far yet.

Chapman. What I'm concerned about is that we see elegance, and this is the part that for a real applied person really gets quite exciting. You come up with something reasonably close and you show elegant results, and I don't know whether the model fits the real world. That's very bothersome to me. It walks close to the applied side, and you say it's the real world, you said a real ship, and I thought, "Oh, God, I'm going to see real data." But we're back to the model thing and there's no comparison to see whether the model's really correct. You've got to watch that when you're trying to push these chaotic ideas into the applications area, because that's the thing that quite often gives it the bad reputation. You study an elegant model, but you haven't applied it to the real problem, the real world.

McRobie. I think that's a very valid comment. We have been given data for a large number of real ships. We are going through this not just for the Duffing equation and the polynomials, but for a very large set of these potential wells, and at the moment the sort of things like optimal escape, our researchers can predict where that's going to be quite accurately, just from this sort of analysis. It's not fitting a real physical model, I know, but it's a robustness between different potential wells. It doesn't matter whether you have polynomials or piecewise linear .... .

Chapman. What I'm concerned about is that I think I know about this problem, and the potential well is being very strongly modulated by another degree of freedom.

Dorning. Well, I mean that's a secondary issue but I think the lack of higher-order terms in the expansion of the well is probably not all that important. What happens when you've already gone on the excursion really isn't all that important; you've already gone.

Thompson. There's a hierarchy of problems. First we've established that it's robust against variations in the shape of the potential. Secondly, we've been looking at the variation of the damping level. There's an awful lot of variables, you see, and I agree, coupling in with another degree of freedom requires looking at.

Farmer. Just a couple of things. What it does sound like though, that ship designers actually think in these terms, I mean is this, using this kind of equation, something that ship designers have already thought of or are you guys the first to use this kind of equation?

Thompson. Ship researchers use this type of equation for one-degree-of-freedom roll analysis. But designers don't do dynamics at all. There's a static criterion about what's going to happen in waves. So, to be quite honest, we're open to criticism, of course we are, but this is a far better analysis than any naval designers would use.

Chapman. Well, a naval architect doesn't rely on this. He takes it out and puts it in the ship basin and tries to do with a close model.

Thompson. The naval architect, the designer, actually has a static criterion. He looks at the shape of the total potential energy curve and makes an estimate about how it's going to respond to waves. So the wave frequency doesn't even enter his discussion. He does nothing about dynamics. Now researchers do. If you look at the research literature in naval architecture, yes, they do dynamics and, by and large, dynamics in one-degree-of-freedom systems.

Chapman. Well, I'm thinking of how I've seen the work done at David Taylor model basin. They'll start out with the design criteria, but before they submit to construction, they'll take it into the ship basin with some kind of a model, to get the Froude number and a few things like that correct, and find out whether it's reasonably within that criteria. They don't build these on the basis of static designs.

Dorning. Yes, actually a comment on ships. I don't think people build ships very different from previous ships. I mean, I think every ship is very much like the previous one. The bulbous bow, for example, was on the drawing boards, for I guess a half a century before it was actually used.

Gorman. As an experimentalist who works on idealized systems, I just wonder whether experiments on idealized systems, you know, make little models and tanks, would be at all useful.

Thompson. Yes, perhaps I can answer these two comments in one. Of course they do model tests, but what do they do with the model tests? They wait for the boat to get into steady-state rolling. This is what they do, and it's absolutely insane. They wait and they wait and they wait.

Now we've said "Look, forget about that—you should be looking at transient conditions," and that, I think, is one of the very important changes in paradigm that we've been able to offer. Why don't you just have the boat sitting there under static conditions and send a short pulse of waves at it? So in a sense, although of course, we're not going to get numbers that teach the naval architects anything, we can maybe tell them that they ought to adopt a shift in paradigm. They ought to look at transient behavior: and for goodness sake, if you're going to test models, why not test the models under transient conditions? I have to say we tried to do this once and we couldn't make the boat capsize because there wasn't enough wave power in the tank! This was very embarrassing!

Gorman. Oh, they did try it on an idealized system?

Thompson. We have tried this on a model, a big model of a surface ship, in a wave tank, but unfortunately we couldn't capsize it.

Linsay. I just wondered—it seems like the sort of metaquestion behind this that we really want to answer is, first of all to see whether you can optimize the potential to withstand capsize and then go back from the potential to the actual design, and then the second thing is I wonder whether people have actually gone through narratives of ships that capsized where they interviewed the captain later on and said "Well what did the ship do just before it capsized?" You would think you might learn a lot from that.

Unidentified Speaker. You don't come back from that!

Linsay. Well, the lieutenant, whoever survived!

Thompson. Capsize, as they say, is a problem in small vessels, by and large. I mean although we have a grant from the Navy to look at capsize, one has to admit that no frigate of the Royal Navy has ever capsized. No chimney designed by Allan has yet to topple, either.

Dorning. Perhaps the Royal Navy is better organized than tramp steamers, but they have capsized.

Thompson. Small fishing vessels capsize fairly regularly. There is concern. In fact we got involved with this, because the Canadian transport authorities asked W. S. Atkins Consulting Engineers to look at the capsize of small fishing vessels, which do go over quite regularly. But of course, not necessarily in these nice beam sea conditions.

Chapman. Normally I know in small boats the well potential is a lot shallower.

Thompson. I don't know what you mean by shallow. You see, the curious thing is, there's a lot of scaling in here. You can always scale the equation so the height of the well is unity if you like.

Chapman. It always comes out to a real value, and small ships have characteristically smaller, shallower wells.

Thompson. Yes, but compared with what, when you say smaller. An interesting question would be what would—let's forget about ships for a moment—what would the best shape of a well be, in order to keep something in it for as long as possible, given that you were only allowed to have a certain height? I mean, suppose you've got a potential well with a minimum here, a certain fixed distance here, you've got a fixed height; what would be the best shape of that well, to keep a ball in (or conversely to get a ball out, since some people of course want to get things out of wells)? That is an interesting sort of theoretical problem.

Yim. When naval architects design a ship I'm sure they're more interested in the probability of failure, like you put down, one wave, two waves, three waves, four waves, and so on. Can you put some properties of structure on it so that the designers can say, "We are still pretty safe with a high probability that it won't overturn"?

McRobie. That's one of those things we're aiming towards with the lobe dynamics that I was talking about. You can define some very formal mathematics to quantify integrity based on lobe dynamics of intersections between particular sets and the probabilities of escape; you can then formulate in terms of lobe dynamics. So again this is a kind of underlying mathematical formalism where you can define mean escape times from, via stepping through particular lobes. This is a step towards hopefully being able to predict those probabilities. Yes, we can certainly measure them, in fact it's a trivial matter to actually measure them numerically for a system: the very simple forward numerical time integration that sort of anyone can do on a personal computer. And then a bit of invariant manifold analysis to back it up, provides an underlying, mathematical formalism with which to try and understand where the probabilities may suddenly increase or suddenly change.

Yim. How would small random noise affect your calculations, or your methodology?

McRobie. The whole point is, I would contend a small random noise wouldn't change anything at all. The small random noise to some extent implies that it's a *long-term* ensemble, whereas we are talking about a few short waves. All that really does, that random noise, is maybe just move things to a slightly different point in the parameter space. That's another way of thinking about it—and so, I don't think it affects the transient escape diagram and those sort of things to any great extent. It would move things around a little, but we are criticized by the people who do stochastic work, that, in a sense we don't think in their terms. In the case of chimney design, for example, there is a very comprehensive stochastic design process which you go through to design a chimney. It takes account of all the coherence functions, cross correlations, the spectrum of the turbulence of the wind, but at the same time is based upon this assumption of material linearity—and one of the things that worries me is the material linearity assumption. There's two load cases to consider: one's the cross-wind resonance, and the other's the long-wind resonance. The cross wind's due to the vortex shedding which we heard about yesterday, involving nonlinear van der Pol–type oscillations. The two problems are separated into a cross wind and a long wind. For the cross-wind ones, there's very sophisticated stochastic dynamics included with this van der Pol–type oscillator. You do a finite-element analysis of the chimney and then you insert the modes that come out of that into the van der Pol–type oscillator with the stochasticity thrown in, and it takes account of mode locking and that sort of thing. But those have a zero-mean displacement, so basically you're going through this constant stiffness range. It's only when you get in a very large vortex-shedding you go into the nonlinear material regimes, whereas in the long wind response, the mean forces push you straight over to the edge of this large drop in your stiffness, and then you apply the fluctuations due to turbulence and so all this comprehensive random dynamics that's done in the chimney design is all done in a very dramatically nonlinear stiffness regime. It probably didn't really come through on that schematic that I drew, but for reinforced concrete sections there's a dramatic decrease in stiffness and that's where you're operating all this comprehensive random dynamics. And my own feeling is on the basis of this sort of work, that maybe we should think in terms of hitting it with a few bad gusts and just integrate the thing forward and see what happens—and you can work out what the probabilities of those gusts would be—and to go back to a sort of a discrete gust approach which is conceptually very very much simpler. When one perceives it on this level it may be a sounder way of doing it than the random dynamics with the nonlinearities ignored. In which case, it ties in to the approach to ship stability as well.

Thompson. Let me just elaborate on that. All the naval architects say, "Well, you should be putting in ocean data as your driving excitation," you see. Now that's a crazy thing to do if you're interested in capsize. That's all right for operating conditions, but capsize is an extreme event. If you put a model in a wave tank and start sending random ocean waves at it, you wait all day for it to capsize and then you perhaps have one capsize. So what have you learned? I mean, you have an enormous problem. It's much quicker and much more efficient (you get through more models and make more parameter explorations) by sending a short pulse of regular waves, rather than waiting for a short pulse of regular waves to emerge in a big statistical piece of data. So one should distinguish between operating conditions and an extreme event like capsizing.

Dorning. I'd like to make a comment about that. It seems to me it would be prudent to analyze random ocean data and then superimpose occasionally on your regular wave train, not some small scale random noise, but a random amplitude in the regular sequence. This, I think, is what tends to happen. One gets maybe three, four, or more likely six regular waves, then a wave that's very much like the others except it has a particularly large amplitude arrives as the famous "seventh." Then, depending on when that wave—that one with the larger amplitude—comes, and where the ship is with respect to the phase of the forcing function, some dramatic events can occur. I've experienced some of them!

McRobie. That's exactly the way we are thinking; in terms of, "you have a sequence of fairly moderate waves which puts you somewhere in the basin, you don't really know where you're starting, then hit it with one bigger wave"; well, if it's the seventh one, okay, the first six put you somewhere in the basin, so we've got that far along your time series. Now we come to the seventh. Now if that basin is heavily eroded then there's going to be a big chance this one big wave is going to hit you over.

Thompson. One of the interesting things, actually, is that because these fractal figures sweep across the whole central area, and they move of course, it means you can do one crucial test: you could always start your test on a model ship in ambient conditions. In other words, let's start at $x = 0$, $t = 0$. Let's always test a ship in ambient conditions, it's still, it's vertical, and send a short pulse at it. This procedure is going to locate this cliff rather accurately because the incursion has swept across the central domain. You see, you've got to get down to a practical test. I mean, on a computer you can do 10,000 runs for each picture. In a wave tank, you are really very limited on time, and so you've got to think of a test that is really very quick, and the very quick one we would suggest is to put the ship at initially at rest.

Yim. At OSU, a colleague of mine has developed a transform procedure that can incorporate a particular deterministic wave form in there. He can generate some random waves and then, depending on where you want it, how you want it, what shape you want it, he can embed it into the random waves. So that way it doesn't take that long to have a random condition at first and a particular wave form that you choose.

Harrison. What we do when we test mechanical structures with impact techniques, such as a wave hitting a structure, is look at the response of the structure knowing the input, and we do individual transfer function measurements. If you look at the coherence function to see if there are new frequencies present that are generated as a result of the impact that aren't in the frequency of the wave train, then we know there are nonlinear effects going on, and that's our initial test for nonlinear behavior. So if there are dips in the coherence function, then we look for other nonlinear effects, but we must rely on fairly large curvatures and displacements that actually excite nonlinearity, in a lot of real-world mechanical structures. The traditional coherence function is a very good test to use.

Discussion: Dynamic Instabilities and Chaos in Running Belts and Their Cleaning Devices. By A. Harrison

Thompson. As a philosophical point in these sort of belt studies, do you find on the whole that chaotic motions are a good thing or a bad thing? In some of the early work we did on impacting systems, we used to find resonance regions, which of course were dangerous because the amplitudes were large, but between the resonances were the chaotic regions in which the amplitudes were actually rather small—and in that sense, for the system we were studying, chaos didn't matter too much. They were actually rather small vibrations. I just wonder whether in your experience (just as an overview as it were), are chaotic vibrations on the whole bad things or are they actually rather good in the sense that they keep the amplitudes low?

Harrison. The answer is, they're good things if one can keep the amplitudes low. If you consider the example of the forced belt, the chaotic vibrations were of much lower amplitude than those vibrations going into resonance. Just before chaos, large resonance amplitudes occur and these are very dangerous to structures.

Thompson. So you confirm the concept that in this case chaos is not in this sense a dangerous thing; it's actually rather a good thing to have.

Harrison. Yes, unless you are operating very close to a resonant boundary, where you've got a very high Q system; then you're in big trouble. Because the energy, which is locked up in chaotic flutter of the system, can all of a sudden go into the system. Suppose the initial conditions change coming into a resonant cavity, in the case of a belt, where mass per length might change, or the width might change a little bit (through manufacturing change). All of a sudden, the belt will be absolutely on resonance whereas if it's chaotic, a bit off, you get much reduced amplitudes—and those things you can't predict, and they do vary. Viscoelastic properties vary with temperature too, and they vary with tension in the belt.

Thompson. Yes; so that's one of the elements of chaos that can be in a sense dangerous: Although the amplitudes are small, the risk is that one can easily go into resonance.

Harrison. My personal view is that we are trying to rewrite the books to design the systems right away from resonances altogether, and that can cost money because you've got to put more support structure in it someplace to kill the vibration.

Gorman. Who designs these belt systems? Are they all over all the countries or is there one country that predominates?

Harrison. No, there are engineers in Germany, America, Australia, and England; mainly those four countries of the world are leaders in conveyer construction.

Gorman. Right. And are they aware of ideas of chaos?

Harrison. No, this is all new to them. We're in "cloud cuckoo land" compared to them. They build a system; when it doesn't work, they call us. Basically, they design them traditionally. They pull out a filing cabinet design from last year and modify it. They increase the tonnage, they increase the strength, they increase the requirement on the belt, they put four motors instead of two, all of a sudden the thing doesn't work. What they're doing, in effect, is reducing safety factors slowly. They reduce structure strengths, for example, to cut costs, and all of a sudden the dynamic loads in starting and stopping occur, and the belt snaps. And that's not particularly chaotic, but what they're doing, in effect, is changing the boundary from what mankind knows as a zone of stability. For example, in the underground mining industry traditionally belts have only been 500 meters long, fabric-reinforced, and it's worked for 20 years. Now with long wall mining we want to make those belts 9000 feet long. So the long wall systems are built, the Germans built the long wall systems, the English built the long wall systems, the building engineers who have been happy with belts for 50 years literally, say "well, we'll just make it a little bit longer, a little bit longer, a bit longer," and all of a sudden it breaks; it snaps and big chunks completely eradicate it. Unstable resonances also now exist whereas before they did not.

Gorman. So, are you testifying in lawsuits, also?

Harrison. Well, yes, in one case for defense of the design which was on the verge of being below an acceptable safety factor. But the operators were playing with knobs and buttons and changing loads at the loading point. But essentially we've always operated in a zone of stability. Now we're sort of attacking that zone of stability which is now a zone of instability, and I guess that's the cause of the problems we're starting to see.

Maulbetsch. Assume that the mass of the stuff on the belt per unit length is larger. Are the variations in the mass along it also large compared to the belt weight?

Harrison. One of the belts we've done work on was 130 kilograms per meter, that's maybe 300 pounds of weight, and the loading was about 100 kilograms per meter (less than the belt weight).

Maulbetsch. Okay, so that's not a good assumption. The belt is the large mass.

Harrison. It can be. I mean, there's steel cords inside the rubber belt. Hundreds of them, side to side on a plane, and it's very heavy stuff. The one in Chile we worked on, for example, had a running tension of 1400 kilonewtons. It failed, it broke, and it caused a catastrophe—and it was a very heavy belt. It moved 5000 tons of copper up a 1 in 4 incline, and it broke because of the snap action on stopping; it threw the takeup winch 100 meters. Like a big sling shot, and it was a 60-ton takeup. The belt had 6 megawatts of power when starting. They're big systems with big forces. But the problem is occurring in the design and engineering; the engineers to cut costs are coming up against this boundary where you get instabilities, starting and stopping. You get forces which are so large you can't contain them anymore. They can be unpredictable by classical methods.

Yim. So, now that chaos can occur in your system, how are you going to incorporate it into your design of the conveyor belts?

Harrison. The whole reason for doing that modeling on the elastic wave stuff is to try and find at what length (length is one of the parameters we have not tested very much yet) there is an effect on the solutions to the wave equation—and I believe that at some point you'll get stable vibration; solve for period-one or period-two, at least, vibrations. If you can predict the type of vibration you're going to get in a system, you've got a handle on what stresses are going to be in the system. If you can do that, then you can work out what spacing your idlers have to be to get to minimum sag so you don't have problems when you start and stop —and you can also space the idlers to stop the resonance.

Yim. So are you trying to design it so that you avoid chaotic response?

Harrison. Yes, I think man's evolving to the point where he's made these things and they've always flapped and they've always done fine in bits and pieces but nobody's really cared about it, they haven't been expensive systems. Now through trying to cut costs they've put the idler spacing further apart instead of closer and therefore the potential for resonance rises, and we have to go back and look at the fundamental design. We might have to change belt stiffness, for example, to fit it, just from a straight cost point of view.

Yim. Do you imagine, in the future, that designs might actually encourage chaotic response, and take them to the chaotic response region as well?

Harrison. I think we're seeing that now. Here's an example (see slide) where the motors that come from Japan are what we call very stiff—electrical people will understand what I'm saying—it's got a very sharp torque–speed curve. Traditionally they've been sloppy, so if you change the load line torque, then you go up the torque–speed curve and the torque goes up, but it's not a big deal, it's quite spongy. But when you've got a sharp motor, a little change in load will put us right up here at a very high torque and then you fall over at the pullout point on the motor—and it's an upside down parabola; it's got two stable states of speed. So if that's occurring, it might be only 200 rpm difference in motor speed, you reduce that down through gearbox, so you've got a speed variation in the belt. Now that speed variation at this point affects what the next drive will see, in this case, and therefore it affects the tensions going in that drive, and the whole things hunt, and when it starts to pull harder, the next motor goes back up over the hill and you've got an elastic oscillation between two drive sources and that's a very complex problem to solve. Because we can't define the way the belts behave at any instant; so we have had to evolve ways of designing stability into it, so we operate away from those boundaries of unpredictability.

Discussion: Atmospheric Flight Dynamics and Chaos: Some Issues in Modeling and Dimensionality. By G. T. Chapman, L. Y. Yates, and M. J. Szady

Dorning. Actually, I had a couple of questions. The first is in connection with the data you showed from the experiment that was done in San Diego, where there was a small spike in the power spectrum at a very low frequency, and your time series also looked as though it had beats in it. It was oscillatory, but it had an envelope that also was oscillatory, which suggests quasiperiodicity of the motion on the torus. Then when you let the wire out, I guess when you increased the velocity, the wire started to vibrate as well.

Chapman. Well, actually they took the damper off.

Dorning. Oh, you took the damper off, I'm sorry. So, that then allows the third frequency into the motion, and then I think you observed something that looked quite aperiodic. Is that right?

Chapman. They showed a wide range of stuff. What I was really trying to show here is the coupling that comes into play; in this particular case I didn't get any feedback directly, but it isn't immediately obvious the types of feedback you get. But the motion of the fluid mechanics seems to come in and give you a nice spike in the motion of the wire, the fluid mechanics now look chaotic, but the motion dependence of the wire still look pretty decent. So we've got feed in one direction but they don't seem to have feedback.

Dorning. When you had the clamp on the wire you had two frequencies, not one, in the system—and adding a third frequency, in fact, very typically does lead to chaotic behavior.

Ottino. When we all fly back home, most likely we will all encounter some turbulence. In a wind tunnel you produce nearly uniform flow. At some point in the real world you encounter fluctuations that are of the length of scale comparable to the airplane or something like that. How do you handle that?

Chapman. Well, when most of that occurs, the control systems are pretty robust and will handle it; I mean, you're flying at 35,000 feet and you encounter it, you get a pretty severe shaking. It's only when you get into bad conditions, when you get into what they call a microburst in the atmosphere, and you have a large region of downflow which now takes you and runs you way out of the regular flight regime that you normally fly these vehicles. Luckily, most of these kinds of things have occurred at 35,000 feet. Now you may get beat around the aircraft a little bit, but 35,000 feet is a fair amount of time for a good pilot to guide himself loose of it. That's why most of the flight testing in military airplanes all takes place at around 20,000 feet when you do this; if you do encounter any of these that you've got plenty of time to recover from it. But the accident that occurred at Dallas airport several years ago was a wind shear, a megaburst. That occurred too close to the ground and there was no chance to recover, but it literally does cause severe problems and mostly what we do is try to put radar on board and try to find out if you can find it. If you've noticed, if you're ever flying at high altitude the pilot once in a while makes small turns. What basically has happened, his radar has picked up the remnants of these and he just tries to fly around, provided he doesn't have to go too far. If you've ever flown in a light, small airplane through the thunderclouds down in the Florida area, at about 8 to 10 thousand feet, where you look out the window there's a thunderhead here and a thunderhead there you'll see the pilot forever flying down valleys trying to avoid exactly that type of thing. Most of the time you can't design an airplane to meet all of those. It's just impossible. No matter who's interested, all the nonlinear dynamics says—and that's partially what happens again in this, we never see so many steady states. We never see asymptotic solutions; we see all transients. It's—whenever a control system comes back, the control system has a function, "I want it to keep on this course at this condition," and the control systems are all going to bring them back; besides that they don't really stay in what we would call a nice control condition. As soon as they immediately start to lose altitude, a lot of other control parameters change, density starts to change—literally, you have to come up with potential energy gained by the vehicle dropping in altitude—and the flight mechanics becomes much more important. Everything I talked about here is level—not level, but constant-altitude flight mechanics questions. They really become, there's one mode that even compares linear aerodynamics to what they call the high frequency, the regular mode is just a pitch oscillation mode, that's quite easily damped. There's a very long period called the Phugoid mode which is an interaction between the angle of attack and the fact that it is continuing to lose altitude. Those two will fight each other and give you a long period of oscillation. Those are all simply nonlinear as you understand it. They go clear back to the turn of the century, with a group in England, when we started out in flight mechanics.

Sen. You've put all this in the context of flight mechanics and rightfully so, but there are industrial applications of flow around heat exchanger tubes and things of that sort. I don't know if you could comment on that.

Chapman. Well, it's just this: It's a fluid, is a fluid, is a fluid. It permeates the society in more ways than you can imagine. The circular cylinder problem, which we're spending a lot of time on, we're only using it because to some extent we see it as a model problem for a lot of our flight mechanics problems, and yet it's a little bit more amenable for testing in small-scale wind tunnels so you can get at it much more, and we use that to identify about three or four what we call special problems that illustrate these basic features and not try to study complete vehicles because that's a difficult problem; if you study complete vehicles they're complex. We're going to keep flying them, but we've got to study the simpler ones, to understand some of these things.

Kim. I'd like to comment on your comment. Indeed this kind of phenomenon is observed in one of EPRI's projects: We are sponsoring a small project. This involves the row of cylinders in cross flow, and we observe the kind of phenomenon that you observed in the airplane wings; we are calling it fluid–elastic vibration, because there's a coupling between the structure and the fluid and (what we found out is at the end of the row) the flow behavior is extremely sensitive to the conditions upstream.

Chapman. I'm always hoping that somehow, if it is chaotic, that somehow the attractor has some of the same characteristics even though the details aren't the same.

Kim. The applications of this kind of phenomenon occur, where this particular cylinder geometry is like in a heat exchanger, where there is some cross flow, so we are studying that. I have also one question, if you may. When you observed this interesting phenomenon, it seems like it is very difficult to do anything about it. Do you have any special idea as to how to control this?

Chapman. Yes, there is in fact a big flight test and wind tunnel test that is just being started with NASA/Ames, Langley and Air Force are all involved, and basically they'll take an F18 aircraft and flight test it but also run it in the huge wind tunnel at Ames. Good airplane; the same airplane they flight test will be wind tunnel tested, and they're looking at, particularly, this asymmetric vortex flow that appears off the nose when you get to these high angles of attack; they would like to find out if they can control them. Because that's an important thing; it gives side forces, you could end up with at least the beginning of an out-of-control maneuver, and so what they're doing is they're trying several mechanical devices when you just pull a little on one side of the vehicle and you feel it coming out on one side and you stick out a little flap on the other side and see if you can bring it back. Or by jet blowing. Jet blowing is preferred because most military airplanes like to have radomes up in front and they don't like to have any kind of mechanical devices up there because they want the full portion of that vehicle to their sole utilization. In fact, aerodynamic flight engineers will see the electronics people up there all the time, but they are looking at that, and there's a fairly extensive program. What's bothersome to me about it is they're looking for fixes, and if they don't understand it they don't care—as long as they get a fix. They're not approaching it in what I would call a systematic scientific way. You're going to put an awful lot of money into very expensive experiment and when we're all done we may find a fix for that vehicle, but if we don't know well enough generalities about it to see whether it applies, there's a good chance we won't be able to apply it to another vehicle. I buy that and I fight with the program offices on this. There's not good fundamental support behind doing work to understand these phenomena so that when you do a large scale one that you know enough about how you're doing it and making the fix of it you extract some general information. But there are programs going on; Notre Dame is doing a lot of work on wing rock, for example. Very extensive, a lot of work on fixes. But they are having an actually full-scale flight test; it's a joint program between NASA and DOD, and, well, like you say, that's strictly engineering. The guys would kill me, because they think they're doing real good science, but they won't touch the idea of chaotic processes at all. They're going to go on, and if they measure unsteady forces, they'll measure mean and rms. Very traditional. Maybe a power spectrum.

On the nose region they don't know how to scale very well. On delta wings, where you have a sharp edge, the scaling is a little simpler there because the sharp edge sets up the location of the vortex for them. On a curved body it depends highly on the Reynolds number. You've got, in that respect interactions with rudders and so it really fouls up the scale. As it turns out, if you do a full-scale flight test and look at the vortex patterns that come off the wings, where you got relatively sharp edges, and compare those to even what you do in a water tunnel, there are subtle differences between them but they're—it's amazing correlation in terms of the global structures. Now there's details in terms of the higher frequency components, but that portion works. Soon as you have a body—remember the one I showed, the store body with the rounded corners on it and how it had hysteresis—it turns out most vehicles fly at the fairly high Reynolds numbers so it's in its well-behaved state. Most wind tunnel tests are done right in the middle of the region where all the damage occurs. In fact, what we spend a lot of time doing when we do wind tunnel tests at high angles of attack is trying to find kluges, for lack of a better word. We just go in and put a tremendous amount of roughness in the nose region to make it look like high-Reynolds-number flow, because what we test in wind tunnels under some of these high-angle-attack conditions is not the same as what we see in flight, and that comes back to the scaling question. It's mostly, it's more of a body question than it is a wing question.

Dorning. I think it was the case when you were talking about the rocking, and you had what I think was a subcritical pitchfork bifurcation followed by a couple of turning points. I think you had something clamped then, and then when you unclamped it you had a Hopf bifurcation that appeared at a lower value of the parameter?

Chapman. Yes, well if you release it, allow it to roll, there's a fold in the moment curve; it would be a fold catastrophe, if you come to think about it, in the sense that you have a solution sheet that's folded. Actually it turns out that if you take the two parameters, the angle of attack parameter and the roll parameter, that maps out a whole space. But it's got a solution space on it that does this, it's two sheets.

Dorning. But, you said, as I recall, in what this was a physically important case—I thought it was the case of the left–right roll or the clockwise–counterclockwise roll—you said you had a Hopf bifurcation at a low value of the parameter. It looked like you had a subcritical pitchfork, a static bifurcation, and then you unscrewed a clamp of some kind . . .

Chapman. I see what you're talking about. When you're free to roll when you're below where there's any asymmetries going on and you move it over to say 10 degrees roll, it will start to roll and will build up to an amplitude. There is a limit cycle there and that really comes from a Duffing-like behavior; there's a damping–undamping going on that occurs in that particular problem.

Dorning. And then you said as the parameter varied, I guess the amplitude of the limit cycle grew and then it crashed into the branches born in the turning-points bifurcations?

Chapman. Into the turning point of the other one and then started off on another. You could see it was definitely this curve coming along here and all of a sudden you could see that it looked like it took on another curve. Of course, when you do experiments you can never find these [unstable branches].

Dorning. In that case, did the period of the limit cycle get extremely long?

Chapman. Not much different. What's so, it's a little bit surprising to me. At that point there's probably an awful lot of interaction going on because this has to do with how they're moving up and down relative to the length the points are going back and forth. So, I'm not sure which controls which at that, . . . Again, so many of these things, I see the phenomena, very little of these have really been attacked from a strictly nonlinear mechanics, very few of the flight mechanics problems I have dealt with have people started out with that as objective. I get artifacts that I keep scraping out of the literature on this. One of the reasons why I took my retirement when I did with NASA was I wanted to have some time to go off and look at these from a fundamental standpoint and unfortunately, within NASA they want big programs, and I wasn't prepared to do those. I would rather work myself out of a job and solve one of these problems, rather than finding fixes—and, unfortunately, well, engineering is a lot of times finding fixes. You've got to make something work today. It's out there flying in the fleet and you're losing some of them, and the Air Force or even NASA likes to find fixes. But you've got to get at this along on your dynamic side because I think it contains some of the real roots of understanding. Again the matches between simple one-dimensional maps and some of this stuff is a big stretch in the imagination. But this is an open system, it has a lot of convective instabilities and it's harder to put them in the same ballpark.

Harrison. Have you done any work on the reentry vehicles?

Chapman. When they stay within boundaries, where things are pretty well behaved, but there have been conditions on some of the early flight tests where they came in, that—again it comes back to where they try to build very strong, robust control systems, and so, when they say how much . . . they build a control system that will handle what we expect, then they probably double the amount of control authority that you've got. In one of the early flight tests when they came back into the atmosphere, they landed safely and all of that, but when they went back over and reviewed all of the flight records, it turned out that they used up every bit of their control authority—they were right up against the limit. Pilots sort of knew it because they knew they were taking the controls much further than what they were doing in normal simulation. It turned out that had to do with some real gas effects and some separation that was occurring over some of the flap control systems, but it does apply—it goes into those. I've studied flight mechanics of vehicles going into planetary atmospheres—the Jovian atmosphere is the one because it loses so much mass—an interesting problem. In those we never have a pilot, so the vehicle has to have an onboard, automatic control system, and what you try to do in that case is to try to make the body inherently stable. Unfortunately, blunt bodies—you can make them statically quite stable; in other words the nose comes down, but dynamically they have very little damping, very, very little damping. So you have to have onboard systems that normally consist of little rocket jets to provide active yaw damping. That's another thing—I've got a whole story about yaw dampers that describe military aircraft. You put devices on that are supposed to keep pilots out of problems, and they get into real combat—this happened both in Korea and South Vietnam—they put something onto vehicles that were meant to keep from getting into these difficulties, but if you've got a Mig on your tail, you don't care if you lose control. That's one of your better options! So you don't always want to be in that condition. So, in Korea, what specifically happened, basically, the F106 people, when they got back after the first few sorties, they told the electricians to go in and find the wire that controls the yaw damper and cut it. Now we're smarter about that. We build them all into the system, but we have a switch on the dashboard—if you get into combat, you take 'em out.

Discussion: New Applications of Chaos in Chemical Engineering: Intuition versus Prediction. By J. M. Ottino

Chang. Your prediction of the drop size is a linear model?

Ottino. Yes. For capturing satellite and subsatellites, we have to zoom in or to do more sophisticated computations (boundary integral techniques). We have movies of two big droplets leading, upon breakup, up to something like seven in between.

Chang. (Inaudible.)

Ottino. In the first comparison that we did we ignored the satellites. However, when you mix the things for 40 periods, and you have 25,000 droplets, which is a reasonable number, you have no idea which one is the satellite and which one is not. However, everything looks like a sphere, satellites or nonsatellites, they all fall under the same scaling curve. They all do. We have a cutoff size, below which we cannot measure things. This is of the order of 10 micrometers. So there is a part of the PDF that we don't see, but the part that we do see, which contains undoubtedly many satellites, does indeed show scaling behavior.

Chang. (Inaudible.)

Ottino. The main reason that we're inspecting this case is because the stretching is itself self-similar, obeying basically the same scaling description. But why things like satellites should fall under the same curve exactly, I don't know. In two-dimensional flows, according to whether there is complete chaos (global chaos) or partial chaos, there are two families of curves that scale, and there seems to be some crossover mechanism which seems to depend primarily on the viscosity ratio.

Cvitanović. Maybe I shouldn't mention this, but you had some very interesting technique of destroying the islands. You went so fast by it, I didn't even understand whether you were changing the mixing prescription as you went from frame to frame.

Ottino. As long as you have something time-periodic or finite, there will be "islands." No question about that. The size might be very small, but they will be there. As long as there are periodic points there will be tiny regions in there which might be undetectable in the Poincaré section. However, even though we do not know how large these islands are going to be, we know how they are going to be placed. For this particular system we know that they will be, for example, vertically placed or horizontally placed, and then what we do rather than continue forever is to change gears, and now we start doing something that will place the islands, the potential islands, in a horizontal fashion—and then we don't keep doing that forever either. We change again, and pick another flow that will place islands also vertically. But in general the islands, the symmetry will be the same as the first flow but the island sizes and locations in general will be different. That's the idea.

Cvitanović. You know, I'm fairly sure that only very few islands, if you really wanted to know about all the islands, of which there's infinity, very few of them lie on symmetry lines but of course those are the biggest ones and maybe you destroy those.

Ottino. When you capture the system at the moment in which it displays symmetry, all the periodic points are symmetrically placed. There is no other way around it.

Cvitanović. Yes, but there is a line of symmetry. I mean there is a flip and every point is either self-dual or dual to another orbit under this flip, I agree. But now it doesn't mean that if I change the axis I will destroy all of this symmetry.

Ottino. I agree. I didn't say that you could destroy all of them, I assume you might have an island that is just at the center of the system and that you are not going to destroy. What I said is that for some systems (such as the one that I was talking about), you can take a snapshot of the system and capture all periodic points on the system in such a way that they display a given symmetry. The simplest type of display will be one that is vertical or horizontal. In other systems it would be a 45 degree line; in other systems the symmetry line could be a curve. Given that, the basic idea is to shift symmetries around. There are other things that can be used, and in fact we are looking at other possibilities. One has to do with developing sequences or actions that somehow emanate from, for example, something which is connected with a golden mean, or a Fibonacci sequence. However, you cannot prove that any one of these is the optimal. Conceivably there could be one that given a certain amount of energy produces islands with the smallest size.

Chang. Is this idea equivalent to the frequency modulation of a perturbation?

Ottino. Could be.

Chang. Basically you have a perturbation, not at the peak frequency, but you're modulating it?

Ottino. The systems I talked about here have no inherent frequency, in the sense that, if I was showing you a movie of this, there is no way on earth for you to tell me whether or not I'm speeding the movie up or slowing it down. It all depends on how I move the boundaries, but, yes, you could probably think along those lines, if you wanted to. Another thing that you can do is try to change the period. In what I talked about the basic building block was always left at the same length and we could combine them, but you could conceivably change that as well. Another degree of freedom which is particularly important: everything that I show you from that particular example was always corotated. However, that's not engraved in stone either. You could move one way and then decide to move it the other way. That might lead to better mixing.

Chang. That's phase modulation.

Ottino. Yes.

Cvitanović. In dynamical systems, one also looks at these periodic points that you look at in symmetry. Now the way we use them is when there's a system with symmetry like yours, then we keep dividing the symmetry and reduce it to a fundamental cell where there is no symmetry. Then we work in this little subspace where any trajectory reflected would produce something that would all be inside—and then this has all the islands in it, and there is no symmetry left, and they haven't gone away, so I'm just surprised that you can make them go away.

Ottino. No, I cannot make all the islands go away. If I could make this, that would be a major result. As long as I stop somewhere, which I do, they're still there. They could be small, but they are there. If I knew how to make the islands go away completely, that would constitute a major breakthrough, I think. What we do is to minimize the impact of potential islands; I can make the contents of one of the islands become part of the fluid that now becomes well mixed. That's basically the idea. Part of the region of the island that was poorly mixed now happens to lie in a region of what is good mix.

Dorning. You're essentially switching them, by changing some parameter.

Ottino. Essentially, yes; that's the idea. Probably within that concept there are 10 other intelligent things that one might do to improve things. I think that the most important applications would be for systems in which there is something like a tube and there is something flowing through and now you decide what objects to put in the flow and in such a way that within that fixed length you get the contents to mix the best.

Discussion: Chaotic Mixing for Heat Transfer Enhancement. By H.-C. Chang and M. Sen

Gorman. Do you think as you open up this angle you're changing your amount of your increase, or does it matter? You had these coils that are not wound straight like this but wound like that and therefore you have some angle, right?

Chang. We have done only one geometry . . .

Sen. But if you open it to 180 degrees you obviously don't get anything.

Chang. The nice thing is—it really doesn't have to repeat itself. You can change, as long as you have some kind of perturbation, but that should also cause . . .

Kim. In your eccentric cylinder, or tube case, do you have a fluid flow?

Chang. In this model we're studying we assume it's two-dimensional so there's no through flow. A through flow really doesn't change things because essentially the through flow is decoupled from the transverse flow field, so it should be the same even though we didn't put that in our model.

Dorning. It depends on the rate of rotation. Even when it's concentric, if the rate of rotation is too high that's when you have all the bifurcations in the Taylor–Couette flow.

Chang. I take it back, that's not unconditional.

Cvitanović. I'm just curious. Is ten percent a significant engineering increase? These numbers seem very small.

Chang. Yes, ten percent is significant.

Chapman. Did you try, when you had the coil, to increase the pumping power just slightly? Did you try to ascertain how much?

Chang. Yes, four percent. That's always a trade-off. You have to worry about power input and pressure drops, and that's why it's nice to work under low Reynolds numbers, because the marginal returns are usually higher.

Dorning. I had a question, I don't remember exactly what it was, during your talk, but something came out to be 1. It seemed to be independent of either the rotational velocity or the eccentricity, or the aspect ratio of the cylinders.

Chang. Yes, that was curious. The speed ratio is always maximum 1.

Dorning. Independent of the ratio of the diameters of the cylinders and independent of the eccentricity?

Chang. It's independent of the gap width. You can always normalize the true cylinder radius. There's 1 plus 1 plus delta, say. They seem to be independent of the gap width.

Dorning. Did you expand in delta to do the calculations?

Chang. No, we expanded the eccentricity, not the delta. There are two parameters, one's eccentricity, and the other is the gap width. It's curious that the delta speed ratio is always 1 as an optimum. There must a reason for that. I didn't look into it...

Discussion: Chaotic Transients and Fractal Structures Governing Coupled Swing Dynamics. By Y. Ueda, T. Enomoto, and H. B. Stewart

Linsay. I'd like to point out that the basin structure is not the only question of interest here, that transient times are probably also quite interesting and important in the stability of these kinds of systems. In the case of two coupled oscillators it's probably not terribly important, but when you have a complicated system with many coupled oscillators —I have experiments of that kind—transient times can become extremely long, even though the initial conditions are such that the system will settle down to a stable fixed point. I've experienced cases where it's 10, 20 thousand oscillations before the oscillators will phase-lock again and come to a stable point. So, I think the analysis has to be generalized somewhat.

Stewart. Well, I think we could clarify that, perhaps, by looking again at this picture [Figure 8.2]. What we're trying to say here is that, let's imagine that this is where you want to be. It's clear that you don't want to be in any of the black regions, that's absolutely out. If you do that, you're going to a very bad sort of transient which is going to destroy your system. So the black regions are absolutely out. Now what you're talking about are probably analogous to white regions which are mixed up with black regions. You probably don't want to regard these as good regions just because they're white.

Linsay. I think the point is, this is two oscillators. When you get to several, 10, 15 oscillators, then these things become very complicated, and your perturbation still leaves you in one of what you call white here. But the time it takes to get back to that fixed point can become extremely long.

Grebogi. In the case of two oscillators you have the formation of strange sets that live in the basin of attraction. Those sets are responsible for the existence of long chaotic transients.

Stewart. Okay, well, what we're saying here is that the first focus of attention should be on the geometric structure. Then there's, I agree, there's a second level of detail in which you look at additional complicated sets which may be entirely contained within the white region here. The point is just to focus the engineers' attention on the geometric structure, to point out that the energy integral methods are basically trying to fit some say sphere, or some nice smooth convex set, to fit some nice set inside that region and—the point here is really so simple that we risk losing sight of it. It's just that the geometric approach is necessary; it's not talking about a luxury, it's talking about what the engineer really should be concerned with.

Thompson. It is of course, as Yoshi is well aware, very easy once one is doing these grids of starts, to simply make a note of how long it took either to escape or to settle. It doesn't actually involve a great deal of extra computer effort and, therefore, one might as well do it. One might as well color code the points as to how long it did take to settle. This would cover these points that have been raised: and really, it's no more computational expense anyway. Well, I do agree, of course, that the first step is to get the absolute boundary, but then transient boundaries are very easily mapped out by the same technique.

Dorning. And refine the category of basins: basins of strong attraction, basins of weak attraction?

Stewart. But still, they generate a structure which you can understand.

Discussion: Probabilistic Analysis of a Chaotic Dynamical System.
By S. C. S. Yim and H. Lin

Kim. Actually, ergodic systems will behave just like that. So what you are actually trying to prove is that this kind of system is ergodic? Is that what you're trying to say?

Yim. Yes, based on my definition, I think the chaotic system possesses ergodic properties. I have a very restrictive definition how I sample. So, in the global sense what I really like to determine is how often does the system become chaotic?, and how often does it fall into this parameter space? Suppose I have a displacement and a velocity parameter space, and I have regions of chaos and regions of periodic response so one thing I would like to see is how often does it fall into the chaotic region, if I start rocking the system at a particular angle and a particular initial angular velocity, and if the response is periodic I have no problem because we can analyze periodic responses and just turn it over to the engineers, but if it's chaotic, then the engineers would like to know how they can account for the chaotic behavior in the design. I'm far away from turning it over to the engineers and saying here, you can analyze it. But I'm trying to bridge the gap between interpretation of chaotic theory and engineers who are familiar with probability density functions and spectral density functions and who want to design for practical systems. So even though I picked this particular simple example, the application is actually generic. In fact, this is a very small part of my research project and my main research is on nonlinear compliant systems out in the ocean, so hopefully I can identify chaotic behavior in ocean systems and then I can convince them to support additional research to apply my analysis!

Stewart. Do the equations that you show there allow the cube to bounce up so that both corners are off the table?

Yim. No, it does not. That's one of the limitations. In my discussion with Mike Thompson the other day, he pointed out that in some experiments they do bounce up. But I do not allow for that. This is often a reasonable assumption. In Japan they tried to use the responses to estimate earthquake action. They have a large array of blocks and there are other applications as well that may be interesting to EPRI. The original research was actually stimulated from radiation protection analysis at LBL, where they have small radiation labs, and what they do is they use large concrete blocks to protect people from the outside, from radiation. But since the laboratory lies right on top of the Hayward fault they were worried about the earthquake excitation and the overturning of those concrete labs and endangering the staff working nearby. That was the origination of the problem. I analyzed that problem for earthquake response and I found the system to be very sensitive, and I did not believe the results myself, until today of course when I look at it as a chaotic system it makes perfect sense. That's the definition of chaotic behavior, it's very sensitive to initial conditions. So, back to the question, I don't see any problem if it doesn't bounce, because in a real system you're going to put some rubber pads on it. So that it will absorb a lot of energy and it will just rock back and forth nicely.

Harrison. In real engineering systems we find the fatigue life—let's say of steel cables that are flexing as they're moving—has a variety of numbers and you get a probability distribution of the fatigue life. When you flex a steel cable in a more chaotic way its amplitude might reduce and you find it fatigues faster. Now does that fit with your view of the probability density function? Can engineers use such a tool?

Yim. I thought about that last night; I was anticipating you were going to ask a question about the vibrating belt, but what I have done is just look at a very primitive problem in which there is no flexibility in the structure. The only fatigue that I'm looking at is actually how often it hits the rubber pad, and so we have a design for the rubber pad and for the flexible belt problem. I actually hope that you can extend it to your system.

Harrison. Rubber impacts are nonlinear of course...

Yim. Right.

Lawkins (Comment extensively rewritten after the meeting). We've talked about the generation of information. Consider an initially tightly clustered set of points on an attractor and think of these points as representing a sample of experiments similarly prepared. If the attractor is associated with a chaotic process, then the flow on the attractor, that is, the flow defined by the dynamic process, will cause that initially small radius cluster to be spread out and mixed over the whole attractor. The initial cluster represents a small amount of information about the process because all the states are alike to some precision, the radius of the cluster. The information about the process as a whole represented by that sample grows exponentially with time as that cluster becomes spread out over the attractor. The exponential rate at which that occurs is, by definition, entropy. The entropy of a chaotic process is an important time scale.

Yim. Right, that's exactly what I'm trying to point out. I have one time series and I can infer on the information of nearby trajectories, that's what I'm hoping. Suppose I know the characteristics or the probability distribution of the chaotic behavior. Then I can look at the whole space and cut it into grids, and superimpose on top of that another probability structure. So that I have probability on top of probability and the two of them combined will give me, will give the practicing engineers, a way of designing. That's how I look at it.

Lawkins. Yes. If one is designing a sampling strategy to test a chaotic process, the property of information growth, or entropy, is important and useful.

Discussion: Chaotic Behavior of Coupled Diodes. By H. A. Cerdeira, A. A. Colavita, and T. P. Eggarter

Dorning. Did I understand correctly that when you generate the bifurcation diagrams, you started all the trajectories from the same initial condition? You didn't do a whole set of initial conditions for each point on the bifurcation diagram?

Cerdeira. No, for a very simple reason. As I told you all, I just saw it about Tuesday last week when I started trying to see what was really happening.

Dorning. This gets back to the point that Leon Glass raised: As you vary the parameter you're probably just moving fractal basin boundaries across the point that you're using as a seed for the initial conditions. So the self-similarity in the bifurcation diagram probably really reflects the self-similarity in the basin boundaries.

Cerdeira. Ah, but then you are talking about within the windows. You're not talking about the beginning of the bifurcation diagram.

Dorning. Yes, within the window, because as you zoomed in you saw a structure again, and that really reflects the movement of what is probably a fractal basin boundary across the single initial condition you're using as a parameter.

Esselman. I may have missed it; are all your curves the result of analysis, or have any been compared with experiments?

Cerdeira. At this moment we are finishing the calculations. Now we can move time off designing all this software and hardware, and then we will do the comparison with experiment.

Thompson. I'm interested in what would seem to be a period triple, your trifurcation, if you like, in which one line simply splits into three.

Cerdeira. Yes, but as I told you that's what they think.

Thompson. Now, generically I think there is no such bifurcation, and it does lead me to wonder whether the three lines simply come together and get very close together. This would of course explain the continuity. It might be worth looking very carefully to see if, as the three points come backwards through the "trifurcation," you've simply got a period-three all the time, but the points are very close together.

Kim. What are the consequences of chaotic behavior in diodes? I presume that you will be using this device as part of some other machine, or whatever.

Cerdeira. Well, the idea is that if you want the device to be able to measure something, you want a predictable answer.

Kim. So you'll be using this device for measuring something else?

Cerdeira. Yes, for measuring the passage of charged particles.

Discussion: Real-Time Identification of Flame Dynamics. By M. Gorman and K. A. Robbins

Cerdeira. What's the argument on the polynomial versus exponential fall-off?

Gorman. The argument is this: If you have a deterministic system, it should be able to be differentiated from all orders—that says you have all frequencies present—and then, it should fall off faster than any power law, but I forget the details of why they say that. And that's the argument. There's no proof. I have to emphasize that. There's no proof that deterministic systems should fall off as an exponential. The stochastic systems, the argument is that they're not ever infinitely differentiable. [See Section 11.2, Background.]

Stewart. How about this: Take the Fourier transform of a polynomial. What do you get? Exponential falloff. It's just to do with the smoothness. You can approximate very smooth functions very well by polynomials. It would be an even more persuasive argument in fluid dynamics than it would in some other system.

Gorman. Yes, that's certainly true.

Thompson. At University College we're interested in building fires. Now of course you've been talking exclusively about steady states, but you did just mention that extinction was some sort of boundary crisis, and presumably there's a lot of hysteresis going on in these systems as you vary control parameters. I just wondered if you could say a little bit about some of these control variations and bifurcations etc. with a view to not only extinction but flash-on, for example.

Gorman. Our experiments have to do with premixed flames. Let me just show you what our stability diagram looks like. Now what we vary is the total flow rate and the equivalence ratio. And then we plot out all of these various dynamical regimes. This diagram [Figure 11.8] is very complicated. There are all kinds of dynamical regimes that occur at the boundaries of these regimes that I didn't have time to talk about. Okay, that's the first thing. The extinction process goes on when one crosses a boundary. Now in a building fire I believe that's more of a diffusion flame, where there's fluid dynamical factors going on. I would be hesitant to extend these results to those kinds of processes right up front. It is fairly apparent that in these premixed flames that this is the dominant instability when you go to very low flow rates, and when you go to leaner rich mixtures. Now this is relevant because, for instance, there are pilot lights on stoves. There are a very large number in the world, and what you want to do is go to the lowest flow rate you can and go to the leanest mixture you can. That's also a diffusion flame, so there is some reason for me to think that the same process which governs extinction in pilot lights governs extinction here because one sees an oscillation, a sort of what seems to be a chaotic wandering that's very similar to what I've showed here. What we could do, and what we are gearing up to do, is basically that we've done our extinction over here, we've literally crossed this boundary right here—and if you cross it the exact same way a large number of times, because you can't control the initial conditions precisely, what you get are different degrees of time it spends in the chaotic mode, and we can do this very reproducibly, study it; and we've also done some experiments on what is called transient chaos, which Dr. Yorke's group has worked on.

Thompson. Presumably you always have to light your flames, I mean there's no question of them self-igniting or anything like that.

Gorman. Correct. There should be a different term for this kind of crisis where you go to sort of the penultimate fixed point, like extinction. You can't go back, certainly. In Dr. Yorke's crisis, when you have a crisis you can always say that the attractor has been destroyed if when you return to the original parameters, then the attractor reappears. Okay, well obviously you can't do that. But if you go back into the same fuel rate and you ignite it—there's always funny stuff with igniting burners but as long as you go into a reasonable range of parameters you can always reignite the flame.

Maulbetsch. A lot of the flames that we're interested in are essentially Bunsen burner flames but very large. Plus, they're solid fuel, they're pulverized coal burners, which is always a diffusion flame in some limit. But you know, in some sense if you look at the primary zone inside, it can be considered sort of Bunsen-like, with the premix region. Was the suggestion of what you said that the premix region internally can experience these sorts of behavior and essentially drive the surrounding diffusion region with it, or is the diffusion region sufficient to damp out that thing?

Gorman. Well, I think you can get both kinds of effects. You know, if you put any sort of flame in some sort of acoustical chamber you can couple the acoustic waves back on the flame, but ours is the reverse. It's where the premixed flame front oscillates due to these effects. Let me remind you that it occurs at the rich and lean boundaries. Basically a lot of what we are talking about today is chaos avoidance. Dr. Ottino is going to talk about fluid dynamics where you use the chaos to do the mixing. Here we're talking typically about chaos avoidance, but when you get to those systems where you want to run very lean, for instance, then you are going to encounter these kinds of problems.

Maulbetsch. Precisely what people are moving for in NO*x* control.

Gorman. Yes. So we think we have a very high degree of documentation here. We're going to begin looking at data from other systems; to begin to use this technique, power spectra and other techniques to begin to provide some sort of insight. Let me show you something about power spectra. All these spectra are recorded at successive times and different values of the parameter between something called the drumhead mode and the axial mode, and what you see here, for instance, are broad peaks all the time and when we do a blow-up of an individual one again we see the exponential falloff. But if you go at a slightly lower value of the parameter you can see that these broad peaks only occur for part of the time. So, there's intermittent chaos. And then as you go to here it becomes more periodic and then as you go up that way it's more periodic. So, the intermittency we've seen, I think it's near the bifurcation where you can get behavior that looks polynomial.

Kim. You mentioned about the Landau–Ginsberg instability. You said that it depends on the frame of reference.

Gorman. As to whether or not you measure the system to be a chaotic attractor or not, Deissler [*Physica D* **25** 233 (1986)] looked at a Landau–Ginsberg equation in a moving frame of reference and showed that it has chaotic solutions which are deterministic chaos. I don't know what technique he used to decide whether it was deterministic or not. But then he looked in the laboratory frame. This is relevant because the falling vertical film of fluid, which is something that is used in the colored-film industry, where they lay down strips of liquid over acetate, has this exact problem; it is a classical problem in engineering. That system is described by a Kuramoto–Sivashinsky equation in a reference frame moving with the waves as they fall down your window pane. This is a very relevant, real-world kind of problem, and what Deissler's point is, is that, in the specific example he chose, he measured a qualitatively different kind of dynamical behavior in the moving frame and in the fixed frame. [*Note added in proof:* Deissler showed that, in the moving frame, nearby trajectories diverge exponentially; but, in the laboratory frame, nearby trajectories converge.]

Tam. What's the physical argument for it?

Gorman. The physical reason is that it's convectively unstable, that is, the system gets more and more complicated as it goes down. You can see this on a window pane because the system breaks up into smaller and smaller drops as it goes down, so it's much more complicated down the pane. So the dynamical system is moving, is evolving as you go down the stream.

Unidentified Speaker. But that's further evolution in the system as you go down the window pane, and so the dimensionality of the system is getting larger, I guess, as the parameter associating the system with the height on the window pane has changed. But in your flames where you have some steady flow is there some moving frame?

Gorman. I'm not talking about the flames, I'm talking about the general issue. The people who do experiments in open flows, they sit at a point and they measure what the turbulence is; they have to invoke Taylor's frozen turbulence hypothesis. They have to say that at a given point, you know, it is as if they had an ensemble of equivalent systems that is equal to, you know, one system as a function of time.

Sen. You've talked about the temporal characteristics of the signal that you have. However, your videos show quite dramatic spatial characteristics. How would one study those?

Gorman. What we've done is we've taken an array of photodiodes—these are out latest kinds of results—and in the periodic regime we can correlate two different channels and what one sees is that these are two different kinds of behavior: This is a radial mode where there is perfect correlation and the system remains; this is the correlation function, this is basically $U(t)$, $U(t + t)$ as a function of t and it oscillates nicely and periodically as one would expect for a periodic oscillation. There are two kinds of modes here: cellular modes, where you have those cells, and pulsating modes, where the system is basically functioning as an oscillator. In the pulsating modes, it seems as if you do not have spatiotemporal chaos, that is, all points on the flame front were correlated. There is a definite correlation which maintains itself. Whereas, when you have these cells, the motion of these cells is in some way uncorrelated. I didn't bring the correlation function for that. It decays right away to zero. It doesn't oscillate at all. So, we are very interested in studying the spatial characteristics again. One of the things we want to do most, though, is make differential videos, where we take a videotape, run it into a computer, subtract the sequential frame of videotape, and display that in real time. That's the way to characterize spatial measurements because—so what you can do is then you have a dial and a monitor you can look in and see, have a better idea of what spatial mode you're in. As you can tell right here, in some sets, especially from those square burner pictures, it's very difficult, in the most general case, to tell what the heck is going on in this flame front. As I said, I can't emphasize enough that I've shown you absolutely the best stuff.

Chang. It looks like there are only a few spatial modes.

Gorman. Right. What is happening in these regimes here is that you have the radial mode and the axial mode which are two relatively simple modes and they're interacting all along this line here [see Figure 11.8]. Okay, now I can show power spectra to demonstrate that they interact and there are no broad components [see Figure 11.9]. They interact and only produce sharp components. But as you move along this line from axial to spiral, those two modes interact and produce broad components [see Figure 11.10]. It is also true for cellular flames. I think the fact that there is a relatively small number of modes is exciting, and simply by changing the pressure you can adjust the number of cells. So, I think we have a system that basically studies a whole bunch of coupled chaotic oscillators. But here we have a relatively simple system. We have enough trouble with this, measuring the temporal characteristics. But you are right, it does seem that there is a relatively small number of modes, which makes them all relatively tractable theoretically. There are other variables here, pressure and type of fuel. If you just went to an arbitrary point in parameter space by turning knobs, things would look rather complicated.

Dorning. You may have already answered this, but I'll admit I don't remember. Do you have capability for creating cross-correlation functions?

Gorman. Yes. Those correlation functions I showed you were cross correlated. There was one of them, I threw it up there very fast, I apologize, with both autocorrelation functions and cross correlation—well, channel 1 and 2, that's hardly very good. Yes, we have done that, we have looked at that, and the point again is that there's no difference in the cross correlation in the pulsating regime, but in the cellular regime different points are very uncorrelated whereas here they're correlated for a while. They are as correlated spatially as they are with themselves.

Dorning. In the pulsating mode?

Gorman. Right. And that just says it's temporal chaos, as Chia was saying, with a relatively few number of spatial modes interacting.

Dorning. And in the other case, are you able to deduce anything in the cellular mode?

Gorman. I could spend another hour talking about the cellular flame. They're interesting from all sorts of points of view. There are arrays of cusps and folds, and I apologize we did not get to that low-angle shot. And they form ordered states, and then there are all those other states we didn't get back to.

Esselman. When you are approaching one of these spatial changes, is that a very sharp indication or is there a broad set of parameters? What I'm thinking of is, is there something you can see in the measurements that tells you you're approaching a new bifurcation?

Gorman. This is what happens we go from the axial mode to the radial mode. What we begin to see is the emergence of other components. For instance, this component at B is the frequency of this mode right here pulled up. This component A is the frequency pulled down. Frequency pulling is characteristic of mode–mode interaction. One begins to see this right here. If one looks at the flame front, it is relatively difficult to detect the slight radial motion, because when the flame moves up it looks like it moves in a little bit anyway. So the answer is we use power spectra. We sit there with a real-time spectrum analysis. We can't spend time computing with it. We have to do things in real time. Now when we go from the axial mode to the spiral mode, the spiral mode is a two frequency mode and when it interacts with this frequency up here what one gets is a broad component. If I were to show you video tape of the two you wouldn't be able to tell them apart. You wouldn't be able to tell which is chaotic and which is periodic. The answer is we use the power spectra; we'd like to develop tools for doing spatial characteristics.

Thompson. While you are discussing this bifurcation diagram, am I right in assuming that there is no hysteresis as you cross any of those boundaries?

Gorman. That is correct for pulsating flames.

Thompson. Rather interesting, actually, because, as you know, in dynamical systems hysteresis phenomena are really quite common.

Gorman. I can't say the same thing about the cellular flames. This is very much like Taylor–Couette flow where there is a lot of hysteresis in terms of the number of Taylor vortices. We have not yet gone through and done an extensive study of hysteretic effects, but all those dynamical modes you saw in the videotape depend on how many outer cells you have and how many inner cells you have. I can't speak about cellular flames, but I can speak about these pulsating flames, which are extremely reproducible, to the accuracy of the flow control devices.

Discussion: A Quantitative Assessment of Three Metal-Passivation Models Based on Linear Stability Theory and Bifurcation Analysis. By A. J. Markworth, J. K. McCoy, R. W. Rollins, and P. Parmananda

Stringer. These are so far the results of the manipulations of the equations. What would you expect to see experimentally?

McCoy. I guess that remains to be seen. Punit Parmananda is doing experiments and we'll see whether under perfect conditions he is getting chaos, these types of bifurcations.

Ottino. I think that indeed you are going to see chaos. There are many people doing things somewhat similar to this, and they find that chaos is very pervasive. But the question that I have is, when you put a piece of metal to be corroded, don't you expect to see some spatial dependence? In fact, unless you operate with single crystals or something like that, there are going to be locations that for one reason or another corrode earlier than another. So, what's the current mode of thinking about this? Do you have to go to PDEs to model some of these things, or can you get away with ODEs?

McCoy. I would suppose that for localized corrosion you would have to go to PDEs to model the transport of material out of the pit, for example.

Ottino. But are people doing it?

McCoy. I'm not aware of it, but that doesn't say that they aren't.

Stewart. I have two comments. One, the limit cycles that you showed in the last model looked very much like relaxation oscillations. Is that correct, that they're very quickly attracted onto the limit cycle. (Yes.) And you mentioned that as you change the parameters, the unstable limit cycle can grow very quickly to annihilate stable limit cycle. Typically, in relaxation oscillation systems the growth is actually so fast that it's virtually impossible to track it, the parameter changes are just incredibly small. It grows smoothly for a while and then the last bit of growth is virtually an explosion that becomes numerically impossible to follow. That's a nongeneric but still very commonly occurring case of dynamical bifurcations. There's a theory of this for what are called slow–fast systems—things are called canards. There's a survey article in the *Mathematical Intelligencer*, a couple of years ago, by Marc Diener about these kinds of bifurcations. Also, I believe, I'm not completely sure about this, but I believe there's some experimental work showing chaotic oscillations in the corrosion problem by Jack Hudson at the University of Virginia.

Dorning. Yes, there certainly is. He and his students are doing some very lovely work!

Ottino. He has even looked at two different parts within the same electrode, and how they interact with each other. So it's really very easy to produce behavior that is very, very, weird.

McCoy. Whether the behavior can be described by this simple model is another matter, of course.

Ottino. Yes, but if you put two of these together and you make them play with each other, you will see all kinds of exotic behavior.

Dorning. The chaotic dynamics of the system that you're describing were described by fairly simple equations, actually, with a reasonable ability to reproduce the experimental results. Not the details of the patterns on the plates, of course, but the overall dynamics in the well-mixed reactive system. A very low-dimensional model was successful.

Discussion: Dynamical Signatures in Electrocardiographic Data.
By R. de Paola, W. L. Norwood, and L. Glass

Gorman. Is the reason you choose babies because they basically have all the external parameters fixed, I mean as opposed to [adult] human beings where they might be doing different things?

de Paola. That's one reason to choose them, but the reason I choose them is because that is what's available. I have complete access to all babies which pass through University of Pennsylvania.

Gorman. Obviously, all their parameters aren't fixed, are they?

de Paola. I think there are advantages to choosing babies. The biggest advantage probably is that they don't have damaged hearts due to smoking or due to myocardial infarction. So, I think that there is a better control.

Linsay. Have you tried to construct any kind of strange attractors or anything like that from the data?

de Paola. No.

Dorning. I have a few questions. First, pigs versus dogs. Actually, we were planning on doing something similar at the University of Virginia Medical School with a dog. Maybe we shouldn't. We've delayed in doing it, and now that we've been delayed, is there strong advantage to a pig over a dog in studying the sinus node? In fact, we had in mind precisely what you did using the atropine to counter the effect of the acetylcholine.

de Paola. There are two advantages that I know of. Given the animal rights movement in the states, dogs are about 10 times the cost of pigs. The second reason in the case of cardiac function is that the anatomy of a pig is more like human anatomy, although there are difficulties in doing surgery on pigs, which make it less desirable than they might be as models.

Dorning. Okay, another question. The ectopic signal that you were talking about—is that from just some sort of general area in the ventricular tissue or is it from the AV node?

de Paola. Not necessarily, but it could be: I was using a generic term. As a matter of fact there are certainly a subset of arrhythmias, perhaps a very large subset that are from an external pacer, a secondary pacer for example. So I was just using that as coin for another oscillator.

Dorning. I see; so it could include the AV node as a special case?

de Paola. Yes, and there's a wonderful dynamics associated with AV and SA cellular interactions.

Dorning. Another question. How do you distinguish between the sinus and the ectopic signals in your data processing? I see the amplitude's different, but, do you distinguish them on the basis of the amplitudes?

de Paola. Yes. Leon Glass and I had a lot of arguments about collecting the actual ECGs. I don't collect the ECGs. All those measurements are made in real time. If one considers an ideal case of a premature beat, you'll find that the premature beat in the event that there's no coupling, back to the sinus node, is reduced in duration by the same amount that the pause following the premature beat is increased in duration. That is, if there's no coupling, it's conservative—time conservative. That's a very good way to pick out, for example, premature beats in which the sinus nodes are not perturbed.

Glass. If I may, just one or two comments. I agree with everything you've said, and I enjoyed it very much. The basic hypothesis that Robert was putting forward for the parasystole is a very well defined hypothesis, and it's one reason why people such as Robert and myself who have studied circle maps and dynamics of circle maps have been drawn to this particular mechanism because it's possible to understand in detail the dynamics that will arise from this and make predictions about what the dynamics will be. One of the problems that arises is that when we have actually tried to look at details of dynamics of a record of somebody who has frequent premature ventricular contractions, we have found it very difficult to fit the experimental data to appropriate phase resetting curves or to simulate long sequences using well-defined phase resetting functions. What I expect is that there are probably many different mechanisms of premature ventricular contraction generation, such as reentry or early after depolarization (these are all jargon terms in cardiac electrophysiology). One of the problems that the cardiologists have now, is when they see patients with frequent premature ventricular contractions they don't have very good ways to classify them, into subgroups. It is known that people with frequent premature ventricular contractions can be at higher risk for sudden cardiac death —postmyocardial infarction—but it is only a small percentage. In other words, it only is a small risk factor compared to other risk factors. It's not clear why this should be; why we can't understand better what acts as a risk factor and what doesn't. One possibility, which I think you said, is that things that look superficially the same may be really attributable to many different possible mechanisms, and one potential way to disentangle the mechanisms is to look at a very detailed dynamical analysis. What Robert and his colleagues have set up at the University of Pennsylvania at the moment comes close to being unique. Few hospitals have invested money into this kind of analysis so I think it's still at a very early stage, but it has to be done. The hearts are generating this vast amount of dynamical data but it's still not clear how to interpret it.

Dorning. I can see two possibilities, I guess. One is the ectopic beats being symptomatic of some other pathology that is not known, and the other is them being not due to another pathology but attributable to the actual event that leads to the cardiac arrest.

Glass. First of all, probably most of us here have premature ventricular contractions without pathology. However, in some people it is associated with pathology, and in some it is believed to precede or help contribute to instability or be a marker of instability in the heart.

Dorning. When superimposed on another pathology?

Glass. Right. For example, if you have very frequent premature ventricular contractions right after a heart attack, people will be watching, this may be some cause for concern. Whereas with you, if you're feeling healthy and happy and you have them, and you'll see your physician he'll say don't worry about it.

Unidentified Speaker. Do you see any slow drift? In the babies, these are in incubators, do you see any drift over a period of six hours or so? We've looked at some babies in incubators and there's a very slow drift before really doing anything we wanted to see, make sure that we understood that or could extract that confidently; and I don't know whether it's due to the temperature controls on the incubators or something physiological in that baby and not due to an external effect.

de Paola. We don't measure the babies in incubators.

Same Speaker. Oh, you don't?

de Paola. Our typical patient is on a mechanical ventilator in a bed. I haven't looked for that.

Same Speaker. The ones that are on the mechanical ventilators are ill for some reason?

de Paola. They're recovering from open heart surgery.

Same Speaker. Oh, I see.

de Paola. They've been anesthetized, they've had tubes stuck down their throats, and they need to be assisted with their breathing until they wake up.

Same Speaker. Are these typically normal babies, they're neonatal, but are they typical premature or not?

de Paola. They're not typical premature, but this is a somewhat unusual patient sample in that the great bulk of these babies have had congenital defects and their surgery was to repair congenital defects. The extent to which that affects clinical analysis is not clear. Certainly many of those defects do not alter the electrical conductivity of the heart, some of them perhaps... but I should have mentioned, with the exception of one or two slides, we were focusing on children and they may have specific symptoms that are really unique to them.

Discussion: Nonlinear Modeling of Chaotic Time Series: Theory and Applications. By M. Casdagli, D. Des Jardins, S. Eubank, J. D. Farmer, J. Gibson, N. Hunter, and J. Theiler

Dorning. There were some methods developed some years ago, about which I know very little other than the name, based on maximum likelihood parameter estimations and Kalman filtering. I assume they just went backward in time and not forward in time and didn't take the advantage of unstable manifolds. I would certainly assume that were true. Are the kind of techniques you are talking about, other than that, related at all to those?

Farmer (Comment extensively rewritten after the meeting). There are relationships between our methods and Kalman filters, but there are also significant differences. Kalman filters are used both for prediction and for noise reduction. In the simplest case, a Kalman filter assumes that the dynamics can be modeled as a linear dynamical system with some noise; the difference between a Kalman filter and an ordinary Weiner-style linear model is that the parameters of the model are updated dynamically, through a simple and elegant recursive least-squares method. One of the advantages of this approach is that because the parameters are updated continuously, it is relatively easy to track a nonstationary dynamical system. In this case the Kalman filter uses different linear models at different times, so in this sense it is nonlinear.

In contrast, our assumption is that the nonlinearities occur in state space rather than in time. As for a stationary Kalman filter, we use different linear models, but the selection of the model is based on the current state rather than the recent time history of the system. (Kalman filters also have been extended to the case where the models are nonlinear, but with a fixed parametric form, such as a polynomial. However, we have tested polynomial models, and have generally found that local models are superior.)

There has also been some work coming out of the nonparametric statistics and time series modeling community, as discussed in Section 15.6.

Glass. In one of the early examples you showed how bad the linear models were, and you showed how good the nonlinear models were. When I look at that I say, if I took the first five components of a Fourier transform and superimposed them and took them out into the future, that I would be able to do something at least halfway decent, better than what I think you're saying you're doing with your linear model. So, I just wonder, since you're someone who believes in the nonlinear models, if you are really doing the linear models as well as you can possibly be doing them.

Farmer. In 1941 Norbert Wiener proved that it is possible to make an optimal model for a linear random process by using a least-squares algorithm to construct a linear map from past and present lags to a future value.

Glass. So that's what you mean by a linear model!

Farmer (Comment extensively rewritten after the meeting). Yes, this is a standard linear time series model (technically called a linear autoregressive model). In principle, in the limit where there is an infinite amount of data, such a model captures all the linear structure, i.e., all the information in the correlation function, or equivalently in the power spectrum—such a model is equivalent to one based on a Fourier transform. In practice, however, with a finite amount of data this may not be the case—a model based on a Fourier transform might be better at capturing complicated oscillations than one based on the standard time series least-squares procedures. In our case we have a lot of data, much more than in many applications where this creates problems, so we believe the model is fairly well specified; also we gave the linear model its best shot by varying the number of lags and picking the number that gave the best result. However, since we haven't tried a model based on a Fourier transform, I cannot say with certainty that such a model could not do better. That's something we plan to investigate in the future, although it is on a long list with many other issues.

Unidentified speaker. Have you given any thought to using more than one probe?

Farmer (Comment extensively rewritten after the meeting). It is clear that using more probes can be more efficient. Again, there is a difference between what is true in principle and what is true in practice. In our original paper in 1980 we showed that for low-dimensional, simple systems it is possible to construct a state space from a single variable, which had the same dimension, Lyapunov exponent, etcetera, as the "original" state space. Takens independently proved that this was true *in the absence of noise.* Thus, in principle, it is always possible to reconstruct a state space, and one probe is as good as many probes.

However, in practice, there is always noise. We addressed this problem in a recent paper called "State space reconstruction in the presence of noise" to appear in *Physica D*—some of the results are also summarized in Section 15.2. We showed that with only one probe, if the dimension is sufficiently high or the largest Lyapunov exponent is sufficiently large, the noise is amplified so much that the system becomes indistinguishable from a random process, even if the initial noise levels are quite low. Although the same problem can still happen, the use of multiple probes clearly provides a method of improving things. Thus, in practice, your intuition is quite good; whenever possible it is worth at least trying to use more than one probe.

Whether or not multiple probes will actually improve the model depends on whether or not the extra probes have useful information. For example, suppose you want to predict the behavior measured by a given probe. If you enhance the model by adding another time series based on the output of a nearby probe, the information may be largely redundant, and so cause only a small improvement; on the other hand, if you use a *distant* probe, the behavior at a far distant point may be effectively causally disconnected, and so may be irrelevant and may make things even worse, since from the point of view of the behavior of interest it can be regarded as noise. Thus, whether or not the use of multiple probes improves the model depends on the positioning of the probes and the nature of the underlying dynamics.

Tam. In the near beginning of your talk you introduced the idea of a noise amplification factor. Is this related to this mutual information?

Farmer (Comment extensively rewritten after the meeting). The noise amplification as we have defined it is based on a conditional root-mean-square variance rather than on mutual information. I have always been very interested in information theory, and we spent a lot of time looking at measures related to mutual information and entropy, but in the end we rejected them. The reason is as follows:

The presence of noise means that it is no longer possible to think in terms of points—it becomes necessary to think in terms of probability densities. A highly "deterministic" dynamical system is one in which states are described by probability densities that are sharply peaked around a single point. A bad embedding, in contrast, can generate a density function that is either not sharply peaked at all, or is sharply peaked around more than one point.

The entropy is a functional that assigns a number to a probability density function. The entropy of a probability density that is sharply peaked around two points (which might be far away from each other) can be the same as the entropy of a probability density that has one peak (which is as broad as the other two peaks put together). The first case corresponds to a very bad embedding, while the second case to a good embedding. Entropy can be regarded as a measure of the "peakiness" of a probability density, however, it is indifferent to the location of the peaks. In contrast, variance-based measures, such as the one we use for noise amplification, are quite sensitive to the location of peaks; for the preceding example, the case with two distant peaks generates a large variance and therefore a large noise amplification. Another way to see this is that entropy measures treat the lowest-order bit as though it had as much information as the highest-order bit. In practice, if one cares about highest-order bits much more than lower-order bits, then entropy is a bad measure.

Mutual information is based on entropy, and similar remarks apply. Thus, our conclusion is that mutual information is not a very good measure of the level of determinism of a dynamical system.

The drawback of using conditional variance is that it does not have the nice invariance properties to coordinate transformations associated with entropy. However, this may be inherent in the problem: the central question is, Which coordinate transformation has the best noise properties? In this context it is not surprising that we end up with measures that depend on coordinates. However, it does make the problem more subjective—to find the "best" coordinates, we must first define what we mean by "best."

All these issues are discussed in more detail in the paper mentioned earlier.

Tam. When you estimate your amplification factor do you just use the original scalar time series or do you do embedding in different dimensions?

Farmer (Comment extensively rewritten after the meeting). If the noise level is known, the noise amplification can be estimated either from a single time series or from a multivariate time series. With a fixed noise level, the noise amplification is proportional to the root-mean-square noise in the reconstructed model; or, equivalently, it is proportional to the rms errors of a model based on nearest neighbors. (To predict the future behavior of a reconstructed state, find its nearest neighbor and predict that its behavior will be the same as that of its neighbor the same time increment in its "future.") It depends on the state space reconstruction, which can be based either on a univariate or on a multivariate times series.

Noise amplification differs from root-mean-square error because it is divided by the noise level, in the limit as the noise level goes to zero. You might ask, why bother to make this normalization? The reason has to do with our desire from a good theoretical framework—noise amplification is independent of noise level. Thus, it can be used to characterize the "quality" of a particular set of coordinates, without reference to a particular noise level. This allowed us to prove several facts about reconstructing state spaces in the presence of noise.

Esselman. Consider the following case: I have a noisy set of sensor signals and I'm interested in knowing whether there's any change in either the calibration or the response time of a sensor. Is this a useful method to look for that?

Farmer (Comment extensively rewritten after the meeting). This is a promising application that we have not yet investigated fully. The basic idea is as follows: Assume that we build a model while all the sensors are operating properly, which allows us to make reasonable predictions about the behavior of the system. Suppose that we monitor the prediction errors, and we know what they are typically supposed to be. Now, suppose the errors suddenly become much larger than normal: One possibility is that the system has moved into an entirely new regime, where its behavior is significantly different from what it was when we built the model. Another possibility is that one of the sensors is broken and is outputting screwy numbers. In either case this seems like a useful thing to know. Thus, I think we can use these methods to monitor whether a system is behaving "typically." This is something that would be good to investigate in more detail.

Kim. Going back to the noise amplification factors, in your presentation you showed a system of dimension 20, and even for that small number of dimensions it just went out of sight. Now suppose we deal with a problem with flow which has dimensions many orders higher, for instance. Would you comment on whether it is just as useful to adopt a conventional way of analyzing this kind of flow using supercomputers and grinding out numbers and so forth, because the noise will be just tremendous for that kind of prediction. So my question is, would you care to comment on whether this kind of conventional way of doing the business is futile or not?

Farmer (Comment extensively rewritten after the meeting). The plot that I showed assumed that it is necessary to reconstruct the entire multidimensional dynamics from a single time series. If it is possible to measure all the necessary variables on a finely sampled grid, then the situation becomes quite different. True, if there are large positive Lyapunov exponents the details of the simulation will rapidly diverge from the real flow. On the other hand, qualitative and statistical properties of the simulation may still be quite realistic. Thus, the usual supercomputer spatiotemporal simulation may still be quite useful for understanding qualitative properties, although it may not be very helpful for detailed predictions. The latter is particularly true because it is almost always impossible to measure everything that is needed on a finely sampled grid.

A good example of this occurs in the global change area. Global circulation models may be quite useful for qualitative simulations, but it will be a long time before they are useful for prediction of detailed behavior, such as whether or not the Earth will be warmer 100 years from now. In contrast, I think nonlinear time series models have a lot of potential for detailed prediction. The ice age predictions that we have made show promise in this direction; by analyzing more time series, for example, ice records on shorter time scales, temperature records, CO_2 and methane records, volcanic activity records, and other related time series, I think we could construct a decent temperature prediction model for variations based on natural fluctuations. If nothing else this would help to pin down whether or not current temperature variations are natural or artficial. Unfortunately, I have not been able to convince the global change funding establishment of the value of such a project.

Dorning. Now I had two questions. Well, first a comment on something that just came up, and that is the prediction of ice ages; you mentioned that your methods shed some light on the relative effect of the tilt versus the eccentricity. I guess that the net heat flow into the Earth really doesn't change too much whether it's tilted or not other than the reflectivity of the ice cap, whereas if you change the eccentricity of its orbit by a small amount, you're actually changing the distance from the Sun, in which case I guess the heat flow would change.

Farmer (Comment extensively rewritten after the meeting). The net *inflow* is not affected by the tilt of the axis, but the *outflow* is affected. The placement of land around the poles is quite different in the two hemispheres. The phase and magnitude of the eccentricity of the Earth's orbit relative to the tilt determines whether summers are hotter or colder in one hemisphere versus the other—for example, if the Earth is further away from the Sun during the northern hemisphere summer, then there may be little melting in the north and ice may accumulate more in the north than in the south. The total buildup of ice in turn has a large effect on the albedo of the Earth, which affects the outflow of heat.

Dorning. Yes, I'm assuming there's sufficient conduction from one pole to the other over the year that, if the heat flux is fixed, it doesn't affect things too much.

Farmer. I think the conduction is much slower than one might think.

Dorning. And now the question I wanted to ask you. Let me make sure I understand intuitively this business of the error amplification. If you are going from the single time series to the time-shifted phase space construction and if, because of that, perhaps there is some other variable in the real phase space, that had a large value when you were sampling the time series you actually are getting, has a small value where the noise is a very large fraction of the signal and then, that gets expanded with the expansion of the information that's in that signal in the other coordinate. Is that basically what's happening?

Farmer (Comment extensively rewritten after the meeting). Yes, I think you have the right idea. It helps to think about a concrete example, such as the Lorenz equations:

$$\dot{x} = 10(y - x)$$
$$\dot{y} = -xz + 28x - y$$
$$\dot{z} = xy - 10z/3$$

Assume that you observe x. Since $\dot{x}$ does not depend on z directly, information about z depends on the flow of information through y; when z changes it causes $\dot{y}$ to change, which causes y and hence $\dot{x}$ to change. When $x \sim 0$, since the only coupling to z is through the xz term, a large change in z causes only a small change in x. Equivalently, a small change in x corresponds to a large change in z. Thus noise in the determination of z from measurements of x is blown up a lot when $x \sim 0$. This gives the basic idea.

Dorning. Can you just avoid that by maybe not using sample points near the minima? Except of course when the maxima in the other function is correlated with these minima.

Farmer. That's true, but there's a limit to how much this is possible. If you make an embedding based on data that occurred a long time ago, then because the system is chaotic, errors are exponentially amplified in extrapolating to the present. Older data is less relevant to current behavior. You're stuck between a rock and a hard spot. For a more detailed discussion I recommend our paper on state space reconstruction.

Unidentified speaker. Related to the question that Dorning was asking, you showed a plot of noise amplification versus the Lyapunov exponent cross-plotted versus the order of the system. You were talking about just the largest Lyapunov exponent. What about a system which has many many positive exponents, a large number of them, which, I mean it's a distribution which falls very gradually?

Farmer (Comment extensively rewritten after the meeting). The noise amplification depends on both the dimension of the system and on the size of the largest Lyapunov exponent. The number of positive exponents doesn't affect things directly, but it does indirectly affect the dimension —according to the Kaplan–Yorke formula more positive exponents implies a larger dimension.

Discussion: A Study of Fluidized Bed Dynamical Behavior—A Chaos Perspective. By S. W. Tam and M. K. Devine

Ottino. Could you please tell me again how the measurement was made?

Tam. Okay, in this particular case the measurement is done by an oxygen probe which is yttrium-stabilized zirconia. That is stuck in several locations within the bed, and in fact in this particular case the probe is contained in a tube which is similar to the cooling tubes in the bed, so the probe does not perturb the geometry. So that is how the p_{O_2} data is obtained. Now once we got that we've got a time series, then we try to evaluate which is the best time delay by analyzing the distribution of the maxima and minima of the mutual information as a function of the time delay.

Ottino. A fluidized bed is basically like a liquid. Your curve on statistical mechanics, that looks suspiciously similar to a pair correlation function for a liquid.

Tam. Mutual information has a lot of connection with correlation function.

Ottino. Yes, but the peaks are completely expected if one thinks in terms of a liquid.

Tam. You're thinking in terms of correlation functions in terms of position; this is not.

Ottino. But the two are related, aren't they? For example, by means of a Taylor frozen-flow hypothesis as in turbulence. So couldn't you think of your curve as taking a picture, a continuous picture, so that the curve shows part of the spatial structure of the fluidized bed?

Tam. This one as I said directly measures the temporal correlation. Mutual information measures the degree of predictability on a variable at time T given a measurement of another variable x at time zero. The measurements on x and y are presumed to be carried out at the same spatial location or else they represent spatial averages over the region of interest.

Farmer. For a Gaussian signal, mutual information and the correlation coefficient can be directly related to each other, but for non-Gaussian probability distributions the two things are substantially different. In particular, the mutual information could be used as a measure of statistical independence rather than linear independence. That's the difference between this and a correlation function. If the data are sufficiently nonlinear there can be a substantial difference.

Ottino. A pair correlation function basically tells you if you're sitting in a bubble or on a particle and what happens at different radii as you move away from the particle.

Farmer. Yes, but it's not the same as the correlation function, that's important. The second point is that this analysis can get dangerous, because first of all, when you're computing mutual information you're just looking at mutual information through the signal at one time and the signal at another time; so what you're really doing is looking at a projection of the system down onto a two-dimensional state space and then looking at the mutual information in this two-dimensional space. The idea originally proposed by Shaw is that the mutual information in the full-dimensional space is really going to tell you but the causal links between the system at one time and the system at another time. The problem is, it's infeasible to compute that in a higher-dimensional space, which is why Fraser and Swinney computed this; and in fact; even with the stuff Fraser did subsequently, they were never really able to compute the full-blown mutual information. So, there's an inherent problem in the technique.

Tam. There is an inherent issue there and we have tried partially to address that. The results that we have shown are all done by embedding in two dimensions, as you said. However, we have also done three- and up to four-dimensional embedding. In our data we have seen that there is a certain dependance of this on the embedding dimensions, in the following way. What we have seen is that the characteristic oscillatory behavior becomes more damped as one tries to embed the data in higher dimension. In three dimensions, the oscillatory amplitude of the mutual information is smaller than in two dimensions. In fact, in four dimensions the oscillatory amplitude has decreased to such a point that the long-range oscillatory behavior becomes hardly discernible. But, this is a key point, the locations of the maxima and minima still occur in reasonably the same places for the two- and three-dimensional embeddings in which the oscillation is still clearly observable. For the purpose of extracting the heat time delay one needs only an approximate location of the first minimum. High precision is not necessary. So the effect of the embedding dimension on the mutual information does not seem to be important for our purpose. Now we did not go beyond four-dimensions because I'm sure you know that it's extremely time-consuming to calculate this mutual information quantity in high-dimensional space. Basically one needs to do counting of points to extract the structure. But the trend indicates that as long as the oscillations are there one can still recognize the minimum and maximum structures to remain in reasonably the same place.

Stewart. I would just like to make the simple observation, which is sometimes overlooked, that the requirements for sensible data set size of course increase geometrically with the number of dimensions that you're considering in your embedding. If you think a hundred points are necessary to reconstruct a two-dimensional attractor, then you would need a thousand for a three-dimensional, 10,000 for four-dimensional, 1,000,000 for a six-dimensional. So really, in a sense, unless you have billions of data points it probably would be wiser not to even try to estimate the dimension of the attractor but simply to pick an embedding dimension that makes sense based on the data set size that you're given, do your best reconstruction, and see if you can learn something from the dynamical structure, which is better than just purely statistical assumption.

Tam. That is an interesting comment, and, in fact, basically that's what we do—we go to the data set we have, we try to extract as much information as we can, and try to draw some lessons from this. The other thing is, about your earlier comment about the relationship between dimensions and data set sizes, I think that's a bound. In other words, there has been a number of similar estimates, but in practice what happens that a lot of people find by experience that frequently one can get away with much smaller data set sizes. These criteria represent closer to a worst-case scenario.

Stewart. Well, I just want to comment that I think the reason you do that is because what you get away with is much larger error bars.

Discussion: Forecasting Catastrophe by Exploiting Chaotic Dynamics. By H. B. Stewart and A. N. Lansbury

Tam. You have two branches near your bisector. How do you know which one to choose?

Stewart. Well, this is the threshold of the bifurcation, so this one moved up a little bit, but it was really this one which was shooting in. And that one is the . . .

Tam. Did you know it beforehand, without further computing?

Stewart. Well, if you don't think about it too much. You can just take a couple of successive parameter values in here and estimate that this thing is indeed moving much faster than this one is. Or you can, as Predrag Cvitanović said, think about it a little bit more carefully, actually watch the dynamics of successive iterates, and then you will see that as this point moves in toward the bisectrix you've got a slowing down of the dynamics when they come into this region, and that slowing down is a typical warning sign that you're approaching a saddle-type object.

Ottino. Have you given any thought to what would be the dream experiment? The kind of practical examples to which this approach could be applied?

Stewart. Any system in which you do observe very clean one-dimensional dynamics, this trick is going to work.

Ottino. Yes, but also the issue that the time variation and all the information needed regarding whatever the dynamic variable you measure has to occur in your lifetime. So what kinds of experiments would be the best?

Stewart. Electrical circuits. Fast oscillation in electrical circuits would be the easiest way to demonstrate this. You probably could do something for mechanical vibrations if you had a low-noise environment.

Cvitanović. I just want to say Christiansen in Copenhagen has taken Rössler flow and also extrapolated this point, and when you reconstruct the flow you get this high accuracy for the unstable cycles which are close to the flow, and it's wildly off for the periodic points that are far away. Reconstruction has a hard time capturing a point unless you get very close to it.

Stewart. There's certainly plenty of cases in which it's not a long range forecast. The most favorable cases would have an external saddle with fast dynamics along the stable directions and slow dynamics along the unstable direction.

Cvitanović. You see, if a region of the space is not explored by a trajectory, the fits are very poor.

Stewart. So it might ultimately go back to being something more like the noisy precursors thing; it's really just the earliest warning that you have.

Dorning. I had a question, but actually maybe Predrag Cvitanović already has answered it. Would this be useful at all, given you knew something about the attractor, in trying to locate the unstable orbits—because you can get the ones that are embedded in the attractor this way, but you can locate the others for the purpose simply of enumerating them. Then they could be used for other purposes such as expansions in the unstable orbits.

Stewart. Well, those expansions hopefully would be useful in describing the structure of the flow.

Dorning. Right, but do you see this could be used in locating the orbits. Rather than knowing that an unstable orbit isn't going to collide with the attractor, instead using the collision of the attractor with the unstable orbit and observing the changes in the stability as a way of detecting the existence and nature of the unstable orbits.

Stewart. Yes. I don't know, I just couldn't immediately think of a convincing way in which that might have an engineering significance, but it might, I suppose. But this—I think, at least I can make a plausible case that there might be some engineering significance to this—that you might have a situation in which this whole thing is something that you like, and that getting to this point which would cause you to fly somewhere else is something that you don't want.

Dorning. It's of direct engineering value. Well, actually in connection with the talk that Predrag Cvitanović gave the other day in the pinball machine example that he considered, he was able to construct the orbits just from common sense, whereas here if you want to use ideas like that to get accurate values for the parameters such as dimension and the Lyapunov exponents from the expansions in the orbits, it is essential to know the orbits. Of course, this would be a long tedious way to locate all the orbits! But, one might be able to locate the first few low-period orbits.

Stewart. I should conclude by saying I'd be very happy if you'd more or less ignored what I said and take a look at the work of Wiesenfeld and of Thompson and Virgin because not enough people have looked at the possibilities of these methods for forecasting bifurcations, the only thing I want to say is chaos is particularly nice; in this case (Wiesenfeld) you exploit the noise, you may not get a true forecast, but you exploit the noise. In this case (Thompson and Virgin) you may get a true forecasting, but you've got to tickle the system to get your information. When it's chaotic you just let it speak. If you know how to listen it'll tell you.

Dorning. I have one comment about tickling the system. Actually, you can use the noise to tickle the system in order to measure correlations. I think this is done in many cases. A spike in the noise corresponds to a pulse. So instead of knowing when you have the pulse, or inducing a pulse that is tickling the system and measuring the subsequent evolution, one simply measures correlations in the noise while the noise itself is providing the initial pulse. This is done fairly routinely in engineering systems.

Discussion: Controlling the Dynamics of Chaotic Convective Flows.
By J. J. Dorning, W. J. Decker, and J. P. Holloway

Gorman. In real engineering systems, where they use the closed-loop thermosyphons, is there any indication that the operating points are anywhere near the chaotic regimes?

Dorning. In this particular case, you notice I drew in the thermosyphon in red; in fact, in the specific engineering system design, it isn't there. It was, however, in an alternative design which in fact has been dropped, for some other reason, not because it had a thermosyphon in it—And, in any case, I really don't know the answer. These things have fairly broad parameter ranges over which they operate. So, I really would expect that one is going to see these kinds of behaviors. Really, in fact, as far as that particular engineering system is concerned, I think a point that is more relevant than perhaps the one I addressed today is that outside of these big vessels that hold all the sodium, the removal of the heat from the heat source when everything is shut down is to be done by an open thermosyphone. It is comprised of thin cylindrical annuli of air; there are two successive ones connected at the bottom and the air comes down one. These are really fairly narrow and extremely tall, and anybody who's ever tried to start a fire in their fireplace when it's kind of calm out and realized it's a little hard to get things going because there isn't a good draft up above, might worry a little bit—And when one looks at these, I feel I would just be awfully surprised if there was simple flow like this and it didn't have lots of little vortices developed, as one sees, for example, in the motion of fluids in a Taylor–Couette cell. I would be awfully surprised if that didn't happen, and one of the results of that is then there occur successive hot and cold layers along this hot wall at the stagnation points.

Gorman. In our experiments on the circular loop, we see that kind of secondary flows. In the convection loop, in the experiments, you know, when you heat the bottom and cool the top then as it comes out of the hot section it cools near the wall and then you get these secondary rolls, within the convection loop.

Maulbetsch. Most of the closed systems and the systems you are referring to are two-phase systems.

Dorning. No, not in what I've just now been commenting on. These are not thermal reactors; these are advanced fast reactor designs with sodium pools and thermosyphons of sodium as well. They're all single-phase.

Sen. I'm not aware of any work in which nontoroidal loops have been shown to be chaotic, at least in experiments; I don't know the reason for that.

Gorman. They have too much trouble getting around the corner, I think.

Sen. That perhaps is related to what you were saying.

Dorning. Too much trouble?

Gorman. Getting around the corner. In the convection loop, what makes things go is you have the buoyant force driving and the friction force opposing it.

Dorning. The friction effect at the corners is...actually...

Gorman. It's equivalent to having a very narrow tube.

Sen. If you take the nontoroidal loop problem and you expand the gravity function of the Fourier series, the toroidal loop is the first order. So as you change, as you increase the number of terms, you ought to be able to get your chaos, but somehow experimentally you don't get that.

Dorning. It's surely not just due to breaking the symmetry though, because there are some experiments where the loop is at an angle, it's heated not at the bottom, so the symmetry is broken.

Sen. Yes, but that changes the stability boundary.

Dorning. Right, but one still gets the same kind of dynamics. We all think essentially that the hot goes up and the cold comes down and they just do that faster, but, of course, when you put too much heat in, Hopf bifurcations occur, etcetera.

Sen. In fact, you can make it go down the hot side and come up the cold.

Dorning. Yes, indeed. Even when the heat source is not symmetrically located. There is some work in a noncircular loop which is kind of a thermosyphon with two-phase flow in it. It's not yet published. The people who are doing it are quite certain they've observed chaotic behavior, but I don't know, I haven't seen any of the data.

Glass. Sometimes in looking at problems that involve periodic forcing it's helpful to systematically search the parameter space of frequencies and amplitudes of the periodic forcing. Sometimes structures that seem very complicated when you just look at random points become much clearer. I wonder if you've tried to do this kind of systematic numerical analysis?

Dorning. Oh, I guess the answer is no, we haven't tried seriously yet. We've done a little bit of that. For example, one sequence I didn't show you actually started from the same point as the previous one I did show, the period-doubling sequence. We saw the period-doubling as a function of the amplitude of the forcing function, and we saw a similar period-doubling sequence as a function of the frequency, and in fact there was a similar period-doubling, in that neighborhood of parameter space as a function of the mean value of the forcing function of the Rayleigh number—and that suggests a clearer and bigger picture. We explored this a little bit, but certainly not enough to say it's there, but our studies suggest that, if we just limited ourselves to the two-dimensional parameter space, this period-doubling path would lie along some line that is perpendicular to neither of the parameters. So if we fix one of the parameters, we cut through this period-doubling sequence, and if we fix the other parameter to go the other way, we also cut through it. So instead of at these points, these period-doublings really probably occur on lines, bifurcation sets in the two-dimensional parameter space —and the same thing happened with respect to the third dimension; so one can think if one put the three parameters ρ_0, ρ_1, and ω, one could then have some kind of closed surfaces that are isomorphic to surfaces of spheres, a sequence of concentric spherical surfaces, defining the bifurcation sets in the three-dimensional parameter space. But we just kind of looked around a little bit there, and what we saw suggests that. But there could be holes in those surfaces, and no matter how much we do, one cannot argue that there are not holes in them. But indeed, if you have some specific things that you would suggest looking for, I should be delighted to hear about them.

Glass. Well, the thing that immediately comes to mind is if you're periodically forcing an oscillation, then you expect to see phase-locking at different ratios and so forth. You have seen some phase-locking, but I would expect that you should be able to get other ratios between your periodic forcing frequency and the oscillator as you change the frequency of the periodic forcing. Then the question arises as to how all those are organized in space, and it's hard to picture that because you're in such a funny system.

Dorning. Yes, the system's so rich as an autonomous system; start forcing it, and there has to be lots there. Everything from the autonomous system, plus lots more! So as far as studying it as a dynamical system, other than the motivation for understanding the ultimate application, well I guess it's worth doing. I think, well, there's an awful lot one could do, but just doing these things, we were kind of networked into everybody's Sun around, we each had six humming along.

Lawkins. But there is a practical aspect to that, too. As Alex Harrison said, you're worried about the design of the system, and you're worried about resonance in the system, often with the objective of safeguarding it from external sources of potentially dangerous frequencies. So if you have, for example, a system like a mixer or something like that where you need to have turbulence, these can perhaps, for the sake of argument, be called chaotic structures. The design would be assisted if there were an available tool, call it a statistical tool, for characterizing broadband processes.

Dorning. Yes, in fact that brings me to the point of the other application. The one application was the idea of oscillating the sink temperature, essentially the cool-side temperature, to exhibit some kind of control. Originally we started to look into this problem, not for the purpose of controlling the system, rather, we just were worried about the case where the heat sources had some time-dependence—some type of oscillatory behavior, for example. Then if the heat source oscillates at some frequency and we design the layer of a height so that in fact the frequencies are phase-locked and we get some kind of phase-locking, that might be very undesirable. Maybe one would want to know about that possibility in advance of the design effort. In fact, that is something that we wouldn't have control over, except in design. We wouldn't have control over that during operations. Yes, that fact suggests it would be very worthwhile doing.

Maulbetsch. Have you looked for, are you expecting to look at nonperiodic forcing, which might be easier to achieve?

Dorning. Yes, but of course one has to worry a little bit about the underlying theory. For example, if one wanted to represent a sort of bang-bang control where you punch this thing, one would want to model those impulses, maybe, in fact, the way they really are—as spikes in some function that's close to zero for a long time then goes back up again—so that at least one had a forcing function that in the continuous limit would be differentiable. I think if you start talking about these systems and you destroy the differentiability of the continuous dynamical system, you start to run into strange objects or strange trajectories in your phase phase. Your trajectories don't have to be smooth; they have to be continuous but not smooth, and so I think one really has to worry about the difference between the numerical artefacts and the properties of the system whose differentiability you've destroyed. But that certainly would be something worth doing; there's lots of different forcing functions that would be very worthwhile to consider from a practical standpoint. Of course, getting back to the big problem, doing all this in the context of a temperature-modulated boundary in the Navier–Stokes equations and more realistic geometry is the ultimate objective.

Discussion: The Power of Chaos. By P. Cvitanović

Cerdeira. Could you use these types of arguments in explaining quantum equations in macroscopic systems? For example, if you imagine a solid in two dimensions, with each atom as a cylinder, and then you look at a system of a million atoms, how would you go about calculating the conductivity of the system?

Cvitanović. I should make the following disclaimer. This is a modest low-dimensional physics. The systems whose classical dynamics can be described in a small number of degrees of freedom, small meaning three or four, not much more, can be run through this kind of analysis because we understand enough about their topology; we can visualize it. We don't have anything substantial to say about larger structures, the things that have very complicated geometry. The problem is that in large systems it is very hard to describe allowed trajectories; I mean, we don't even know how to describe diffusion very well. The periodic orbit theory just happens to cover almost everything that has been done in so-called chaos. All the averages that have been computed seem to be of this form.

Tam. Is the reason why this method may not be able to tackle high dimensions because you use high order for your orbits?

Cvitanović. You know, our successes so far have been sort of Swiss watchmaker successes. Period doubling was a very detailed first understanding of topology, and all of the examples are of that nature; one sets several people to stare at the topology, the stable–unstable manifolds, and then if you know the symbolic dynamics, if you know the topology, you can feed it into this machinery—and we have no clue how to really attack high-dimensional problems, unless they happen to have some regularity. Maybe there's a couple of lattices that could be approximated in this way.

Keefe. Are you saying that if you could elicit all the characteristics of these periodic orbits in a given system, then essentially all of the usual previously used measures could be expressed in terms of them?

Cvitanović. If you're interested in them. It's a little disappointing because now you have them with 20 orders of magnitude more accuracy. Those plots that you plot now, you know them to an accuracy of 10^{-16} instead of 10^{-3}.

Keefe. Okay, let's take something really gross like just what is the average state of the system? What is the centroid of the attractor?

Cvitanović. Well, that particular question depends on particular coordinates. But, the interesting question is what is the fluctuation around the center, and what is the scale of the fluctuation? The theory accomplishes these kinds of things. I mean that's what correlation spectra do. You know, you take the value of some variable, you look at its time evolution and you find out whether there are resonant frequencies in the correlations, etcetera. The Perron–Frobenius operator is precisely designed to do this: produce the correlation spectra.

Thompson. You've been predominantly talking about conservative systems, have you?

Cvitanović. No. The theory requires only hyperbolicity, it doesn't care whether the system is conservative or not. Symplectic structure is an additional nice symmetry of the system. The formalism does not require it.

Dorning. Have you already computed the probability measures and compared them with the measured ones or the inferred computed ones?

Cvitanović. My collaborators like to do this; I've been just computing eigenvalues myself. When you look at eigenfunctions, it's a probability measurement, you're making a coordinate choice; but they like to do this. There are a bunch of these eigenfunctions evaluated for low-dimensional systems, meaning one and two. Mostly one.

Dorning. And how do they compare, I mean the various quantities you talked about like Lyapunov exponents and correlation dimensions? They're integral quantities.

Cvitanović. No, they're also integral, they're eigenvalues. They're built by the union of long stable orbits, so they coat the whole phase space, and describe the particular smearing out due to the dynamics.

Dorning. The integral quantities average over the probability measures. Actually you get a distribution over the attractors.

Cvitanović. Yes, but I view it as a global contribution, because it is asymptotic smearing over the whole thing. The cycle is actually local, it locally probes the probability distributions; the union of cycles probes global averages.

Dorning. Yes, and how do they compare with what have now become traditional computed indices?

Cvitanović. They look okay. Again, they're much more accurate than just numerical simulation. You have to understand convergence, you have to improve the convergence of spectra. The focus has shifted. With simple problems they just work too well. For example, from the period-doubling factor we get Hausdorff dimensions to 30 significant digits using 30 cycles. I mean, this now works extremely well. The hard problem is getting symbolic dynamics in the systems you really want to study. The real problem's shifted from computing to understanding.

General Discussion

The object of the general discussion was to identify potential applications of chaotic dynamics to problems of interest to the electric power industry, to help EPRI in the selection of areas for research. The discussion was introduced by John Stringer, who highlighted points arising out of the previous discussions.

Stringer. There are several different areas which, it seems to me, have arisen from the proceedings of the last few days which deserve further attention. The first of these relates to the matter of control and the wider issue of understanding the range of parameters over which the operation of an engineering system is stable and in line with the expectations of the original design; the matter of the basin of attraction. Some of the things that Michael Thompson had to say were an illustration of that. To develop control strategies to make the system operate in a desired way: several people have commented on that. Another important related issue which has not been addressed is this: the actions that must be taken to recover from an undesirable excursion. This kind of situation arises frequently, and in the worst scenarios can be very damaging—this was the essence of the problem at Three Mile Island—or, indeed, the recovery from the wide-area electrical blackouts that you've heard about. This recovery situation may have important differences from the type of control strategy with which we are more familiar. Often, the recovery strategy can be counterintuitive, and as a result that's what most people get wrong; that's the thing that causes the most damage in engineering systems. Lastly, among these system-control-related topics is the ability to predict the future behavior of a system. It can be the future health of a person or the future of the weather or the future of an electrical generation system. It seems to me that these are the kinds of things that an engineer or the user of a system wishes to do, and it's things like that that I would like you to focus on for this discussion.

I want to emphasize that there are two possible scenarios in all this. The first one is that chaos is undesirable, for whatever reason. In that case we are probably more interested in the routes to chaos—things like bifurcations and catastrophes of one sort or another—and it seems to me that in that case we are probably not particularly interested in the chaotic behavior itself. I suspect that several of you may feel otherwise, and I am particularly interested to hear your views. The second alternative, however, is that chaos is desirable. Often, papers on the topic of chaotic behavior start out with an opening sentence that says something like "obviously chaotic behavior is undesirable"; but of course that's not necessarily so: Chaos is highly desirable if you're dealing with chemical reactors or mixing systems, for example. In those cases where chaotic behavior is desirable, it seems to me much more likely that one may be interested in the form of the chaos, the topology of the strange attractor and so forth. Again, for the sake of stimulating discussion let me suggest that even in these cases it may well be that detailed topological studies may not have any practical importance: It may be sufficient to know that the system is chaotic and what the boundaries of the basin of attraction are. I suspect that Julio Ottino might well disagree with this statement on the basis of your work.

In connection with the detailed structure of the chaotic behavior there was a question during the meeting about periodic windows, although there was very little discussion. A periodic window is very significant if you are trying to operate a chemical reactor or a mixer in a chaotic region. When you are operating a system of this kind you can tell when it's operating well because everything is smooth. You start to worry when the building starts to shake, and that means you are getting a significant periodicity in the behavior. So in these cases, periodic windows may well present a danger.

On the basis of these points, then, the actual behavior of the chaotic system may be not nearly so important as the topology and so forth of the boundaries of the basins of attraction.

As I understood some of the earlier discussion, there seems to be a situation in which we can go up a staircase from low-dimension chaos to high-dimension chaos (however that is defined), to essentially random behavior, and it seemed as though people were saying that only the first of these was accessible to quantitative analysis—and the question then I think is: "Is this actually *an* important or *the* important area for study?" Then there's a number of what Mike Thompson called house-keeping things:

How large the data sets have to be?

How can one best deal with noise?

Is the inhomogeneity of an engineering system so important as to destroy the usefulness of some of the projects we've got?

What I mean by this last question is, if I've got a system and it's going to fail somewhere by fatigue, I can't make global statements about that system that will tell me anything about the fatigue failure. That is just a plain fact of life. I have to do some very detailed finite-element analysis to find out which is the part or component most at risk or I have to wait and see which one breaks first. So there is an inhomogeneity in engineering systems and this may be a nuisance. Some of the model systems, even several of the model engineering systems that were dealt with in this particular meeting that assumed (or appeared to assume) homogeneity in the overall system. It may be, for example, that a ship that's capsizing is, from this point of view, a relatively homogeneous problem; but it is not so obvious to me that some of the other examples are.

We need to deal with the insertion of some probability aspects into the overall design of engineering systems, and I would be interested in your views as to how would one do that.

Lastly in this summary of questions and comments for discussion is a topic that is again getting down to the very fine scale of what we at EPRI have been doing in the area of nonlinear dynamics and chaos. As you heard, we picked out three small problems for an initial study, and our principal objective at this stage was to gain some familiarity with the field and to determine whether or not there seemed some useful applications of this field of research to our industry. You understand, these are very small studies. I'm glad that these knowledgeable speakers told you exactly how much money they were getting for these projects! We're interested in your views on whether the research topics we have selected are good areas to pick. If the answer is no, what would be better? If the answer is yes, what else could we reasonably do? And what sort of critical experiments should we undertake? For example, I have been giving sets of data from some of our other work to people and asking them to develop time series to explore for possible low-dimension chaotic behavior, and now I'm not at all sure that on the basis of the discussion here that that's the right thing we should be doing.

Gorman. What about the topic you mentioned of recovery from an undesirable excursion? Can you give me an example of that?

Stringer (Comment rewritten after the meeting). Yes, one recent example involved the fluidized-bed combustion of coal in which there was a loss of fluidization air, due, I believe, to the inadvertent shutdown of a fan. The coal feed (overbed, in this case), however, continued. Now actually the coal doesn't burn very well under those circumstances, so the situation ended up with a layer of coal on top of a hot, but largely defluidized, bed. The response of the operator was, when the fan came back on again, to put a lot of air through the bed to refluidize it, in attempt to return to the original operating condition. The coal that had accumulated on top was rapidly mixed into the freshly fluidized bed. What then happened was a large positive temperature excursion because the heat release from this relatively large amount of fuel was way above the capability for removing heat from the system, with the result that the bed material (principally coal ash and a sulfur-capturing material) fused. The whole thing essentially collapsed and the operator had to go in with jackhammers and take the whole solid bed out. This particular case is interesting because, as you have heard, we believe that a fluidized bed combustor may well be a chaotic system.

The point is that in many of these situations your normal response, the way you think you should do things, turns out to be wrong. According to what I read it appears that the Three Mile Island incident was a case where the method adopted to try and recover once things had gone bad turned out to be the wrong thing to do. So sometimes the recovery from a way-off situation is not simply doing the same thing you do to keep it on line, only a bit more so.

Dorning. Yes, I don't have any idea how to do this, but I certainly realize you have to go beyond intuition. When something has moved away from an equilibrium point, your intuition switches back. Of course if it resembles some kind of a cyclic behavior and it's gone to a maximum and is headed back toward the general vicinity of the equilibrium, but is not there, and you push it, you are just going to increase the amplitude. The kid on the swing is the classic example. If you push him when you're trying to stop him and you're a little out of phase, why you'll send him around the loop; and I think, when I first read about Lasalle 2, for example, I was very happy to see that the operators took no action. Some actions that would have been appropriate under different circumstances might have been catastrophic.

Stringer. Right, very often that's not a bad thing to do, but it's very difficult to keep your hands off the controls when things are going wrong, and all the bells are going off.

Chapman. An example of that is an airplane loss of control. There's a very good case for, if you lose control, take your hands off the controls and the airplane; at least a stable airplane will recover itself. If you fight, if you keep tight in the loop, you'll destroy it.

Stringer. Yes. I think that the recovery situation could perhaps be described as, "you've fallen out of your basin of attraction, how do you get back into it?" On the other hand, the usual control problem is a case of, "if you're wandering around in the basin, how do you get back to where you want to be within it?" That's really the difference, I suppose.

I would like to ask you now to address some of the specific questions I raised. The first one is, it appeared from some of the statements made on the first day that people felt that the situation with high-dimension attractors was intractable or useless, if I understood correctly.

Gorman. It's just that the apparatus that has been developed to date has concentrated on the low-dimensional chaotic systems. I think there's been a lot of work on systems like coupled lattice maps and things like that, things where people try to simulate spatial characteristics, and those aren't as far along and I would like to think that it would be very helpful to have experiments on such systems, to be able to do detailed comparisons of the theory and the experiment—and this is where we think the flames can come in because those cellular flames are basically coupled chaotic oscillators, so that we can begin to make contact between theory and experiment.

Harrison. With signal processing equipment the important factor is its capability to be able to take higher-order signals and smooth them out, get rid of that noise, not just to filter it using classical filtering techniques. That might allow us to go to higher orders. Whereas traditionally to get a good chaotic representation of a process you need a very clean signal in the first place.

Farmer (Comment extensively rewritten after the meeting). I don't think the high-dimensional situation is necessarily useless, although in order to make this clear, it is necessary to be precise about what we mean. It's important to keep in mind that the dimension is just a single number and may or may not do a good job of describing the full picture—other factors may play a role as well, for example, a flow might have some behaviors with a large dimension, but others with a low dimension. What the "dimension" of a system is depends on what you care about. For example, consider a fluid. From a strict point of view, it is composed of molecules that act more or less like a hard-sphere gas, which we know is always highly chaotic and large-dimensional. On the other hand, the bulk properties can be quite simple—if the fluid is at rest, at the level of description of the Navier–Stokes equations it has a fixed-point attractor and its dimension is zero. I am just saying this to illustrate how the same physical system can be viewed in different ways, and the dimensions associated with each view may be quite different.

Things can get even more complicated. Suppose the behavior of the fluid can be roughly decoupled into two parts—some small-scale eddies and a large-scale flow. Even if there is some interaction between them, for the big picture it may be valid to treat the small-scale stuff as noise, like the molecules in the first example, and concentrate on the large-scale part. Even if there is some coupling between the two scales, so that the "low-dimensional" behavior of the large-scale motion is quite noisy, the realization that such low-dimensional structure exists may be quite useful. There may be many situations where such behavior happens.

Stringer. You're telling me that in fact quite a lot of attractors in real situations are of low dimensions over quite a wide range of parameters...

Farmer. Well, yes and no. There are certainly many experiments, in many different areas, including fluid flow, chemical reactors, mechanical and control systems, and others, where it is quite clear that low-dimensional chaos exists. When I think back to 1978, when no one really knew whether or not chaos was widespread in nature, I would say that the results so far have exceeded the expectations of even the most ardent believers. In the *real* world, outside of controlled experiments, it is a bit harder to say. Nonetheless, there are many areas where dimensional analysis and/or prediction show lots of nonlinear structure, and it is suggestive that this is due to low-dimensional behavior, at least in the loose sense I was just describing.

Stringer. As we heard this morning, one needs a geometrically increasing number of data points to characterize high-dimensional attractors and numbers like a million data points for six dimensions were mentioned, I think.

Tam. I've got a question about high dimensionality. Doyne Farmer, do you think the difficulty of dealing with high dimensionality, or at least one of the difficulties, is that we are trying to reconstruct the attractor from scalar time series? Can this be helped substantially by experiment, that is, do we simultaneously measure two state variables? Would that help a lot?

Farmer (Comment extensively rewritten after the meeting). At least *part* of the difficulty of dealing with high-dimensional chaos comes from noise amplification and the reconstruction problem, but this is not the only difficulty. Let me back up a bit: Whenever a model is fitted to data, the model is usually not perfect, that is, there is always some estimation error in determining the values of the parameters. All other things being equal, the estimation error grows rapidly with the dimension. This is sometimes called the curse of dimensionality. The problem is obvious if you think about fitting, say, an mth order polynomial in d dimensions. The number of parameters required is roughly d^m, which says that the amount of data needed to get a fit of similar quality increases rapidly with the dimension. This is a big problem, and it is not clear whether there is any way round it. Whether or not it is possible to overcome the curse of dimensionality is a problem of considerable mathematical and practical interest, perhaps comparable to some of the Hilbert problems. Note that *in principle* it is always possible that someone might *guess* the correct form of the model, in which case the curse has been lifted. In a sense, this is what happens when someone has a good fundamental understanding, and derives the "true" model. However, this is quite different from an automated procedure that works directly from the data.

Noise amplifiction is important because, when it is necessary to reconstruct a state space, it complicates the issue *even when the correct model is known*. In this sense it is a problem that is even more fundamental than the curse of dimensionality.

Dorning. Kind of in connection with, I'll say, high-dimensional physical systems and low-dimensional phenomena within them, it seems that if you wanted to study complex problems in complex systems, a program of study would be to start with a complex system and try to create a simple problem. There are examples from problems in convective flow; one has an infinite-dimensional system and in some cases three-dimensional equations do fairly well. In fact, there's an example in one of Michael Gorman's papers about the thermal syphon. I think he cites an earlier paper where at very, very high Reynolds number the friction factor became important, and they had to extend I guess to a four-, possibly five-, I think a four-dimensional Lorenz model based on some earlier work. I don't remember who the earlier author was. So as they were moving into these other parameter regimes, the model had to be extended. Hence, I see a few possible directions. One is a patchwork of models for different parameter regimes for a complex system and another is a hierarchy of models for successively more complex phenomena in a specific complex system. And superimposed on all that, to bring in the mathematicians, I think studying the structural stability of the models as you simplify the models, the structural stability with respect to the robustness in predicting the nonlinear phenomena, would be crucial—if it's possible to do all this—and a lot that might have to be done numerically.

Ottino. Can I make a comment? When I gave my talk, I forgot the last two slides, so in order to save time I'm going to show those here [these are the last two figures in Chapter 5]. I want to make two or three points. My first point is concerned with what people do. Some researchers start with some physics, make assumptions and then have equations that could be PDEs or ODEs. Although this is kind of wishful thinking, in many cases you will try to reduce the dimension of the problem to something manageable, and you try to find out whether or not there is chaos. Then you try to characterize it. The classic example would be the Lorenz system, which everybody talks about. There are people here who started somewhere in here, with the equations, and the idea is that you got something that is simple and you try to find what are the routes to complexity in all of this. So, a mathematician would jump right in at the level of a map, or something like that, but you have to remember the physics in here. Then the other situation you might have is that you have something in which you have no idea about the physics, but my point is that if you have some information on the physics, you should try to use it. For example, if you know that in a fluidized bed there are bubbles that are 10 times the size of the particles, somehow that piece of information should be used. Thus in the latter case you have the time signal and then you do lots of things to it and you try to find out whether or not there are attractors, and possibly at the end you will try to make some forecasting or to build some possible model. One example in here would be the Rayleigh–Bénard problem, or Taylor, or whatever. So we can do these two things. Another point worth making is that there are many models that without assumptions lead to low-dimensional systems. In fact, the main reason I have been working in mixing is that without any need of making too many assumptions, essentially none, I get equations that are two- or three-dimensional. Finally, another thing that we have to also remember is that chaos per se is not going to solve anything. Let me illustrate this point by means of a figure from my paper [Figure 5.14]. This slide more refers to my view concerning my problem, but I think many people probably would feel the same way about their problems as well. I have two boxes that are roughly the same size—fluid mechanics and chaos theory—and somehow the merging of those two is going to tell me something about mixing in general, but I cannot possibly understand anything with one box alone—and then I have to have in mind who are going to be the end users of the results. The people who will consume the results (and I think most people here as well) belong to two classes. For example, a geophysicist cannot really touch the mantle of the Earth; engineers, on the other hand would like to "touch" their systems and make them do something they have in mind. We have to have in mind the end users. Chaos theory by itself, without the other building blocks, like physical chemistry or genetics or whatever contributes to these boxes, is not going to solve anything. So those are basically the points that I wanted to make. Let me finish by addressing some previous remarks. The comment was made here that somehow simple equations are giving chaos a bad name. In many cases that's the only thing that you can do, but there were many talks in this meeting that were basically equation-free, and some (like mixing) that naturally lead to simple equations. And it's also very likely that in two or three years we could see a merge between these two columns in there. For example, nothing prevents me from thinking that we could have a little ship and we match the Reynolds number and the Froude number and we build a model at the same time that we are gathering data from the little ship floating in water. So, I'm really hopeful that interesting applications will emerge, and not all of them will require that the system be infinite-dimensional. There are many that are very concrete applications involving low-dimensional systems.

Linsay. I think another point about the low-dimensional systems is that although they may not be realistic they often give you a lot of qualitative information that is applicable to the high-dimensional systems. You won't get the details correct but you get a good understanding of what's going on and you can then fill in the details by some less analytic techniques such as, say, some sort of empirical techniques.

Lawkins. There is nothing new about constructing approximations by projecting a system with a huge number of degrees of freedom onto a small-dimensional approximation subspace. Yes, some precision will be lost, but there is still good reason to believe that such a projection will contain useful information.

Stringer. Let me ask a question. Before we got involved in chaos we were involved in examining possible applications of fractals, and my question really derives from that kind of thinking. Quite a lot of attractors published in the literature have fractal dimensions that are very close to an integer. We saw a couple this week, for example, where the fractal dimension was 2.04 or something like that. Just what's involved in regarding that as being two-dimensional? I understand that there's a qualitative change there because if it's got dimension 2 it isn't fractal, whereas if the dimension is 2.04, it is. But in practical engineering terms, to what extent can one approximate the attractor's dimensions down to smaller spaces?

Stewart. Could you repeat that question?

Stringer. Let me put it this way. If we're looking at getting a dimension of something in the way that Shiu Wing Tam was showing today, about trying to embed the thing in the traditional Shaw-type approach—if I have a shape which is a purely fractal shape (this is a statement about fractals, it has nothing to do with chaos), I can easily envisage a shape that formally has very high dimensionality because of the small deviations of the shape into different dimensions. It can easily have a dimension of 7 or 12 or 20, it doesn't matter. But nevertheless, for most practical purposes, in a purely topological sense, that thing will appear to have a dimension which is virtually two; it will have little wiggles into the other dimensions. Now, in a purely formal topological sense, that thing has a very high dimensionality. And yet, if you look at it in the way an engineer looks at things (as I think you were just going to say about mapping things down onto lower dimensions), it will appear to have a dimension of only two. Some of the things that Bruce Stewart was talking about had little jiggles along the edges, which he said were irrelevant. Well, if they are irrelevant, to what extent can I use low-dimensional approximations as accurate descriptions of things that are genuinely high?

Farmer. The point you are making here is precisely the one I was making earlier about "effective" dimensions as compared with "true" dimensions. From an engineering point of view the effective dimension is probably the most relevant.

Ottino. Everything depends on what kind of magnifying glass you use and what question you really try to answer. It depends on whether or not you want to do averages way at the beginning or somewhere at the end. I mean if you don't want to average out anything, in almost every physical problem that I can imagine, you'll have to deal with PDEs and essentially infinite-dimensional spaces. But if you make reasonable assumptions, like something is well mixed (averaging, essentially) you can effectively reduce the dimensions right at the beginning to some number that is not going to be unrealistically high. So complexity depends on the magnification, I think.

Dorning. I'd like to make a few comments in connection with modeling. Of course, we all learned as students—and suggest to our students—that in order to get a handle on something new you develop a simple physical model that recovers the salient physics and press on and do that calculation, then face the real calculation and do it. Of course in the face of nonlinearity you want to capture the salient physics and the salient nonlinearity as well, which may not obviously be connected with the salient physics to start with. So that's one point on simple models. And the other is—in fact Lorenz's original work is perhaps a perfect example of what I'm about to say. He had read an earlier paper, I guess by Salsman, who indicated that in rather high-order expansions almost all of the coefficients died away in time except for a certain three which Lorenz then called x, y, and z. So, for the problems in which he was interested he thought the system evolved onto a three-dimensional manifold at long times and therefore studying that would be a useful thing to do. And if one has a complex model and one carries out enough analysis, either analytical or numerical, one puts oneself in a position where one can establish the manifolds onto which the flows evolve at long times. I think that's very important. Because direct numerical simulations, depending on the details of the numerical scheme, often have different bifurcation or nonlinear phenomena, typically bifurcation, or for example more bifurcation points, than the original physical system does. So a straightforward numerical attack, even though one may be investing a lot of money in computer effects and developing difference schemes and the like, may in fact yield results that have qualitatively different properties than the original continuous physical system has. So, the idea of developing models that have the same nonlinear structure, in my eyes, is crucial. In fact, Bruce Stewart talked about something like this at a meeting a year or two ago in connection with modeling. I don't remember the details, but let me give you a simple example. A few years ago a student, Yousry Azmy, and I were looking at numerical calculations for driven cavity problems, and we found that when we used too coarse a grid in these numerical methods that we had developed, which in fact were very capable of getting very accurate solutions on coarse grids to many problems, there were actually bifurcations that were not in the physical problem. When we refined the grid, of course they went away—and if one did some sort of a convergence study just on the basis of looking at the numerical convergence of some set of these grids, a sequence of the grids, they would have looked like they were converging to the solution to a problem, with several turning points. If one went beyond that, all the turning points went away, and then the finer grid sequence converged to what in fact was the correct solution. So, a simple model that captures the right physics, orchestrated with complex numerical methods is essential, I think, to a correct development. Numerical methods alone can be very misleading.

Thompson. If I can make a comment as well, while we're talking about models—I think what's happened in the past (and this can be particularly true since industrial designers have been used to linear systems, they were brought up with linear systems, they were taught linear systems) is that designers have tended to put enormous effort into building a model. I mean, typically, the offshore oil people will have a finite-element model of an enormous structure with thousands of degrees of freedom: but it is very expensive to run. So since they expect it to settle down to a unique solution, they do a rather short run. It settles down—"Hurrah! That's the answer, close the discussion." Now I think what they'll have to realize is that, in the future, dealing with nonlinear systems they'll at least have to put in an equal weight of effort into exploring the response of the model. And here you have a problem. If you're going to make big models with too many degrees of freedom you simply aren't going to be able to spend the computer time to explore it, and this again points to the fact that you are actually in many cases better off with a rather low-dimensional model that you can then do a decent job on, rather than making a very complex multidegree of freedom model that you just cannot afford to explore computationally. I know some cases exactly like this, where a beautiful sophisticated model of a boat, for example, has the boat surface triangulated. The fluid forces are calculated, every step in time from the waves; but you can't afford to run it. Even if you've got a Cray for a fortnight, you can't really do the sort of explorations that we would do to explore the basin boundaries. So in a sense you are forced back to more simple models. I think it's a very important point.

Farmer. I wanted to make a comment on one of the points that John Stringer raised in his initial discussion, concerning "inhomogeneity." When a system is highly inhomogeneous, with lots of complicated parts, this obviously enormously complicates the problem of modeling from first principles. On the other hand, for time series modeling, inhomogeneities are not necessarily all that bad—a time series is a time series is a time series, and the fact that one is produced by an inhomogeneous system does not necessarily make it harder to model, unless the inhomogeneities increase the dimension,

Stringer. A lot of situations that we are interested in, we meaning EPRI, are intrinsically very noisy, where often the noise has nothing to do with the subset of the system that we're actually wanting to look at. We just get noise from fans and we get noise from this and that. We just have noisy systems.

Farmer. The methods that we have developed for eliminating noise are only as good as the underlying prediction model. The key is to eliminate "irrelevant" influences and concentrate attention only on the "relevant" ones, to ignore the noise at the outset. We have been working on methods to do this, which I believe could give significant improvements in noisy situations.

Stringer. Okay, to be specific, for example, Alan McRobie presented his work on the concrete stack this morning. You know, the properties of a concrete stack that lead to its eventual failure are not well defined, they're defined statistically. There's a thing called Weibull index which determines the probability of the material failing at any particular stress. Howe would you build a failure criterion into your model, for example? You've already discussed the failure, or the potential failure, of a large stack. The model that you presented really talks about the perturbations that are taking place in the stack as a function of the elastic or nonlinear properties of the material as it behaves in a condition where it hasn't failed. So, if you're interested in whether it's going to break, you have to have in there somewhere a criterion for the failure of the concrete itself, the actual breaking of the concrete, that has to be in the experiment, in your overall analysis, right?

McRobie (Comment rewritten after the meeting). In the simple model that I have been talking about, there are geometric nonlinearities as well as material nonlinearities. These provide a smooth hump on the potential well, so loosely speaking it "buckles" first; it undergoes the geometric instability, it starts to fall, and then it reaches the material failure some time later, out to the right of the potential maximum. In that case then, the exact nature of the material failure criterion is not so important. In a lot of cases, though, the model should be different, and the material failure would come before any geometric instability. One of the things this would do is to give a potential well with a nonsmooth hump, and there would be no hilltop saddle cycle, so we can't use its invariant manifolds to look at the flow. I would still argue though that the way to look at the problem would be in phase space, to look at the organization of transient motions, and look at those that lead to failure

Stringer. Yes, I think I understand what you're saying. Probably the question which I asked referring to the introduction of the variability of the materials properties to the statistical function might have better been applied to say, the fatigue failure of a turbine blade. One might be using exactly the same kind of approach to describe the mechanical fluctuations that take place as a result of shedding within the turbine, which has formally quite similar dynamics; but the fatigue criterion is based on the usual description of the variation of the fatigue properties of the material. In this case, you could not possibly avoid quoting the statistical properties of the material.

Kim. I think we need to think about probabilistic techniques in the deterministic system, because in the energy systems, in plant systems, there are too many variables that we have to take into account. Invariably, we have to reduce them to very simple models. Therefore, we lose certain things that may become important during certain operations, and we are dropping them because it becomes very complicated. I wonder if there is any way of getting a little bit of a handle on the uncertainty in the modeling itself by incorporating probabilistic techniques. I think the amalgamation of the probability and the determinism may be quite useful for very complex systems.

Stringer. The next question relates to what research is it appropriate to do. Are we interested in the forms of the attractors?

Thompson. It should be emphasized that chaos is not necessarily undesirable. In lots of resonance problems chaos is desirable because it means that the system hasn't got into a synchronized motion and the energy's not being pumped in—and therefore it defuses, in a sense, the amplitudes of vibrations. So in that sense it is desirable.

Lawkins. I would like to make two points. First, nonlinear dynamics theory, including chaos theory, provides a basis for developing statistical tools to empirically study time-dependent, chaotic nonlinear processes. Second, if chaotic structure is an inherent component of a process, like a chemical reactor, for example, then we need tools to analyze it.

Dorning. Actually, if this is the case, should we be interested in the routes to chaos? I think the answer to that is yes, whether it's the case or not. If it's desirable, you want to know how to get there; if it's undesirable, you want to know where to make the right turn to not go there.

Gorman. But, then you don't care about the details of the chaotic behavior.

Stringer. What I really mean is the routes to chaos as opposed to the chaotic behavior itself. It seems to me that you could make a case that it's the routes to chaos that are important and not the chaotic behavior.

Dorning. I can certainly think of cases where you'd be interested in the chaotic behavior. If it's chaotic but the whole attractor is in some bounded region that is very tolerable, you don't really care just because it goes on long excursions.

Gorman. I wanted to pick up on your point which was on the very first slide. The only time one might be interested in the chaotic behavior of such a system is if one wants to get out of it. And I think that was very stimulating because I would like to see if I could think of an experiment that could raise and test the kind of problem that question five on the first slide posed. It's only then I think that one is interested in the details of the chaotic behavior. But I agree with you, I think that this is astute, it's exactly right in terms of the flames. One is more interested in how one gets there, the details of the attractor in most of those cases aren't important. But, until we do this other class of experiments that you suggested, that you're in the chaotic state and how to get out, you can't rule that out.

Stewart. On point one there, I'd just like to say that in a sense I agree with you. I think the really interesting thing is to focus on geometric methods for nonlinear dynamics problems, and chaos is an interesting part of that, but because of our increase in attention to just the geometric methods themselves we also learn—have made a lot of progress recently in understanding regular phenomena in bifurcations—so it really is nonlinear dynamics and chaos geometrical methods, for me. On the second point, wouldn't necessarily downplay the importance of the topology of the strange attractor. For example, if you're worried about the periodic windows as being a danger you may be able to look within the structure of the attractor and see whether the collapse through a periodic orbit is incipient. Another example where topology might be important: We're used to, we're maybe used to, the idea that topology of a strange attractor is a very subtle thing because we only have a few low-dimensional examples where we say a lot about the topology, but it might be something much more dramatic, like the difference between a three-dimensional and a five-dimensional attractor, which could have a serious consequence.

Stringer. That's true. I suppose I was thinking of some of the very elegant work that Chua has done for example on the topology of his attractor—which is an extremely complex attractor—and I think the work he's done is extremely elegant; but I just don't know what I would use it for.

Unidentified Speaker. I think one case where you do want to know the structure is in the work Celso Grebogi was talking about yesterday on trying to control a system that's gone chaotic—and there you actually want to understand the details of the chaotic attractor itself, so that you can implement this kind of control system.

Stringer. Why would I want to control in that sort of way?

Same Speaker. You might not have any choice in your parameters and you're forced into a situation where...

Grebogi. In our work, practically any parameter can be used for control, provided it obeys some technical condition.

Stringer. I was excited by what you said, but supposing, for example, you're in a chaotic system—as we think we are in a fluidized bed—and I want to keep it there. I think I need a control paradigm which enables me to keep things chaotic and away from periodic windows so it's not only in the chaotic regime but in a rather special place in that chaotic regime. So, I mean, if the things you are suggesting would enable us to do that, that needs to be considered.

Grebogi (Comment rewritten after the meeting). Indeed, we should be able to control chaos without having to drive the system hard. This kind of forcing with the sole purpose to quench chaos, as Hubbler and co-workers do, takes the system to other regions of the phase space with a considerably different dynamics, which is typically not desirable. We, on the other hand, control chaos by keeping the dynamics restricted to the phase space of the original chaotic attractor, as you indicated this to be a desirable feature. In addition, Hubbler's control technique requires the knowledge of the differential equations governing the system. In our control technique, the relevant quantities necessary for control can be extracted from a time series of some experimental scalar quantity.

Cvitanović. I'll just make Grebogi's point again. In Grebogi and Yorke's paper they have this nice conceptual example of how you can get to the end of the solar system. You do it not by controlling stability but by controlling instability. Send *Voyager* along an unstable orbit, it gets close to something, then you give it a kick and send it elsewhere. So one conceivable use of understanding the topology extremely well is as kind of Maxwell demons; you align the things so you get a series of kicks, and there's also the idea of increasing the mixing, increasing diffusion. You go and look at the topology, and the topology can really shove the stuff around. For this you really have to know how the things work; you have to know that you place it here and you place it there and then you have to adjust it, and the detailed knowledge might just pay off. Now that I'm talking I might say something else: On your wish list, maybe, there is sort of an engineer's wish list and scientist's wish list. For example, quantum mechanics was motivated by trying to understand spectra, it was not motivated by trying to build a transistor. But everything of interest that came out of this thinking was not conceived at the time the thinking was done—and yet I think, optimistically, the nonlinear science is actually an interesting development in science. I hope that nothing that's on your list will 20 years hence be remembered as an important contribution. There are actually applications which by definition are engineering except are inconceivable at this moment. For example you are interested in energy but maybe the nonlinear advances don't fit the intended applications; advances in science almost never solve the intended problems. They always create a bunch of new problems and unexpected solutions. So I hope this will happen here as well.

Stringer. I agree with you, I'm glad you said that.

Cvitanović. Then you should put ten percent of your money into that!

Stringer. I will write letters to NSF on your behalf! What I mean by that is the different funding agencies have different targets and I don't want either you or anybody else to think that because I've been concentrating in this discussion on practical applications from an EPRI point of view that I don't think well of basic science.

Unidentified Speaker. Back to controlling chaos. A practical system is a tokamak plasma, and there you have this enormous engineering problem: How do you keep this plasma stable? Nobody knows really the equations of motion, and if you can show it's chaotic, then you could possibly implement some of these schemes and keep the plasma stable, rather than having it blow up in these weird and wonderful ways that it does. And then you might actually have a machine that EPRI would be interested in.

Dorning. Yes, I was just going to make a comment. I guess it could even be extended to chaotic, but I was thinking more of just nonlinear. A part of the area I reviewed the other morning was this business with limit cycles in reactors; and of course one does in existing commercial reactors try to avoid them. I happened to be speaking with some people who were doing some kind of "Star Wars" work and they were concerned about getting pulse power out of power sources. If one goes to some kind of a gas-core reactor then, rather than taking this steady-state energy and converting it through some kind of energy conversion system to make it pulsed, one could run the reactor at a power where it was in a limit-cycle mode. Then one could even have nonlinear waves which have very narrow troughs and very high, very wide maxima or vice versa. So one could get out wide peaks with narrow spaces or narrow peaks with wide spaces. So the idea of understanding some of these nonlinear systems and using that understanding in engineering—to engineer the nonlinear denominant in the designs—makes a lot of sense to me, and one might even be able to do that. This is a great leap of faith, but suppose one did have a tokamak system or some other magnetically confined system, and one were really stuck with chaotic output of the energy. I mean, separate from the structure of the field lines and the stochasticity in the tokamak, suppose what you got out in terms of energy was chaotic and that was about the only way you could easily confine the plasma. Well, knowing that, knowing the bounds on the size of the amplitude of the oscillations that will come out in the chaotic time series at various times, then one could engineer the energy conversion system—the blanket and so forth—to accommodate the originally chaotic energy output. In such a system one wants to understand the attractor and to adapt the rest of the engineering system to that attractor's behavior, if one can't control it.

Lawkins. Coming back to the attractor, and the fluidized-bed example, and the engineering application, it's operating and you want to know if it's operating right. Well, what do you mean by right? Let us suppose that the process has been analyzed, to the point where we can identify the attractor and have a set of parameters characterizing the process. Now I don't know whether it's going to be 6 parameters or 76; it depends on the system, and just how well we need to identify it from an engineering point of view. Now, first of all, we've got to get the system operating, and we will sense those parameters over time. Before we worry about control, the first thing we want to know is where we do want the system to be and how can we identify if it drifts away? One can think of a box representing the acceptable ranges of the parameters: if the system drifts out of the box, some action is called for, e.g., shutdown for maintenance. But we'll never know what the parameters are until we do the detailed analysis.

Stringer. The last objective, let's see, let me pick your brains. Within EPRI I would like to have something that represented a successful engineering demonstration for our sort of industry of the usefulness of chaos. This is perhaps a redundant question because obviously the work on the stack is a fine example. We don't have many boats, but we have an awful lot of stacks! But I'm asking the question not of you as engineers but of you as chaos people. What do you feel would be a system which would show the virtues and benefits of this branch of science to a pragmatic and skeptical audience?

Thompson. I'd like to ask just a little bit about the vibrations of turbine blades, I mean what sort of problem is involved there?

Stringer. One of the troubles of interest is a sort of soliton-wave-type problem. One finds that when one tries to tune the turbine blades, and obviously there's a whole set of criteria for tuning a turbine blade; what one occasionally finds is that one blade in a row will start to vibrate with a very considerable amplitude and will eventually fail if something isn't done quickly about it—and there seems to be no very obvious reason why that should be so. When one looks at that blade it doesn't seem to be significantly different from all the others. In fact one of the methods of correcting this problem at the moment is to disorder it—in other words, to broaden the band of properties that one has in the turbine blades with the idea that, somehow or another, one is coupling it in a way which is not very clear to produce this extreme vibration and that by decoupling them by making them just a little bit more different from each other, although still within the criteria of avoiding the many different potential resonant frequencies that one has to avoid ... That's one particular blade-failure process which we don't really understand. It seems to be a nonlinear sort of process.

Dorning. Don't you have shrouds on those blades to keep them from vibrating?

Stringer. On some of them, yes; but not on others. Not all turbine blades have shrouds, and the shroud may or may not be a help. Shrouds also have that problem too.

Dorning. They couple the oscillators!

Glass. How many first-rate mathematicians are you funding rather than engineers? I really liked what Predrag [Cvitanović] said. I think that you folks are very focused on very narrow issues and very narrow problems, and you're asking what areas should we fund. Maybe you should be saying what people should we be funding?

Stringer. Yes, I agree with that.

Glass. And I think that sometimes there are strong results from areas in which one doesn't expect results. The people who are trying to get a lot of money are not necessarily the people who are going to make the most creative applications to what your major problems are. You should realize that some small percentage of your total funds is still a lot of money to some people.

Stringer. We do, actually, within the Exploratory Research effort, which I hope was explained to you at the beginning of the meeting, have a section whose specific objective is to try and develop mathematics work. However, I think it is fair to say that we do not currently have a top-flight mathematician on our staff to help select the topics and the people. One of the things we do, of course, is hold workshops such as this to which mathematicians come along; and our intention is to define some work that's regarded as high-quality by the professionals, done by first-quality mathematicians. But I take your point and to the best of our ability we're trying to evolve towards doing something like that. I quite agree we certainly need it. I also agree with your point about funding the person, not the project.

Gorman. I'd like to echo what they said, in that the only place to go in the U.S. for chaos funding is with Mike Sluschinger at ONR. Funding is very very tough. This field has attracted the best and the brightest in all the disciplines of the physical sciences and it's still very difficult to grow and compete with the better-funded disciplines and I appreciate the fact that you guys had the foresight to begin to move into these areas. And I think if anything, it seems to me the nineteenth century was characteristic of Newtonian dynamics, where things were somehow predictable and one thing was connected to another in a very simple way, but if there's anything that characterizes the twentieth century it's chaotic dynamics, where things are connected in very complicated ways.

Kim. What was the objective of this workshop? So let me show you the overhead that I showed you the first day. I think it might be worthwhile just to take a look at it again. We started out saying that the workshop would focus on aspects of chaos theory with clear and direct applicability, and how chaos science can lead to chaos technology. I feel very gratified that most of this part has been really satisfying, indeed most of the people gave some practical examples and most of the papers have addressed how the science part can lead to technology. So thus far, I think the workshop was reasonably successful. We also wanted to define and determine the potential (research areas) for practical applications of chaotic dynamics, and overall I think we got something out of this workshop on that one too. I think this is really the first attempt to gather the experts to focus on practical applications of chaotic dynamics.

INDEX